SIMON FRASER UNIVERSITY
W.A.C. BENNETT LIBRARY

RA 644 H38 C76 2007

HANDBOOK
OF

Helminthiasis
for
Public
Health

HANDBOOK
OF

Helminthiasis
for
Public
Health

D.W. T. CROMPTON
LORENZO SAVIOLI

Taylor & Francis
Taylor & Francis Group
Boca Raton   London   New York

CRC is an imprint of the Taylor & Francis Group,
an informa business

CRC Press
Taylor & Francis Group
6000 Broken Sound Parkway NW, Suite 300
Boca Raton, FL 33487-2742

© 2007 by Taylor & Francis Group, LLC
CRC Press is an imprint of Taylor & Francis Group, an Informa business

International Standard Book Number-10: 0-8493-3328-8 (Hardcover)
International Standard Book Number-13: 978-0-8493-3328-6 (Hardcover)

Library of Congress Cataloging-in-Publication Data

Crompton, D. W. T. (David William Thomasson), 1937-
   Handbook of helminthiasis for public health / D.W.T. Cromptom and Lorenzo Savioli.
      p. ; cm.
   Includes bibliographical references and index.
   ISBN 0-8493-3328-8 (alk. paper)
   1. Helminthiasis--Handbooks, manuals, etc.   I. Savioli, Lorenzo. II. Title.
   [DNLM: 1. Helminthiasis. 2. Intestinal Diseases, Parasitic. 3. Developing Countries. 4. Health Promotion. 5. Helminths--parasitology. 6. Public Health. WC 800 C945h 2006]

   RA644.H38C76 2006
   616.9'62--dc22   2006007997

Visit the Taylor & Francis Web site at
http://www.taylorandfrancis.com

and the CRC Press Web site at
http://www.crcpress.com

Preface and Purpose

Public health attends to the health of communities rather than individuals. Public health professionals apply biomedical knowledge and environmental technology to improve and sustain the health of the communities they serve. The effectiveness of public health interventions is measured by decline in human incapacity and increase in human well-being, self-esteem, and prosperity. In September 2000, the member states of the United Nations adopted an unprecedented global commitment to achieve the Millennium Development Goals. Public health is a major item in this agenda for development and resources from wealthy nations will be needed if the justifiable demands from impoverished nations for the implementation of public health interventions are to be met. Helminthiasis, the assemblage of diseases that accompany helminth infections, affects the health and productivity of millions of poor people in developing countries. The control of helminthiasis falls unquestionably into the purview of the Millennium Development Goals.

Major efforts, based on partnerships between those in need and those with resources, are already dealing with dracunculiasis, lymphatic filariasis, and onchocerciasis. New partnerships are emerging to tackle schistosomiasis and soil-transmitted helminthiasis. Experience shows, however, that global health problems such as smallpox and poliomyelitis are not solved overnight. Years of work by dedicated public health workers and supportive communities are essential if lasting success is to be secured.

The purpose of *Handbook of Helminthiasis for Public Health* is to assist public health workers now and in the future with the daunting task of reducing the burden of helminthiasis. We also hope that the book will be a useful source of information to all teachers and students of helminthology. A special challenge has been to draw attention to the uniqueness of the human host. This is not simply a matter of human DNA and the information it carries. Human-helminth interactions are determined to a large extent by social, cultural, economic, and political influences in addition to the biological principles that are considered to underpin host-helminth relationships elsewhere in the natural world. Interventions to control helminthiasis have to be adapted to the lifestyles of people in communities where helminth infections are entrenched.

The Internet provides many of us with access to journals on line and facilities to search for sources of information. During one search for information about a particular theme, the computer screen displayed 20 citations and carried a message to say that 835 additional pages of citations were awaiting inspection. Despite much work, we know that the literature we have cited in this book is but a smattering of what is available about human-helminth relationships and how to control helminthiasis. We hope our choice will serve as a guide to the knowledge awaiting application for the benefit of those in need.

D.W.T. Crompton and Lorenzo Savioli
Glasgow and Geneva

Acknowledgments

We thank Dr. Judith Spiegel and the staff of CRC Press-Taylor & Francis Group LLC for the opportunity to write this book and for seeing it through production and distribution. We are particularly indebted to Kari Budyk, our production co-ordinator, for patiently and sympathetically guiding us through the maze of details that required our attention. We are equally grateful to Jay Margolis, our project editor, for his support, expertise and attention to detail.

We thank the copyright holders who granted us permission to use their property; acknowledgment can be found in figure captions and table titles, and traced through the reference lists at the end of the chapters. We have tried to contact all relevant copyright owners and comply with their conditions. Changes in the publishing industry seem to have been accompanied by changes in copyright ownership and so we apologize for any oversights or omissions that have unintentionally occurred in our acknowledgments. We also wish to acknowledge the excellent facilities, comprehensive resources, and helpful staff in the libraries of the University of Glasgow, the University of Nebraska-Lincoln, the World Health Organization in Geneva, and in the Wellcome Library for the History and Understanding of Medicine. PubMed has proved to be a marvellous source of information.

Writing a book is an enterprise that very much depends on the help of friends and colleagues. It is our pleasure to acknowledge and thank the following people for their generous support, advice, constructive criticism, and help, and for sending us information: Adebayo Adeloye; Marco Albonico; Samuel Ore Asaolu; Stephen Attwood; Boakye Boatin; Dannai Bunnag; Jong-Yil Chai; Isabel Coombs; Tessa Crompton; Dirk Engels; Ann Lackie; Duncan Mara; Edwin Peters; Santiago Mas-Coma; David Molyneux; Antonio Montresor; Darwin Murrell; Malden Nesheim; Marion Nestle; Brent Nickol; Edwin Peters; Claudio Politi; Edoardo Pozio; Mary Hanson Pritchard; Francesco Rio; Han-Jong Rim; Paiboon Sithithaworn; Vaughan Southgate; Russell Stothard; John Swaffield.

We are most grateful to Ian Ramsden, Effie Crompton, and Patricia Peters. Ian Ramsden, sometime Head of the Department of Medical Illustration at the University of Glasgow, has exceeded our hopes with the style and clarity of the illustrations. Effie Crompton has prepared graphical data for Ian, managed files, tested websites, and most importantly of all rescued material and unravelled tangles generated by DWTC's idiosyncratic word processing. Patricia Peters of Kirriemuir Secretarial has unfailingly provided dedicated and effective support for every aspect of this project.

About the Authors

D.W.T. Crompton has held a lectureship in parasitology at the University of Cambridge, the John Graham Kerr Professorship of Zoology at the University of Glasgow, and an adjunct professorship in the Division of Nutritional Sciences at Cornell University. He has worked with colleagues at the World Health Organization since 1981. In 1977 he was awarded the Scientific Medal of the Zoological Society of London. In 1999 he was appointed OBE by HM Queen Elizabeth II for services to the development of health care in West Africa. He is a Fellow of the Royal Society of Edinburgh and an honorary member of the American Society of Parasitologists, the Helminthological Society of Washington, and the Slovak Society for Parasitology.

Lorenzo Savioli is the director of the Department of Control of Neglected Tropical Diseases of the World Health Organization. He graduated in Rome in medicine in 1977, in tropical medicine in 1979, and in infectious diseases in 1985 in Rome. He has an MSc in parasitology from the London School of Hygiene and Tropical Medicine and the DTM&H from the Royal College of Physicians of London. In the 1970s he developed a fascination for classical clinical semeiology and tropical medicine and in 1979 decided to go to Zanzibar to work in the small district Hospital of Chake Chake on the island of Pemba. In 1986 he started the Pemba Island Schistosomiasis Control Programme that a few years later was extended to include the control of soil-transmitted helminthiasis. In 1991 he joined WHO in Geneva as the medical officer in charge of the Programme on Intestinal Parasitic Infections and in 1996 was appointed chief of the schistosomiasis and intestinal parasites unit. He is senior associate in the Department of International Health of Bloomberg School of Public Health, Baltimore, Maryland, and a Fellow of the Islamic Academy of Sciences, Amman, Jordan. In 1986 he received the 1st Prize of the Rorer Foundation for Medical Science for *Italian Medicine for Developing Countries*.

List of Tables

Table of Contents

Part I

Human Health and Helminth Infection

Absolute poverty is a condition of life so characterized by malnutrition, illiteracy, disease, high infant mortality and low life expectancy as to be beneath any reasonable definition of human decency.

Robert McNamara, President of the World Bank, 1978

Poor people in low-income countries have little option but to share their lives and meager resources with helminth infections. Interventions and resources are available to reduce this chronic burden of disease and much experience has been gained from major control programs. Nevertheless we all still live in a wormy world. Helminthiasis has to be included in the development agenda and be given a higher place in the list of public health priorities.

World Bank. 1993. *Disease Control Priorities in Developing Countries* (Edited by DT Jamison, WH Mosley, AR Measham and JL Bobadilla). A World Bank Book. Published for the World Bank by Oxford University Press.

World Health Organization. 2003. *Global Defence Against the Infectious Disease Threat* (Edited by Mary Kay Kinderhauser). Geneva: World Health Organization.

World Health Organization. 2004. *Comparative Quantification of Heath Risks. Global and Regional Burden of Disease Attributable to Selected Major Risk Factors,* Volumes 1 and 2. (Edited by M Ezzati, AD Lopez, A Rodgers and CJL Murray). Geneva: World Health Organization.

There is little joy in life nor any kind of justice for a child condemned to disease or early death because of the accident of birth in a developing country.

Halfdan Mahler, Director-General of the World Health Organization, 1980

1 Helminthiasis — A Challenge to Health and Development

There is now wide agreement in the clinical and biomedical science communities that helminthiasis has greater public health significance than was previously thought and that action to relieve and reduce its debilitating effects on the health and livelihoods of millions of poor people deserves high priority. We have written this book to support all who are concerned to confront and overcome helminthiasis. Although we may enjoy explaining how helminths travel from host to host and we may marvel at how these remarkable animals obtain food and shelter in an immuno-competent environment, we have sought to concentrate on the fact that helminths are agents of human disease.

Although many species of worm have been retrieved from humans, we consider that species listed in Table 1.1 and those whose names are shown in bold type in the list compiled by Coombs and Crompton (1991) merit most attention in the context of public health. Coombs and Crompton's list as revised by Isabel Coombs is shown at the end of the chapter. Readers are also referred to the annotated checklist of parasites published by Ashford and Crewe (1998).

The worms that challenge health and development thrive and persist in econom-ically impoverished countries. Millions of people are infected and millions live with the risk of infection. In 1978, Willy Brandt chaired an independent commission that reported on global development issues. The commission divided our world into the affluent North and the impoverished South (Brandt Report, 1980). In this handbook we have retained the terms North and South as convenient if simplistic shorthand to stress where affluence and poverty prevail (Figure 1.1). The North includes member states of the G8, the OECD (Organization for Economic Co-operation and Development), the EU (European Union), the Paris Club and the dominant members of the WTO (World Trade Organization). The South includes the territories occupied by major forms of helminthiasis (Table 1.1). Despite economic growth in some big countries of the South, this region remains the arena of malnutrition, malaria, TB, HIV/AIDS and the constellation of infectious disease.

The South contains the lower income countries once described as the Third World. There has been an increase in national and individual prosperity for some countries in the South since the days of the Brandt Report but not for many of those in the continent of Africa. Prosperity has largely passed by Africa. This continent is the only region to have suffered further deprivation and poverty instead of enjoying some measure of prosperity (van der Veen, 2004). The average GNI *per capita* for

TABLE 1.1
Estimates of the Numbers of Human Helminth Infections (Millions)[a]

Helminth (Chapter)	Stoll (1947)	Sturchler (1988)	Peters and Gilles (1989)	Hopkins[b] (1992)	Crompton (1999)	Specific Reference (Millions)
Cestodiasis (5)						
Echinococcus granulosus	—	0.10	—	—	2.7	2–3 (Craig et al., 1996)
E. multilocularis						0.1–0.3 (Craig et al., 1996)
Hymenolepis nana	20.00	75.00	36.00	—	75.00	4–5 (Pawlowski, 1984)
Taenia saginata	39.00	45–60	76.00	7.0	77.00	See Pawlowski (1982)
T. solium	3.00	6.00	5.00	—	10.00	See Pawlowski (1982)
Dracunculiasis (6)						
Dracunculus medinensis	48.00	4.00	98.00	3	0.08	0.032 (WHO, 2004)
Food-borne trematodiasis (7)						
Clonorchis sinensis[d]	19.00	25.00[e]	28.00	15	7.01	7.00 (WHO, 1995a; Rim, 2005)
Fasciola gigantica and F. hepatica	0.10	0.004	—	17	2.40	2.40 (Mas-Coma, 2004)
Fasciolopsis buski	10.00	10.00	15.00		0.21	
Heterophyes heterophyes[f] and related species	(< 0.5?)	—	—	—	1.05	4.40 Rep Korea (Chai and Lee, 2002)
Opisthorchis viverrini and O. felinus	—	25.00 [e]	11.00	15	10.33	10.5 (Watanapa and Watanapa, 2002)
Paragonimus westermani[f]	> 3.00	20.00	5.00	6	20.68	20.0 (Toscano et al., 1995)

Lymphatic filariasis (8)						
Brugia malayi and B. timori	19.00	9.10	4.50	—	13.00	120.00
Wuchereria bancrofti	170.00	82.00	85.50	90.00	107.00	(Ottesen et al., 1997)
Onchocerciasis (9)						
Onchocerca volvulus	20.00	40.00	40.00	18	17.66	0.14 in Americas (WHO, 1995b)
Loa loa[g]	13.00	2–3	33.00	—	13.00	2–13 (Pinder, 1988)
Schistosomiasis (10)						
Schistosoma haematobium	39.00	90.00	78.00	200	113.88	112.00 (van der Werf et al., 2003)[h]
S. intercalatum	—	—	—	—	1.73	See Doumenge et al. (1987); Tchuem Tchuente et al. (1997)
S. japonicum	46.00	5.00	69.00	—	1.55	0.99 (Hotez et al., 2006)
S. mansoni	29.00	100.00	57.00	—	83.31	54.00 (van der Werf et al., 2003)[h]
S. mekongi	—	—	—	—	0.91	0.14 (Hotez et al., 2006)
Soil-transmitted helminthiasis (11)						
Ancylostoma duodenale	457.00	670.00	900.00	900	1298.00	740.00 (de Silva et al., 2003)
Necator americanus						1221.00 (de Silva et al., 2003)
Ascaris lumbricoides	644.00	800.00	1233.00	1000	1472.00	795.00 (de Silva et al., 2003)
Trichuris trichiura	355.00	520.00	670.00	750	1049.00	
Strongyloidiasis (12)						
Strongyloides stercoralis	35.00	75.00	70.00	80	70.00	See Sato (1986)
Trichinellosis/trichinosis (13)						
Trichinella spiralis[f,i]	27.00	0.30	49.00	11	—	11.00 (Dupouy-Camet, 2000)

TABLE 1.1 (continued)
Estimates of the Numbers of Human Helminth Infections (Millions)[a]

[a] Between 1950 and 2000 the world's population increased from 2,515 million to 6,100 million. That in the South increased from 1,683 million to 5,000 million during the same period (Hewitt and Smyth, 2002).

[b] Hopkins (2002) refers to infections by disease nomenclature, e.g., schistosomiasis and does not subdivide data to species.

[c] About 50,000 deaths annually from neurocysticercosis (Schantz et al., in Macpherson, 2005).

[d] Some 35 million cases of infection are reported by Lun et al. (2005).

[e] Sturchler (1988) combines values for C. sinensis and Opisthorchis spp. at 50 million; Hopkins (2002) combines values at > 30 million.

[f] Taken as representative of a complex/group of species.

[g] Loa loa is included because it may induce serious adverse events when treatment is provided for other filarial infections.

[h] Data published by van der Werf et al. (2003) apply to endemic countries in sub-Saharan Africa.

[i] 25,161 cases with 240 deaths reported since 1964 in China where pigs are common (Liu and Boireau, 2002).

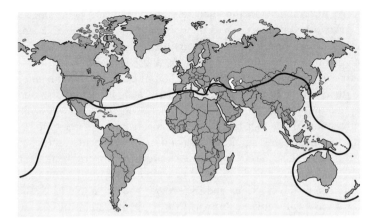

FIGURE 1.1 The economically affluent North lies above the line while the impoverished South, where most helminthiasis prevails, lies below the line (see Brandt Report, 1980). The figure is based on the projection of the world by Dr. Arno Peters.

the 46 countries of sub-Saharan Africa is USD 496 compared with USD 28,337 for the 39 countries of the North (UNICEF, 2004).

Estimates of the extent of helminth infections and the burden on human health and development caused by some of them are offered in Table 1.1 and Table 1.2. Disease can be quantified in terms of morbidity and mortality rates, but these statistics depend on the accuracy and coverage of the systems in place for obtaining

TABLE 1.2
Estimates of the Global Burden of Selected Helminthiases

Disease (Chapter)	Morbidity Cases (Millions)	Annual Death Rates (Thousands)	DALYs (Millions)	References
Lymphatic filariasis (8)	42.81	**	5.64	WHO (2002b); de Silva et al. (2003)
Schistosomiasis (10)	78.5	0.5–15	1.76–4.5	Warren (1989); Chan (1997); van der Werf et al. (2003); Hotez et al. (2006)
Soil-transmitted helminthiasis (11)				Warren (1989); Chan et al. (1994); Chan (1997); WHO (2002a); de Silva et al. (2003); Hotez et al. (2006)
Ascariasis	335–350	4–60	1.18–10.5	
Hookworm disease	159	4–65	1.82–22.1	
Trichuriasis	220	2–10	1.65–6.4	

** Data not found.
DALYs, Disability-adjusted life years.

and recording information. In low-income countries, where helminthiasis is most prevalent, overworked health services may not have the resources to keep complete case records, may be forced to rely on superficial diagnostic procedures and may overlook concurrent ill health (see Pawlowski and Davis, 1989).

Such difficulties led to the creation of a non-monetary economic unit known as the DALY (disability-adjusted life year) to measure the burden of disease. A DALY is a conceptual unit based on "disability weights", ranging from 0 = healthy to 1 = death, that takes into account the socioeconomic context of those to whom it is applied and the epidemiology of the disease or impairment of interest (WHO, 2002a). DALYs are important because they may (1) aid in organizing health services for both curative and preventive priorities, (2) assist in identifying disadvantaged groups and targeting health interventions and (3) provide a comparable assessment of the results of various health interventions (Murray, 1994; Carabin *et al.,* 2005). The controversial aspect of using DALYs concerns deciding about the values to be used in the disability weightings. The closer the weighting is to1, the greater its impact on the DALYs. This has proved to be a particular difficulty when dealing with the burden of disease associated with helminthiasis because the real significance of subtle morbidity (WHO, 2002a) seems not always to be taken into account when calculating DALYs (Table 1.2). There is a subjective problem concerned with the calculation of DALYs. Health professionals who feel that their area of concern deserves a higher place in the list of public health priorities may be tempted to use the higher DALY value when more than one exists.

A person presenting with such afflictions as haematuria, anaemia, blindness and so on is obviously ill and weightings can be agreed and combined with estimates of the number of people suffering in this way. However impaired cognitive perfor-mance, increased absenteeism from school and work, reduced worker productivity and synergistic effects of multiple helminth infections fall into the category of subtle morbidity and need to be included in the DALY calculations. In 2000, the member states of the United Nations and development agencies in the North committed themselves to work with countries where poverty prevails to achieve eight Millen-nium Development Goals (MDGs) by 2015 (World Bank, 2004). There are convinc-ing reasons to accept that reducing the subtle morbidity of helminthiasis in addition to the obvious morbidity will help to achieve five of the MDGs (see WHO, 2005a).

How can we meet the challenge of global helminthiasis to human health, well being and development (Table 1.1 and Table 1.2)? A consensus is now emerging that helminthiasis is one of the interacting processes that are generated and sustained by poverty (see Figure 2.1). A collaborative approach is required to reduce poverty. The North with its bank of technical and financial resources cannot impose or drive poverty reduction. Political decisions and willing co-operation from the sovereign nation states of the South are essential. The obvious need for partnerships between the North and the South, which was pressed home in the Brandt Report (1980), is now gaining support. Interestingly, Henderson (2004), in his list of the top ten scientific breakthroughs of 2004, identified at number nine the public-private part-nerships that have brought important advances against diseases endemic in devel-oping countries.

Global parasite control, which gained attention thanks to the concern of a former prime minister of Japan (Hashimoto, 2000), is advancing because public-private partnerships have been established between governments, UN agencies, universities, national and international NGOs, philanthropic institutions and the pharmaceutical industry (WHO, 2005b). The main thrust of such partnerships in the sphere of helminth control is to provide anthelminthic chemotherapy to reduce morbidity and disrupt transmission. Every partnership accepts that improving health education will increase compliance with control measures. This is a relatively short-term strategy. Long-term, sustainable helminth control will not be achieved until appropriate sanitation and safe drinking water are universally available in the countries where helminthiasis is endemic.

Provision of these secure essentials for healthy life will depend not only on the resources of the North but on the South having stronger economies and full control over their developmental destinies. Countries that are poor and heavily indebted will need more relief. The economies of poor countries should benefit if trade barriers are lowered, allowing a freer exchange of goods and services. Sceptics may still advise that there is little point in tackling helminthiasis until sanitation and access to safe water are in place. A few years ago, those of our persuasion who argued that anthelminthic chemotherapy was a realistic public health intervention would be put down in the public domain and told "they will only get re-infected". We reject that uncaring attitude. In this handbook, we shall examine knowledge of the biology of our choice of helminths that have public health significance (Table 1.1). We shall show how access to modern anthelminthic treatment is improving the health of people and can benefit millions more even in spite of continuous transmission. We will offer practical guidance as to how this knowledge can be applied to counter the suffering caused by helminthiasis. We will not ignore problems and difficulties. We are committed advocates and practitioners of helminth control and deworming. Overcoming the challenge of helminthiasis will improve human health and enhance national development.

REFERENCES

Ashford RW and Crewe W. 1998. The Parasites of Homo sapiens. An Annotated Checklist of the Protozoa, Helminths and Arthropods for which We are Home. Liverpool: Liverpool School of Tropical Medicine.

Brandt Report. 1980. North-South: A Programme for Survival. (Report of the Independent Commission on International Development Issues.) London and Sydney: Pan Books Ltd.

Carabin H, Budke CM, Cowan LD et al. 2005. Methods for assessing the burden of parasitic zoonoses: echinococcosis and cysticercosis. *Trends in Parasitology* **21**, 327–333.

Chai J-Y and Lee S-H. 2002. Food-borne intestinal trematode infections in the Republic of Korea. *Parasitology International* **51**, 129–154.

Chan M-S. 1997. The global burden of intestinal nematode infections — fifty years on. *Parasitology Today* **13**, 438–443.

Chan M-S, Medley GF, Jamison D, and Bundy DA. 1994. The evaluation of potential global morbidity attributable to intestinal helminth infections. *Parasitology* **109**, 373–387.

Coombs I and Crompton DWT. 1991. A Guide to Human Helminths. London and Philadelphia: Taylor & Francis.

Craig PS, Rogan MT, and Allan JC. 1996. Detection, screening and epidemiology of taeniid zoonoses: cystic echinococcosis, alveolar echinococcosis and neurocysticercosis. *Advances in Parasitology* **38**, 169–250.

Crompton DWT. 1999. How much human helminthiasis is there in the world? *Journal of Parasitology* **85**, 397–403.

de Silva NR, Brooker S, Hotez PJ et al. 2003. Soil-transmitted helminth infections: updating the global picture. *Trends in Parasitology* **19**, 547–551.

Doumenge JP, Mott KE, Cheung C et al. 1987. Atlas of the Global Distribution of Schistosomiasis. Bordeaux, France: Presses Universitaires de Bordeaux.

Dupouy-Camet J. 2000. Trichinellosis: a worldwide zoonosis. *Veterinary Parasitology* **93**, 191–200.

Hashimoto R. 2000. Global Parasite Control for the 21st Century. Report of the Hashimoto Initiative Meeting, 27 March 2000.

Henderson M. 2004. Mars probe findings get star billing for advances in science. *The Times*, Friday December 17 2004, p. 28.

Hewitt T and Smyth I. 2000. Is the world overpopulated? In: Poverty and Development for the 21st Century (eds. T Allen and A Thomas). Oxford University Press in association with the Open University. pp. 125–140.

Hopkins DR. 2002. Homing in on helminths. *American Journal of Tropical Medicine and Hygiene* **46**, 34–43.

Hotez PJ, Bundy DAP, Beegle K et al. 2006. Helminth infections: soil-transmitted helminth infections and schistosomiasis In: *Disease Control Priorities in Developing Countries*, 2nd edition (eds. DT Jamison, G Alleyne, J Breman et al.). Oxford University Press. Chapter 24.

Liu M and Boireau P. 2002. Trichinellosis in China: epidemiology and control. *Trends in Parasitology* **18**, 553–556.

Lun ZR, Gasser RB, Lai DH et al. 2005. Clonorchiasis: a key foodborne zoonosis in China. *Lancet Infectious Diseases* **5**, 31–41.

Macpherson CN. 2005. Human behaviour and the epidemiology of parasitic zoonoses. *International Journal for Parasitology* **35**, 1319–1331.

Mas-Coma S. 2004. Human fascioliasis. In: Waterborne Zoonoses (eds. JA Cotruvo, A Dufour, G Rees et al.). London: IWA Publishing on behalf of the World Health Organization. pp. 305–322.

Murray CJL. 1994. Quantifying the burden of disease: the technical basis for disability-adjusted life years. In: Global Comparative Assessments in the Health Sector. (eds. CJL Murray and AD Lopez). Geneva: World Health Organization. pp. 3–19.

Ottesen EA, Duke BOL, Karam M, and Behbehani K. 1997. Strategies and tools for the control/elimination of lymphatic filariasis. *Bulletin of the World Health Organization* **75**, 491–503.

Pawlowski ZS. 1982. Taeniasis and cysticercosis. In: *Handbook Series in Zoonoses Section C: Parasitic Zoonoses* Vol.I (eds. L Jacobs and P Arambulo). Boca Raton, Florida: CRC Press, Inc. pp. 313–348.

Pawlowski ZS. 1984. Cestodiases: taeniasis, diphyllobothriasis, hymenolepiasis and others. In: Tropical and Geographical Medicine (eds. KS Warren and AAF Mahmoud). New York and London: McGraw-Hill Book Company. pp. 471–486.

Pawlowski ZS and Davis A. 1989. Morbidity and mortality in ascariasis. In: Ascariasis and its Prevention and Control (eds. DWT Crompton, MC Nesheim and ZS Pawlowski). London, New York and Philadelphia: Taylor & Francis. pp. 71–86.

Peters W and Gilles HM. 1989. A Colour Atlas of Tropical Medicine and Parasitology, 3rd edition. London: Wolfe Medical Publications Ltd.

Pinder M. 1988. Loa loa — a neglected filaria. *Parasitology Today* **4**, 279–284.

Rim H-J. 2005. Clonorchiasis: an update. *Journal of Helminthology* **79**, 269–281.

Sato Y. 1986. Epidemiology of strongyloidiasis in Okinawa. Collected Papers on the Control of Soil-transmitted Helminthiases Volume III. Tokyo: The Asian Parasite Control Organization. pp. 20–31.

Stoll NR. 1947. This wormy world. *Journal of Parasitology* **33**, 1–18.

Sturchler D. 1988. Endemic Atlas of Tropical Infections, 2nd edition. Toronto, Lewiston NY, Bern, Stuttgart: Hans Huber Publications.

Tchuem Tchuente LA, Southgate VR, Vercruysse J et al. 1997. Epidemiological and genetic observations on human schistosomiasis in Kinshasa, Zaire. *Transactions of the Royal Society of Tropical Medicine and Hygiene* **91**, 263–269.

Toscano C, Yu Sen Hai, Nunn P, and Mott KE. 1995. Paragonimiasis and tuberculosis — diagnostic confusion: a review of the literature. *Tropical Diseases Bulletin* **92**, R1–R26.

UNICEF. 2004. The State of the World's Children 2005. New York: United Nations Children's Fund.

van der Veen R. 2004. What Went Wrong With Africa? Amsterdam: KIT Publishers (English version).

van der Werf MJ, de Vlas SJ, Brooker S et al. 2003. Quantification of clinical morbidity associated with schistosome infection in sub-Saharan Africa. *Acta Tropica* **86**, 125–139.

Warren KS. 1989. Selective primary health care and parasitic diseases. In: New Strategies in Parasitology (ed. KPWJ McAdam). Edinburgh, London, Melbourne and New York: Churchill Livingstone. pp. 217–231.

Watanapa P and Watanapa WB. 2002. Liver fluke-associated cholangiocarcinoma. *British Journal of Surgery* **89**, 962–970.

World Bank. 2004. Mini Atlas of Global Development. Washington, DC: The International Bank for Reconstruction and Development/The World Bank.

WHO. 1995a. Control of foodborne trematode infections. Technical Report Series 849. Geneva: World Health Organization.

WHO. 1995b. Onchocerciasis and its control. Technical Report Series 852. Geneva: World Health Organization.

WHO. 2002a. Prevention and control of schistosomiasis and soil-transmitted helminthiasis. Technical Report Series 912. Geneva: World Health Organization.

WHO. 2002b. Surgical approaches to the urogenital manifestations of lymphatic filariasis. Geneva: World Health Organization. WHO/CDS/CPE/CEE/2002.33.

WHO. 2004. Eradicating Guinea-worm disease. Geneva: World Health Organization. WHO/CDS/CPE/CEE/2004.47.

WHO. 2005a. WHO and the Millennium Development Goals. WHO/CDS/CPE/PVC/2005.12. www.who.int/mdg.

WHO. 2005b. Deworming for health and development. Report of the third global meeting of the partners for parasite control. Geneva: World Health Organization. WHO/CDS/CPE/PVC/2005.14.

HELMINTH SPECIES RECOVERED FROM HUMANS

In *A Guide to Human Helminths,* Coombs and Crompton (1991) published a list of helminth species reported to have been found in humans. Literature was cited by Coombs and Crompton to enable readers to decide whether the helminths were habitual parasites, accidental infections, or even spurious associations. Some of the entries are based on a single case or result from deliberate experimental infections. The list below has been revised by Dr. Isabel Coombs and 38 more species have been added. Addition to the list has depended solely on the existence of a published report. We invite readers to communicate directly with Dr. Coombs about matters concerning the list such as additions, deletions, synonyms, evidence of human infection, and so on (I.Coombs@bio.gla.ac.uk).

A comprehensive checklist covering the protozoa, helminths, and arthropods that have been retrieved from humans has been compiled by Ashford and Crewe (1998). They have indicated which species they consider to be pseudo parasites, which helminth infections may be doubtful or spurious, and where synonymy is involved. The criterion to be satisfied if a helminth is to be registered as a genuine parasite is that it should be shown to complete reproductive activity or development in a human host.

- New entries in the list below are identified by a superscript number that relates to a publication describing the worm's having been recovered from a human.
- The list does not take into account the existence of genotypes identified from the application of DNA technology.
- Worms with names shown in bold are those that are discussed in detail in the chapters of this handbook. We contend that our selection serves for discussion of principles of human host-helminth interactions and approaches to the control of helminth-induced morbidity.
- In subsequent publications we ask that this list should be cited as follows: "Coombs I. 2006. Helminth species recovered from humans. In: *Handbook of Helminthiasis for Public Health* by D.W.T. Crompton and Lorenzo Savioli. Boca Raton, Florida: CRC Press, Taylor & Francis Group. pp. 12–24."

PLATYHELMINTHES (Flatworms)
  Turbellaria
    Tricladida
      Bipaliidae
        Bipalium fuscatum
        Bipalium kewense
        Bipalium venosum

TREMATODA
  Digenea (Flukes)
    Schistosomatidae
      Schistosoma bovis
      Schistosoma guineensis[1]

Schistosoma haematobium
Schistosoma intercalatum
Schistosoma japonicum
Schistosoma malayensis
Schistosoma mansoni
Schistosoma mattheei
Schistosoma mekongi
Schistosoma rodhaini
Schistosoma spindale
Austrobilharzia terrigalensis
Bilharziella polanica
Gigantobilharzia huttoni
Gigantobilharzia sturniae
Heterobilharzia americana
Orientobilharzia turkestanica
Schistosomatium douthitti
Trichobilharzia brevis
Trichobilarzia ocellata
Trichobilharzia paoi[2]
Trichobilharzia stagnicolae
Clinostomatidae
   Clinostomum complanatum
Cathycotylidae
   Prohemistomum vivax
Brachylaimidae
   Brachylaima cribbi[3]
Diplostomatidae
   Alaria americana
   Alaria marcianae
   Diplostomum spathaceum
   Fibricola seoulensis
Paramphistomidae
   Fischoederius elongatus[4]
   Gastrodiscoides hominis
   Watsonius watsoni
Gymnophallidae
   Gymnophalloides seoi[5]
Fasciolidae
   Fasciola gigantica
   Fasciola hepatica
   Fasciola indica
   Fasciolopsis buski
Echinostomatidae
   Acanthoparyphium tyosenense[6]
   Echinochasmus fujianensis[7]
   Echinochasmus japonicus

Echinochasmus jiufoensis
Echinochasmus liliputanus[8]
Echinochasmus perfoliatus
Echinoparyphium recurvatum
Echinostoma angustitestis[9]
Echinostoma cinetorchis
Echinostoma echinatum
Echinostoma hortense
Echinostoma ilocanum
Echinostoma lindoense[10]
Echinostoma macrorchis
Echinostoma malayanum
Echinostoma revolutum
Episthmium caninum
Euparyphium melis
Himasthala muehlensi
Hypoderaeum conoideum
Psilostomatidae
    Paryphostomum sufrartyfex[11]
    Psilorchis hominis
Lecithodendriidae
    Phaneropsolus bonnei
    Phaneropsolus spinicirrus[12]
    Prosthodendrium molenkampi
Plagiorchiidae
    Plagiorchis harinasutai
    Plagiorchis javensis
    Plagiorchis muris
    Plagiorchis philippinensis
Strigeidae
    Cotylurus japonicus[15]
Cathaemasiidae
    Cathaemasia cabrerai
Philophthalmidae
    Philophthalmus lacrymosus
    Philophthalmus sp.
Dicrocoeliidae
    Dicrocoelium dendriticum
    Dicrocoelium hospes
    Eurytrema pancreaticum
Paragonimidae
    Paragonimus africanus
    Paragoniumus bankokensis
    Paragonimus caliensis
    Paragonimus heterotremus

    Paragonimus hueit'ngensis
    Paragonimus kellicotti
    Paragonimus mexicanus
    Paragonimus miyazakii
    Paragonimus ohirai
    Paragonimus philippinensis
    Paragonimus pulmonalis
    Paragonimus sadoensis
    Paragonimus skrjabini
    Paragonimus uterobilateralis
    Paragonimus westermani
    Paragonimus sp.
Achillurbainiidae
    Achillurbainia nouveli
    Achilluraninia recondita
    Poikilorchis congalensis
Nanophyetidae
    Nanophyetus salmincola salmincola
    Nanophyetus salmincola schikhobalowi
Opisthorchidae
    Clonorchis sinensis
    Metorchis albidus
    Metorchis conjunctus
    Opisthorchis felineus
    Opisthorchis guayaquilensis
    Opisthorchis noverca
    Opisthorchis viverrini
    Pseudamphistomum aethiopicum
    Pseudamphistomum truncatum
Heterophyidae
    Apophallus donicus
    Centrocestus armatus
    Centrocestus caninus[14]
    Centrocestus cuspidatus[15]
    Centrocestus formosanus
    Centrocestus kurokawai[16]
    Cryptocotyle lingua
    Diorchitrema formosanus[17]
    Diorchitrema pseudocirratus[18]
    Haplorchis microrchis[19]
    Haplorchis pleurolophocerca[20]
    Haplorchis pumilio
    Haplorchis taichui
    Haplorchis vanissimus
    Haplorchis yokogawai

      Heterophyes dispar
      Heterophyes equalis[21]
      Heterophyes heterophyes
      Heterophyes katsuradai[22]
      Heterophyes nocens
      Heterophyopsis continua
      Metagonimus minutus
      Metagonimus takahashii[23]
      Metagonimus yokogawai
      Phagicola sp.[24]
      Procerovum calderoni
      Procerovum varium (experimental)
      Pygidiopsis summa
      Stellantchasmus amplicaecalis[25]
      Stellantchasmus falcatus
      Stictodora fuscata
      Stictodora lari[26]
    Microphallidae
      Carneophallus brevicaeca
      Microphallus minus
    Isoparorchiidae
      Isoparorchis hypselobargi

CESTODA (Tapeworms)
    Trypanorhyncha
      Tentacularidae
        Nybelinia surmenicola
    Pseudophyllidea
      Diphyllobothriidae
        Diphyllobothrium cameroni
        Diphyllobothrium cordatum
        Diphyllobothrium dalliae
        Diphyllobothrium dendriticum
        Diphyllobothrium elegans
        Diphyllobothrium erinaceieuropaei
        Diphyllobothrium giljacicum
        Diphyllobothrium hians
        Diphyllobothrium klebanovski
        Diphyllobothrium lanceolatum
        Diphyllobothrium latum
        Diphyllobothrium mansoni
        Diphyllobothrium mansonoides
        Diphyllobothrium minus
        Diphyllobothrium nenzi
        Diphyllobothrium nihonkaiense

Diphyllobothrium pacificum
Diphyllobothrium scoticum
Diphyllobothrium skrjabini
Diphyllobothrium theilari
Diphyllobothrium tungussicum
Diphyllobothrium ursi
Diphyllobothrium yonagoensis
Diplogonoporus balaenopterae
Diplogonoporus brauni
Diplogonoporus fukuokaensis
Ligula intestinalis
Pyramicocephals anthocephalus
Schistocephalus solidus

Cyclophyllidea
  Anoplocephalidae
    Bertiella mucronata
    Bertiella studeri
    Inermicapsifer cubensis[27]
    Inermicapsifer madagascariensis
    Mathevotaenia symmetrica
    Monezia expansa
  Davaineidae
    Raillietina (R.) asiatica
    Raillietina (R.) celebensis
    Raillietina (R.) madagascariensis
  Dilepididae
    Dipylidium caninum
  Hymenolepididae
    Drepanidotaenia lanceolata
    Hymenolepis diminuta
    Hymenolepis microstoma[28]
    Hymenolepis nana
  Mesocestoidae
    Mesocestoides lineatus
    Mesocestoides variabilis
  Taeniidae
    Echinococcus granulosus
    Echinococcus multilocularis
    Echinococcus oligarthus
    Echinococcus vogeli
    Multiceps brauni
    Multiceps glomeratus
    Multiceps longihamatus
    Multiceps multiceps
    Multiceps serialis

Taenia crassiceps
Taenia saginata
Taenia saginata asiatica[29]
Taenia solium
Taenia taeniaeformis

NEMATODA (Roundworms)
 Dioctophymatidae
  Dioctophyma renale
  Eustrongylides ignotus[30]
  Eustrongylides sp.
 Trichinellidae
  Trichinella britovi[31]
  Trichinella murrelli[32]
  Trichinella nativa
  Trichinella nelsoni
  Trichinella papuae[33]
  Trichinella pseudospiralis[34]
  Trichinella spiralis
 Trichuridae
  Anatrichosoma cutaneum
  Aoncotheca philippinensis
  Calodium hepaticum
  Eucoleus aerophilus
  Trichuris suis
  Trichuris trichiura
  Trichuris vulpis
 Mermithidae
  Agamomermis homiis oris
  Agamomermis restiformis
  Mermis nigrescens
 Cephalobidae
  Micronema deletrix
  Turbatrix aceti
 Rhabditidae
  Cheilobus quadrilabiatus
  Diploscapter coronata
  Peloderes strongyloides
  Pelodera teres
  Rhabditis axei
  Rhabditis elongata
  Rhabditis inermis
  Rhabditis niellyi
  Rhabditis pellio

Rhabditis taurica
Rhabditis terricola
Rhabditis sp.
Strongyloididae
    Strongyloides canis (experimental)
    Strongyloides cebus (experimental)
    Strongyloides felis (experimental)
    Strongyloides fuelleborni
    Strongyloides fuelleborni kellyi[35]
    Strongyloides myopotami (experimental)
    Strongyloides papillosus
    Strongyloides planiceps (experimental)
    Strongyloides procyonis (experimental)
    Strongyloides ransomi
    Strongyloides simiae (experimental)
    Strongyloides stercoralis
    Strongyloides westeri
Ancylostomatidae
    Ancylostoma braziliense
    Ancylostoma caninum
    Ancylostoma ceylanicum
    Ancylostoma duodenale
    Ancylostoma japonica
    Ancylostoma malayanum
    Ancylostoma tubaeforme
    Bunostomum phlebotomum
    Cyclodontostomum purvisi
    Necator americanus
    Necator argentinus
    Necator suillus
    Uncinaria stenocephala
Angiostrongylidae
    Parastrongylus cantonensis
    Parastrongylus costaricensis
    Parastrongylus mackerrasae
    Parastongylus malaysiensis
Metastrongylidae
    Metastrongylus elongatus
Chabertiidae
    Oesophagostomum aculeatum
    Oesophagostomum apiostomum
    Oesophagostomum bifurcum
    Oesophagostomum stephanostomum
    Ternidens deminutus

Syngamidae
Mammomonogamus laryngeus
Mammomonogamus nasicola
Trichostrongylidae
Haemonchus contortus
Marshallagia marshalli
Mecistocirrus digitatus
Nematodirus abnormalis
Ostertagia circumcincta
Ostertagia ostertagia
Trichostrongylus affinis
Trichostrongylus axei
Trichostrongylus brevis
Trichostrongylus calcaratus
Trichostrongylus capricola
Trichostrongylus colubriformis
Trichostrongylus instabilis[36]
Trichostrongylus lerouxi
Trichostrongylus orientalis
Trichostrongylus probulurus
Trichostrongylus skrjabini
Trichostrongylus vitrinus
Oxyuridae
Enterobius gregorii
Enterobius vermicularis
Syphacia obvelata
Anisakidae
Anisakis physetoris[37]
Anisakis simplex
Contracaecum osculatum
Pseudoterranova decipiens
Ascarididae
Ascaris lumbricoides
Ascaris suum
Baylisascaris procyonis
Lagochilascaris minor
Parascaris equorum
Toxascaris leonina
Toxocara canis
Toxocara cati
Toxocara pteropodis
Toxocara vitulorum
Dracunculidae
Dracunculus medinensis
Philometridae
Philometra sp.

Gnathostomatidae
Gnathostoma doloresi
Gnathostoma hispidum
Gnathostoma nipponicum[38]
Gnathostoma spingerum
Physalopteridae
Physaloptera caucasica
Physaloptera transfuga
Rictulariidae
Rictularia sp.
Thelaziidae
Thelazia californiensis
Thelazia callipaeda
Gongylonematidae
Gongylonema pulchrum
Spirocercidae
Spirocera lupi
Acuariidae
Cheilospirura sp.
Onchocercidae
Brugia beaveri
Brugia guyanensis
Brugia malayi
Brugia pahangi
Brugia timori
Dipetalonema arbuta
Dipetalonema sprenti
Dirofilaria immitis
Dirofilaria repens
Dirofilaria spectans
Dirofilaria striata
Dirofilarai tenuis
Dirofilaria ursi
Loa loa
Mansonella ozzardi
Mansonella perstans
Mansonella semiclarum
Mansonella streptocerca
Meningonema peruzzi
Microfilaria bolivarensis
Microfilaria rodhaini
Onchocerca volvulus
Setaria equina
Wuchereria bancrofti
Wuchereria lewisi

NEMATOMORPHA (Horsehair worms)
    Gordiidae
        Gordius aquaticus
        Gordius chilensis
        Gordius gesneri
        Gordius inesae
        Gordius ogatai
        Gordius peronciti
        Gordius reddyi
        Gordius robustus
        Gordius setiger
        Gordius skorikowi
    Chordodidae
        Chordodes capensis
        Neochordodes colombianus
        Parachordodes alpestris
        Parachordodes pustulosus
        Parachordodes raphaelis
        Parachordodes tolosanus
        Parachordodes violaceus
        Parachordodes wolterstorffii
        Paragordius areolatus
        Paragordius cinctus
        Paragordius esavianus
        Paragordius tanganyikae
        Paragordius tricuspidatus
        Paragordius varius

ACANTHOCEPHALA (Thorny-headed worms)
    Moniliformidae
        Moniliformis moniliformis
    Oligacanthorhynchidae
        Macracanthorhynchus hirudinaceus
        Macracanthorhynchus ingens
    Echinorhynchidae
        Acanthocephalus rauschi
        Pseudoacanthocephalus bufonis
    Polymorphidae
        Bolbosoma sp.
        Corynosoma strumosum

REFERENCES TO ADDITIONS TO THE LIST

[1]Pages JR, Jourdane J, Southgate VR and Tcheum Tchuente LA. 2003. Reconnaissance de deux espèces jumelles au sein du taxon *Schistosoma intercalatum* Fisher, 1934, agent de la schistosomose humaine rectale en Afrique. Description de *Schistosoma guineensis* n. sp. In: *Taxonomie, ècologie et evolution des metazoires parasites.* (Livre homage a Louis Euzet). Tome 2. (eds. C Combes and J Jourdane). PUP Perpignan, France. pp. 139–146.

[2]Hu WQ, Zhou SH and Long ZP. 1994. Investigations for reasons of paddy-field dermatitis in some areas of Guanxi. *Chinese Journal of Zoology* **29**, 1–5.

[3]Butcher AR and Grove DI. 2001. Description of the life-cycle stages of *Brachylaima cribbi* n. sp. (Digenea: Brachylaimidae) derived from eggs recovered from human faeces in Australia. *Systematic Parasitology* **49**, 211–221.

[4]Huang ZQ et al. 1992. First case report on *Fischoederius elongatus*. Proceedings of the Third Symposium on Parasitology, Chinese Society of Zoology, p. 116.

[5]Lee SH, Chai JH and Hong ST. 1993. *Gymnophalloides seoi* n. sp. (Digenea: Gymnophallidae), the first report of human infection by a gymnophallid. *Journal of Parasitology* **79**, 677–680.

[6]Chai JY, Han ET, Park YK, Guk SM and Lee SH. 2001. *Acanthoparyphium tyosenense:* the discovery of human infection and identification of its source. *Journal of Parasitology* **87**, 794–800.

[10,11,15–20,25]Harinasuta T, Bunnag D and Radomyos P. 1987. Intestinal fluke infections. In: *Balliere's Clinical Tropical Medicine and Communicable Diseases. Intestinal Helminthic Infections* (ed. ZS Pawlowski). London and Philadelphia: Balliere Tindall. pp. 605–721.

[7,9]Yu S, Xu L, Jiang Z, Xu S, Han J, Zhu Y, Chang J, Lin J and Xu F. 1994. [Report on the first nationwide survey of the distribution of human parasites in China. 1. Regional distribution of parasite species]. Chung Kuo Chi Sheng Chung Hsueh Yu Chi Sheng Chung Ping Tsa Chih. **12**, 241–247.

[8]Xiao X, Wang TP, Lu DB, Gao JF, Xu LF, Wu WD, Mei JD and Xu FN. 1992. The first record of human infection of *Echinochasmus liliputanus*. *Chinese Journal of Parasitology and Parasitic Diseases* **10**, 132–135.

[12]Kwaekes S, Elkins DB, Haswell-Elkins MR and Sithithaworn P. 1991. *Phaneropsolus spinicirrus* n.sp. (Digenea: Lecithodendriidae), a human parasite in Thailand. *Journal of Parasitology* **77**, 514–516.

[13]Fried B, Graczyk TK and Tamang L. 2004. Food-borne intestinal trematodiases in humans. *Parasitology Research* **93**, 159–170.

[14]Waikagul J, Wongsaroj T, Radomyos P, Meesomboon V, Praewanich R and Jongsuksuntikul P. 1997. Human infection of *Centrocestus caninus* in Thailand. *Southeast Asian Journal of Tropical Medicine and Public Health* **28**, 831–835.

[21]Khalil LF. 1991. Zoonotic helminths of wild and domestic animals in Africa. In: *Parasitic Helminths and Zoonoses in Africa* (eds. CNL MacPherson and PS Craig). London: Unwin Hyman. pp. 260–272.

[22]Cross JH. 1974. Diagnostic methods in intestinal fluke infections: A review. SEAMEO-TROPMED/network Technical Meeting: Diagnostic Methods for Important Helminthiasis and Amoebiasis in Southeast Asia and the Far East. pp. 87–108.

[23]Hong SJ, Chung CK, Lee DH and Woo HC. 1996. One human case of natural infection by *Heterophyopsis continua* and three other species of intestinal trematodes. *Korean Journal of Parasitology* **34**, 87–89.

[24]Chieffi PP, Leite OH, Dias RM, Torres DM and Mangini AC. 1990. Human parasitism by *Phagicola* sp. (Trematoda: Heterophyidae) in Cananeia, Sao Paulo state, Brazil. *Revista Instituto de Medicina Tropical de Sao Paulo* **32**, 285–288.

[26]Chai JY, Han ET, Park YK, Guk SM, Park JH and Lee SH. 2002. *Stictodora lari* (Digenea: Heterophyidae): the discovery of the first human infections. *Journal of Parasitology* **88**, 627–629.

[27]Lloyd S. 1998. Other cestode infections: hymenolepiosis, diphyllobothriosis. Coenurosis and other larval and adult cestodes. In: *Zoonoses* (eds. SR Palmer, Lord Soulsby and DIH Simpson). Oxford University Press. pp. 635–649.

[28]Macnish MG, Ryan UM, Behnke JM and Thompson RC. 2003. Detection of the rodent tapeworm *Rodentolepis* (= *Hymenolepis*) *microstoma* in humans. A new zoonosis. *International Journal of Parasitology* **33**, 1079–1085.

[29]Fan PC, Lin CY, Chen CC and Chung WC. 1995. Morphological description of *Taenia saginata asiatica* (Cyclophyllidea: Taenidae) from man in Asia. *Journal of Helminthology* **68**, 265–266.

[30]Narr LL, O'Donnell JG, Libster B, Alessi P and Abraham D. 1996. Eustrongylidiasis — a parasitic infection acquired by eating live minnows. *Journal of the American Osteopathic Association* **96**, 400–402.

[31]Piergili-Fioretti D, Castagna B, Frongillo RF and Bruschi F. 2005. Re-evaluation of patients involved in a trichinellosis outbreak caused by *Trichinella britovi* 15 years after infection. *Veterinary Parasitology* **132**, 119–123.

[32]Dupouy-Camet J, Paugam A, De Pinieux G, Lavarde V and Vieillefond A. 2001. *Trichinella murrelli*: pathological features in human muscles at different delays after infection. *Parasite* **8**, 176–179.

[33,34]Pozio, E. 2001. Taxonomy of *Trichinella* and the epidemiology of infection in the Southeast Asia and Australian regions. *Southeast Asian Journal of Tropical Medicine and Public Health* **32**, 129–132.

[35]Ashford RW, Barnish G and Viney ME. 1992. *Strongyloides fuelleborni kellyi:* infection and disease in Papua New Guinea. *Parasitology Today* 8, 314–318.

[36-38]Ashford RW and Crewe W. *The Parasites of* Homo sapiens. *An Annotated Checklist of the Protozoa, Helminths and Arthropods for which We are Home.* Liverpool: Liverpool School of Tropical Medicine.

2 One Day in the Life of Mumbua

Those of us in the North concerned with helminthiasis attend meetings and workshops to devise strategies for prevention and control. Our education and professional affiliations have exposed us to the complexities of helminth biology, but not always to the complexities of human life styles. We may learn about human lymphocytes, antibodies, and cytokines, but what do we know about the daily lives of the people we seek to help? Obviously colleagues from the South, where helminthiasis is endemic, know what faces those at home, but we in the North need to remember that our version of helminthology and our plans for control may be insensitive, unsympathetic, unrealistic, and impractical.

In an attempt to draw attention to the real world where human-helminth interactions occur we invite our readers to spend a day with an African lady called Mumbua and the household under her care. A simple genealogy is set out in Figure 2.1 to help with names and relationships. We have invented Mumbua from our observations of African daily life, but there is much published evidence to support the story we tell. She represents millions of others who eke out a subsistence livelihood through sheer hard work. Mumbua is unaware of her ignorance and is unconsciously repressed by traditional gender attitudes. Her limited education has shielded her from understanding health messages. She believes that illness and death come from mysterious and magical influences. She accepts that her place in society is to be subordinate to men and she accepts the role that society expects from her. In Mumbua's world, a man will always make major decisions, be head of the household, and so on. In Mumbua's world, a man will not fetch the water because that is woman's work. We are not critical of her culture; that is how it is.

2.1 MUMBUA'S DAY

Mumbua lives in a village somewhere in tropical Africa. The village is not laid out in any obvious orderly system. The terracotta-colored houses are scattered around on the hillsides with the amount of land belonging to each house determining the distance between them. In the old days big trees grew on the hillsides, but demands for fuel have taken their toll and the heavy rains between March and May have been washing some of the soil away. The rains and the erosion have formed a swamp at the bottom of Mumbua's hillside. The stream that used to flow there has been become a patchwork of shallow pools, banks of silt and clumps of rushes. She wakens before sunrise to the cries of the Hadada ibis that come to feed in the swamp. Perhaps the birds have carried the snails that have brought blood fluke (*Schistosoma haematobium*) to her village, but Mumbua does not know that yet. She gently lifts Kavete,

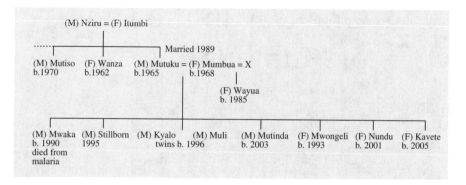

FIGURE 2.1 The occupants of Mumbua's household somewhere in sub-Saharan Africa in 2005.

born three months ago, and begins to feed her. Mumbua's long day has begun well before daylight. Perhaps bearing Kavete will be the last of her eight pregnancies.

Mumbua's place consists of three buildings clustered together in a fenced compound. The largest building, where the family sleeps, has three rooms, open windows with wooden shutters, a wooden door and a new roof made of corrugated iron sheets. The next building is the kitchen, which has a thatched roof and always reeks of wood smoke. Wood is the household energy supply. Any hot water that is needed and all cooking depends on a supply of wood, which is becoming harder to find. There is neither electricity nor gas. Nor is there any piped water. All water must be manhandled from a stream, a task that takes four hours to get just enough for two days. The family's chickens roam in and out of the kitchen. The family's seven goats have to be tethered or the crops would be destroyed. Both buildings are built of bricks made from locally available clay and both have earth floors trodden hard by years of occupancy long before Mumbua arrived in the village. There is neither toilet nor latrine. The third building, which is much smaller than the others, is thatched, has walls and a floor of logs and stands on stilts about a meter above the ground. There are no windows and the door is strong and fits tightly into its frame. This is where the crops are stored to sustain the family between harvests. Every effort is made to protect its precious contents from rats, weevils, and other destructive pests. There is not a scrap of litter to be seen in the buildings and compound; daily sweeping and the goats see to that.

Maize, beans, and cassava are the staple foods for the family. Bananas grow by the big house, over the graves of Mutuku's ancestors. Mutuku married Mumbua in 1989 (Figure 2.1). Mangoes from the family's tree are enjoyed when in season. In a good year, the family gets two maize and bean harvests and one cassava crop. Years ago Nziru, Mutuku's father and his neighbors, planted a patch of coffee bushes. Families with coffee bushes send their beans to a coffee co-operative for roasting, packing, and marketing. Coffee prices fluctuate and planning household improvements on their share of money that the co-operative gets each year is too risky, but every little helps.

Mumbua grew up in the next village. When she married Mutuku 16 years ago she was 17. In accordance with tradition she followed him to his family's home and now she runs the household. Tradition also ensured that Mutuku, the eldest son of Nziru and his wife Itumbi, became the head of the household on his marriage. Mutuku is away from home for most of the time working as a houseboy for a diplomat in one of embassies in the big city. Mutuku has earned enough money to pay for the new roof and for a radio. He is now saving to buy water tanks to collect the water running off the roof during the rainy season. Mutuku also supports Mutiso, his youngest brother, who followed him to the city in search of work. When the diplomat entertains, Mutuku is allowed to employ Mutiso to clean and work in the kitchen. In Mutuku's absence, Mumbua remains subordinate to Nziru. Nziru has been blind for the past 15 years due to an infection with *Onchocerca volvulus*. Mutuku's unmarried sister, Wanza, is also a member of the household. She suffered from polio as a young child and is now crippled and unable to do much physical work.

When Mumbua married Mutuku she already had an infant daughter called Wayua. Mutuku's parents were happy to agree that their son should marry a young woman of proven fertility. In this society, children are economic resources. Children contribute to household chores, harvest coffee beans, forage for firewood, and provide care for younger siblings and grandparents. Mumbua is respected in the community. She has done well first by bringing Wayua to her husband's family and then by giving birth to four boys, Kyalo, Muli, Mwaka, and Mutinda, and three more girls, Mwongeli, Nundu, and Kavete. Sadly, Mwaka died from malaria when he was a year old and another boy was stillborn. Despite Mumbua's best efforts Muli, Mutinda, Mwongeli, and Nundu all show signs of inadequate nutritional status. Weight for age is less than it should be for each of them. Mwongeli shows signs and symptoms of iron deficiency and Nundu is often absent from school because she suffers from a persistent type of dysentery.

All the children are infected with the common roundworm (*Ascaris lumbri-coides*). The presence of *Ascaris* in the gut leads to reduced food intake, feelings of nausea and abdominal pain, disturbed digestion and impaired absorption of nutrients. Mumbua accepts that her daily duties interfere with her ability to offer her children as much food as they need when they need it. Those who go to school this morning may get little more than a cup of tea before they leave home. Undoubtedly *Ascaris* infection retards childhood growth and development. Nundu's predicament is worse because she is concurrently infected with whipworm (*Trichuris trichiura*). During a *Trichuris* infection, the wall of the colon may be damaged to the extent that allows pathogenic bacteria to invade the tissues resulting in chronic dysentery.

A severe *Trichuris* infection also causes bleeding from the wall of the colon. If Nundu's *Trichuris* infection leads to blood loss and if she contracts a hookworm infection she will rapidly become iron deficient and may develop iron deficiency anemia. Mumbua, Wayua, and Mwongeli are already infected with hookworm (*Neca-tor americanus*). Hookworms are the vampires of the gut. They bite into the wall of the small intestine and suck blood. Like all blood feeders they discharge an anticoagulant so that bleeding continues when the worms have finished feeding. Hookworm disease is synonymous with iron deficiency. A confounding problem is that most of the dietary iron available to Mumbua and her children is supplied in vegetable foods. Iron from plants is less readily absorbed than iron from meat. The family rarely eats meat unless Mutuku and Mutiso are at home or relations are visiting. Tradition decrees that the men and boys should have their fill of any meat and then the women and girls can have the leftovers. Such circumstances help to explain why iron deficiency is such a widespread problem in low-income countries, especially for girls and women.

Mumbua is concerned because Kavete, like Mwaka who died in infancy, was very small when she was born. Babies with a low birth weight are more vulnerable to ill health than those falling in the normal or desired range of birth weights. Mumbua's iron deficiency will have contributed to the child's low birth weight and the strain that breast feeding puts on a mother's iron metabolism is likely to move her into a state of anemia. Studies in sub-Saharan Africa indicate that during the time a woman is aged from 18 to 43 years she may spend about 25% of her life being pregnant and about 60% lactating. The physiological demands on her iron status are immense even if she is spared the iron loss from regular menstruation.

Daylight is upon the village, Kavete is now cooing in a sling on Mumbua's back and work begins. We know from careful observations made in rural Uganda that women such as Mumbua will spend at least 15 hours a day looking after the members of the household, preparing and serving meals, tending crops and animals, gathering firewood, doing the housework, and fetching water. Much of this work is physically demanding involving digging, carrying, and chopping. Mumbua's spine is quite deformed because she has been carrying heavy burdens since she was a young girl. On alternate days water must be fetched and on some days surplus crops are carried to market for sale or barter. Fortunately today is not a water day. Mumbua is too weak to walk for two hours to the stream and walk for two hours back to home carrying 15 liters of water as well as Kavete who must go with her food supply. On a water day, Mumbua sets off in the darkness to give her a chance of getting home before the sun is high in the sky.

Mumbua is extremely worried about Mutinda who has a high fever and is listless. Although she cannot read or write she listens eagerly for health messages on the radio. Today a mobile clinic that gives top priority to sick children is due to visit the next village and treatment is free. Mumbua decides that Mwongeli must miss school and sets off at once carrying Mutinda to the clinic for treatment. The radio also tells her that free deworming medicine will be offered to all preschool and school-aged children who either attend or report to the neighborhood school on a named day next week. The deworming program will first be explained to parents and caregivers before any medicine is given. Apparently the medicine is particularly effective against roundworm and hookworm and also reduces the effects of whipworm. Mumbua does not know which worm is which or how the children get them. The radio message asks that children who attend for deworming should bring a urine sample with them on the day. This should be collected when the child gets up. If there is any blood in the urine the child will be given another tablet to treat blood fluke infection. There is often blood in Mwongeli's urine. Sometimes Mumbua sends Mwongeli to the swamp to find some extra water for the household and innocently causes her daughter to be exposed to the cercariae of the blood fluke.

Mumbua's biggest worry concerns Wayua. Two years ago she went to the big city to earn money for herself and the household. For a young woman with so little education prostitution soon proved to be essential for her survival. She has recently returned to the village, five months pregnant and HIV positive. Mumbua knows nothing about viruses, but she knows that an awful affliction may overtake women who are trapped by prostitution or must submit to the demands of promiscuous partners. Everyday the radio is talking about a terrible disease called AIDS that is spreading among women all over Africa. How will Wayua's baby survive if Wayua dies? Mumbua knows nothing about babies becoming HIV positive before they are born. She fears that Mutuku may force the affliction on her when he next comes home from the city. How will her family manage if she is stricken down?

Everything that could have been done today has been done. Nziru has sat in the shade of the mango tree between meals. Itumbi and Wanza have helped greatly by doing the cooking. Wayua has lain on her bed crying. Kyalo and Nundu have been to school. Mwongeli managed to carry Mutinda to the mobile clinic to get treatment for malaria. Muli has been guarding the goats. Mumbua has spent almost all day weeding between the rows of maize in the field and gathering and chopping firewood. She is exhausted and daylight is long gone. Wanza lights the oil lamps and the family has supper of boiled maize, beans, and tea. Mumbua gently releases Kavete from her sling and begins to feed her. Soon the whine of hungry mosquitoes will warn the weary family that the night is dangerous. Mumbua thinks that Mutuku should next buy bednets for his family instead of the water tanks. Every week a man on the radio says that bednets protect children from mosquitoes and the malaria they carry. Tomorrow is another water day.

2.2 COMMENT

Mumbua's story puts helminthiasis and its impact on families into its normal setting. Poverty and inequality dominate every aspect of Mumbua's life and that of her children and household (Lukmanji, 1992; Mwaka, 1993; Allen and Thomas, 2000). Poverty is their reality. Poverty is not simply shortage of money. Poverty denies people choice and the opportunity to fulfil their capabilities. The situation would be essentially the same if Mumbua and the family had tried to make their way in the city. Water might have been easier to get, especially for money, but accidents, violence, and the squalor of the unplanned slums would have ensured that urban poverty was as hard to bear as rural poverty. Illness would have been different, but just as significant in their lives. Mumbua has little income and low spending power so she has little freedom or opportunity to choose what she will do. She is held in the grip of tradition and the gender issues that it sustains. Mumbua and millions like her suffer from having unequal access to safe water, good sanitation, education, and health care. Despite the paid employment of her husband, the family has few resources and remains vulnerable to infection and other risks. The vulnerability to infection of millions like Mumbua ensures that helminth and other infections persist.

A conceptual framework designed to elucidate the synergistic interaction between soil-transmitted helminthiasis and childhood malnutrition is depicted in Figure 2.2. The framework owes much to the ideas of Cravioto and DeLicardie (1976), Sanjur (1982), Sen (1982), Carrin and Politi (1997), and Parker and Wilson (2000). Malnutrition is included because helminthiasis is manifestly a contributor to this chronic problem (WHO, 1972, 1996; Stephenson, 1987; Crompton, 1991;

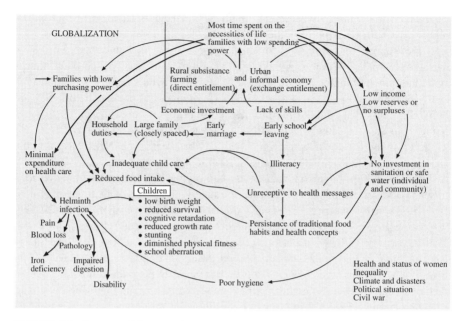

FIGURE 2.2 Conceptual framework of the relationship between helminthiasis and childhood malnutrition. Based on the work of Cravioto and DeLicardie (1976), Sanjur (1982), Sen (1982), and Carrin and Politi (1997).

Viteri, 1994; Crompton, 2000; Crompton and Nesheim, 2002). Helminthiasis is a proxy for poverty. When the grip of poverty weakens helminthiasis declines and health improves. We urge public health workers to note that interventions are needed to relieve Mumbua, and women like her, of having to cope with so much of the burden of helminthiasis (see Bandyopadhyay, 1996). Actions to control helminthiasis, while of unquestionable value in their own right, will have still more benefit if included as part of a holistic effort to reduce poverty and inequality.

REFERENCES

Allen T and Thomas A. (eds.). 2000. *Poverty and Development into the 21st Century*. Milton Keynes: The Open University in association with Oxford University Press.

Bandyopadhyay L. 1996. Lymphatic filariasis and the women of India. *Social Sciences and Medicine* **42,** 1401–1410.

Carrin G and Politi C. 1997. Health Economics. Technical briefing note. Poverty and health: an overview of the basic linkages and public policy measures. Geneva: World health Organization. WHO/TFHE/TBN/97.1.

Cravioto J and DeLicardie E. 1976. Malnutrition in early childhood. *Food and Nutrition* **2,** 2-11.

Crompton DWT. 1991. Nutritional interactions between hosts and parasites. In: *Parasite-Host Associations* (eds. CA Toft, A Aeschlimann and L Bolis). Oxford University Press. pp. 228–257.

Crompton DWT. 2000. The public health importance of hookworm disease. *Parasitology* **121**, S39–S50.

Crompton DWT and Nesheim MC. 2002. Nutritional impact of intestinal helminthiasis during the human life cycle. *Annual Review of Nutrition* **22**, 35–59.

Lukmanji Z. 1992. Women's workload and its impact on their health and nutritional status. *Progress in Food and Nutritional Science* **16,** 163–179.

Mwaka VM. 1993. Agricultural production and women's time budgets in Uganda. In: *Different Places, Different Voices* (eds. JH Momson and V Kinnard). London: Routledge. pp. 46–51.

Parker M and Wilson G. 2000. Diseases of poverty. In: *Poverty and Development into the 21ˢᵗ Century* (eds. T Allen and A Thomas). Milton Keynes: The Open University in association with Oxford University Press. pp. 75–98.

Sanjur D. 1982. *Social and Cultural Perspectives in Nutrition*. Englewood Cliffs, New Jersey: Prentice-Hall Inc.

Sen A. 1982. *Poverty and Famines. An Essay on Entitlement and Deprivation*. Oxford University Press (paperback edition).

Stephenson LS. 1987. *Impact of Helminth Infections on Human Nutrition*. London, New York and Philadelphia: Taylor & Francis.

Viteri FE. 1994. The consequences of iron deficiency and anaemia in pregnancy on maternal health, the foetus and the infant. *SCN News* **11**,14–18.

WHO 1972. Nutritional anaemias. *Technical Report Series* **503**. Geneva: World Health Organization.

WHO 1996. Report of the WHO informal consultation on hookworm infection and anaemia in girls and women. Geneva: World Health Organization. WHO/CTD/SIP/96.1).

3 Public Health Priorities: Decisions for Controlling Helminthiasis

Demands for intervention in health matters always seem to outstrip resources for the provision of health care. Much more money may be available for health care in the North than in the South, but there never seems to be enough. Costs continue to rise as the sophistication of treatment based on the latest research continues to develop. In the North, demographic changes mean that expenditure on health care is being skewed toward the needs of an aging population while in the South the trend is toward care of children and younger adults doomed by the misery of HIV/AIDS (Warren, 1996). Generally, there is a positive correlation between national income and expenditure on health care (Musgrove et al., 2002). As GNI (gross national income) increases so does the amount of public money spent on health care (Carrin and Politi, 1997). There are exceptions to this conclusion. For example, if life expectancy at birth and U5MR are taken as a reliable indicators of national health, then Sri Lanka, with its life expectancy of 73 years and a U5MR of 15, must be allocating a disproportionate amount of revenue toward health care since its GNI is > USD 1000 *per capita* (UNICEF, 2004). A life expectancy of about 62 years and a U5MR of about 90 would seem to be more typical of a country with a GNI similar to that of Sri Lanka. Provision for public health interventions has a political dimension.

Priorities for health care must be set (Walsh and Warren, 1979) and, since resources are required, meeting demands has become an economic problem requiring economic analysis (Carrin and Politi, 1997). In theory, priorities should emerge from comparing the costs of interventions for one disease with the costs of interventions for other diseases. This approach, however, is fraught with difficulty because decision makers find that they may not be comparing like with like. Murray and Lopez (1994) refer to over 80 categories of communicable and noncommunicable diseases and injuries. How can an objective assessment of the public health significance of helminthiasis and appropriate interventions be compared with an assessment of prostate cancer or road traffic accidents and interventions appropriate for them? An attempt to deal with the difficulty led to the concept of the DALY (Chapter 1 and Chapter 4) which in itself may be fraught with difficulties when the information needed for DALY calculation is inadequate or incomplete. Warren (1996) convincingly argued that public health measures based on preconceptions or intuition are invariably unsatisfactory; knowledge based on scientific investigation has to be the basis of setting priorities for public health programs. Nothing would ever happen, however, in the public health arena if decision makers had to wait for perfect or complete information.

There is or ought to be a moral dimension to making decisions about the allocation of resources for public health interventions. Is the golden rule to be applied? A mosaic by Norman Rockwell, entitled the golden rule, is displayed in the headquarters of the United Nations in New York. The rule "do unto others as you would have them do unto you" stands out from the mosaic. Do decision makers reflect on how they would feel if they were living with lymphatic filariasis or schistosomiasis? Does the ghost of Jeremy Bentham join in their deliberations imploring them to adopt his utilitarian philosophy? He would advise "do the greatest good for the greatest number" and would no doubt add "with your limited resources."

3.1 DECISIONS CONCERNING THE CONTROL OF HELMINTHIASIS

A scheme outlining components in the preparation for an analysis to help with assigning public health priority to endemic helminthiasis is shown in Figure 3.1. It is assumed that decision makers will be working with the approval of national and local authorities. It is also assumed that they will have access to knowledge of the biology of the helminth infections and the social, economic, and environmental conditions of the communities of concern. Some information may be available nationally, can be extracted from the literature (Appendix 2), or requested from the World Health Organization. Investigations may be needed to fill gaps, update, or improve the body of knowledge required for the analysis.

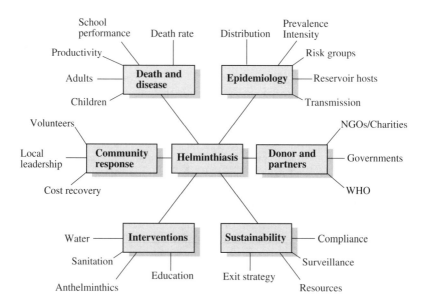

FIGURE 3.1 Components in the decision-making process for the control of helminthiasis. From the work of Walsh and Warren (1979) and Carrin and Politi (1997).

3.1.1 EPIDEMIOLOGY

This phase of the process requires information about the following variables as appropriate:

- Prevalence and intensity in relation to age, gender, occupation, or other characteristics.
- Distribution of infection, noting that patchy distributions are usually observed.
- Mode of transmission including human activities that enhance exposure.
- Population biology and ecology of intermediate hosts, reservoir hosts and vectors.

This body of data should then be interpreted and, if the decision is subsequently taken to establish a control program, should contribute to the baseline for monitoring the effects of the intervention. The data may also be useful for setting targets such as reductions in prevalence or intensity, decline in transmission rates, or numbers of intermediate hosts.

3.1.2 MORBIDITY AND MORTALITY: DISEASE AND DEATH

The most urgent objective of any intervention should be to improve health. If donors are to be recruited to finance the program (Section 3.1.5), the most persuasive advocacy will be to provide them with accurate information about the disease and death that accompanies the infection and to offer a plan indicating how these rates can be reduced. This part of the process should consider the following variables:

- Estimate the numbers of cases of overt and subtle morbidity. There will be overlap with the results of epidemiological surveys. Information may be confirmed and strengthened by the records of dispensaries, clinics, and hospitals.
- Measure the time course and progression of the severity of disease. For example in the case of onchocerciasis, note the time that the infection takes to progress from living microfilariae in the eye to binocular blindness (Chapter 9).
- Identify risk groups classified by age, gender, and occupation.
- Calculate the cost of morbidity on worker productivity (paid employment), childcare, and household maintenance (unpaid work) and subsistence livelihoods including food production. Include physical incapacity and effects on mobility.
- Determine the impact of infection on (1) attendance and performance of children at school and (2) the course and outcomes of pregnancy. Research has demonstrated that both groups are vulnerable to severe morbidity.
- If possible, estimate out-of-pocket, insurance, and public expenditure costs on health care that the infection drains from the community. For example, what proportion of meager incomes are poor people spending

on anthelminthic drugs that may be fake, counterfeit, or dangerous? How many surgical interventions are taking place annually in the hospital service to treat intestinal obstructions due to ascariasis and what does this cost?
- Obtain data about death rates and the section of the population most at risk.

The information obtained from the epidemiology and from the morbidity and mortality phases of the process establish public health significance and provide the basis for the calculation of DALYs (Table 1.2).

3.1.3 Community Response

Decision makers will be encouraged if there is a strong felt need in the community in favor of intervention. The following community responses should be investigated:

- Negotiate the agreement of community leaders such as village elders, imams, and paramount chiefs.
- Experience indicates that local involvement creates a sense of ownership and empowerment. The more bottom-up an intervention is seen to be the more likely it is to succeed and be sustained (Section 3.1.6).
- Review recruitment of volunteers (traditional healers and midwives) to promote and assist with the program. Caring people with the minimum of training can undertake such tasks as washing the legs of people incapacitated by elephantiasis (Chapter 8).
- Explore opportunities for cost recovery or community financing (see Macaluso, 2000). People with a financial commitment, however small, are more likely to support their investment and comply with the program.

Yu et al. (2001) have investigated the problem of revenue generation in support of public health programs by reference to the continuing control of schistosomiasis japonicum in China. Individuals might be willing to pay for their own treatment and that of their families, but there seemed to be less incentive to chip in with money and support public health interventions that would be for the benefit of neighbors. Yu et al. concluded that the cost to a household of a schistosomiasis control program would be equivalent to 0.94% of the household's annual income. Some 30% of heads of household interviewed by Yu et al. were willing to pay toward the cost of control while 70% were willing to volunteer to work for a program.

3.1.4 Interventions

There are four categories of intervention for controlling, eliminating, and eradicating helminthiasis (Figure 3.1). Each requires intersectoral collaboration, not just for eventual implementation but also to ensure that the most realistic costs are obtained:

- The provision of safe water and satisfactory sanitation (Chapter 16) is the policy to adopt if permanent relief from helminthiasis is to be achieved. Costs of installation and maintenance should not be ignored; projections could show that seemingly massive expenditure at the time will prove to be sound investment for subsequent productivity and future economic growth.

- Water developments such as the construction of dams for irrigation or hydroelectric power should include the costs of measures to protect people from exposure to helminth infections.
- The provision, delivery, coverage, and costs of increasing health education should be explored. Education about HIV/AIDS could incorporate education about helminthiasis and other health issues.
- Anthelminthic treatment in the community is proving to be a valuable tool for controlling morbidity. Decision makers need to know the costs of safety-tested quality drugs, drug supply, frequency, and duration of treatment, delivery, and coverage in the community, pharmacovigilance, technical staff, and support services.

3.1.5 DONORS AND PARTNERS

In the South, especially in sub-Saharan Africa, public health interventions to control helminthiasis have invariably needed support from agencies in the North in the form of money, technical help, equipment, and supplies of medicine. If support for helminth control is provided from the North, the governments, agencies, NGOs, charities, universities, philanthropic foundations, and pharmaceutical companies there will be transferring funds, received by taxation, voluntary giving, or carefully regulated commercial activity, to the South. Inevitably some system of transparent accountability will need to be arranged and a degree of fair conditionality negotiated.

Experience gained from the control of dracunculiasis, lymphatic filariasis, and onchocerciasis has demonstrated that formal partnerships (Figure 3.1) between governments in the South and donors in the North offer a constructive solution to both imagined and actual ways in which funds are used. Recently the Partnership for Parasite Control has been established to extend the control of schistosomiasis and soil-transmitted helminthiasis (WHO, 2005). Such a partnership functions because it includes mechanisms for quality drug procurement, schemes for training health workers, funding for essential operational research, and is co-ordinated by a secretariat of impartial staff from the World Health Organization. The secretariat is able to report objectively to all partners about the progress of programs.

3.1.6 SUSTAINABILITY

Support from the partners in the North cannot be expected to continue beyond the point at which a control program's targets have been met. Decision makers should be able to offer the partners a costed exit strategy (Figure 3.1) that will explain how progress is to be sustained. What is meant by sustainability in terms of public health? Applying the opinion of LaFond (1995), a sustainable helminth control program would be one that had the capacity to function effectively over time with a minimum of external input. Sustainability in this context will still require resources, robust procedures for surveillance, and determination to retain community compliance.

Encouraging bottom-up or people-centered responsibilities for those living in remote communities in large countries could be considered as a means of supporting compliance. Elements of sustainability should be costed. They may be beyond the means of the partner in the South if the partners in the North have withdrawn.

3.2   ECONOMIC EVALUATIONS: VALUE FOR MONEY

Once adequate information has been obtained (Figure 3.1), three forms of analysis may be employed by economists involved in decision making about public health intervention: cost minimization analysis (CMA); cost benefit analysis (CBA); cost effectiveness analysis (CEA) (WHO, 2002).

CMA seeks to investigate costs of alternative interventions with demonstrated equivalence in effectiveness or health benefits and then identify the most efficient of them. CBA focuses on both costs and benefits of an intervention where both costs and benefits are expressed in monetary units. For example, what would be the monetary value to an intervention of partial cost recovery from the community? How much in net monetary terms would be gained from the intervention in some area of productivity such as tea plucking, rubber tapping, or road building? The CEA appears to be the most useful economic tool for decision makers to rely on (WHO, 2002). Costs are measured in monetary units. Effectiveness is expressed as positive changes in the values of variables such as blood hemoglobin concentration, birth weight or growth percentiles, and DALYs. The purpose of a CEA is to find the intervention that will give the best value for money or the maximum improvement to health from the minimum of expenditure.

3.3   INFRASTRUCTURE AND INTEGRATION

The costs of controlling helminthiasis can be reduced wherever proposed intervention can be integrated with existing health care programs. Furthermore, it should be noted that minimal infrastructure is needed for helminthiasis control or may exist already. Intervention for soil-transmitted helminthiasis based on anthelminthic treatment of school children needs no new infrastructure. Regular treatment can be delivered at school combined with health awareness programs or other interventions. Regular treatment for all kinds of helminthiasis may be offered through primary health care clinics, dispensaries, faith-based institutions, and so on. In most cases the infrastructure is already there. On a per patient basis some interventions are of extremely low cost compared with the predicted improvements in health for individuals (Chapter 5 to Chapter 13) and the expected economic return to the community (Molyneux, 2004; Table 3.1).

3.4   SUMMARY

The case for sending helminth control up the list of public health priorities has been championed by Evans and Jamison (1994) who examined economic aspects of the

TABLE 3.1
Economic Costs and Returns of Interventions for Various Diseases in Low-Income Countries. From a table on page 23 (WHO, 2005) with permission from the World Health Organization (www.who.int), based on information compiled by Molyneux (2004) with permission from Elsevier.

	Economic Rates of Return	Unit Costs per Treatment	No. Targeted, Interventions and Coverage
Onchocerciasis (Donation of Mectizan®)	17–20%	USD 0.10– USD 0.20	20 million 65–80%
Lymphatic filariasis (Combination of Mectizan and Albendazole)	> 20%	USD 0.03– USD 1.50	120 million (2003) 65–90%
Guinea-worm disease	29%	N/A	All infected villages; filters, health education, temephos case containment, safe water
Soil-transmitted helminthiasis		USD 0.02	36 million (2003) 75% coverage in Cambodia achieved
Chagas disease	30%	N/A	N/A
Trachoma (Donation of Zithromax)	N/A	USD 0.30	2.8 million (2002)
Schistosomiasis		USD 0.20– USD 0.30	10 million (2004)
Vitamin A supplements		USD 0.02	167.8 million children 70% coverage in 44 countries
Total estimated range of chemotherapy package per annual treatment(s) for all above diseases		c. USD 0.90– USD 2.40	

Note: USD, U.S. Dollars.

control of parasitic diseases. In the following quotation from their review, we have changed "parasitic disease" to "helminthiasis," a defensible modification. So Evans and Jamison might well have written with the extent of helminthiasis among poor people in mind "By restricting economic potential, helminthiasis exacerbates existing inequalities in society to a much greater extent than noncommunicable diseases. This is an excellent reason for intervention. However ranking diseases strictly according to the total economic burden they place on society, as has been done in the U.S., would not be of great value in setting priorities for helminthiasis control. It is not the size of the problem which alone should determine the priority of intervention from an economic viewpoint, but the extent to which the problem could be reduced for the available resources."

REFERENCES

Carrin G and Politi C. 1997. Health Economics. Technical Briefing Note. Poverty and health: an overview of the basic linkages and public policy measures. Geneva: World Health Organization. WHO/TFHE/TBN/97.1.

Evans DB and Jamison DT. 1994. Economics and the argument for parasitic disease control. *Science* **264,** 1866–1867.

LaFond A. 1995. *The Sustainability Problem. Sustaining Primary Health Care.* London: Earthwatch Publications.

Macaluso A. 2000. Financing health care systems and reforms in developing countries. Community financing schemes. *Health and Development.* Four-monthly periodical of CUAMM Organization. 2/2000. pp. 18–21.

Molyneux DH. 2004. "Neglected" diseases but unrecognized successes — challenges and opportunities for infectious disease control. *Lancet* **364,** 380–383.

Murray CJL and Lopez AD. 1994. *Global Comparative Assessments in the Health Sector.* Geneva: World Health Organization.

Musgrove P, Zeramdini R and Carrin G. 2002. Basic patterns in national health expenditure. *Bulletin of the World Health Organization* **80** [on line], 134–142.

UNICEF. 2004. *The State of the World's Children 2005.* New York: The United Nations Children's Fund.

Walsh JA and Warren KS. 1979. Selective primary health care. An interim strategy for disease control in developing countries. *New England Journal of Medicine* **301,** 967–974.

Warren KS. 1996. Rationalizing health care in a changing world: the need to know. *Health Transition Review Supplement* **6,** 393–403.

WHO. 2002. WHO-CHOICE (World Health Organization Choosing Interventions that are Cost Effective). www.who.int/whosis/cea.

WHO. 2005. Deworming for health and development. Report of the third global meeting of the partners for parasite control. Geneva: World Health Organization. WHO/CDS/CPE/PVC/2005.14.

Yu D, Manderson L, Yuan L et al. 2001. Is equity being sacrificed? Willingness and ability to pay for schistosomiasis control in China. *Health Policy and Planning* **16,** 292–301.

Part II

Helminthology

Helminthological science makes us acquainted with the forms, habits, structure, development, distribution, and classification of a multitude of evertebrated organisms which take up their abode, at one or more periods of their lifetime, in the bodies of man and animals.

T. Spencer Cobbold, 1864

In an address to the Cambridge Meeting of the British Association for the Advancement of Science in 1862, Cobbold said "As man is infested by a great variety of internal parasites, and some of them prove exceedingly troublesome, it is evident that a large amount of practical good would ensue if we were more perfectly informed respecting the origin and economy of these creatures; for not only are our personal interests directly affected by their intrusion, but also we suffer indirectly from the injury and destruction they occasion amongst our domesticated animals."

Anderson RM and May RM. 1991. *Infectious Diseases of Humans. Dynamics and Control.* Oxford University Press.

Ash LR and Orihel TC. 1990. *Atlas of Human Parasitology,* Third Edition. Chicago: ASCP, American Society of Clinical Pathologists.

Muller R. 2002. *Worms and Human Disease,* Second Edition. Wallingford: Oxon: CABI Publishing. (The book includes contributions and a chapter on immunology of helminths from Derek Wakelin.)

Roberts LS and Janovy J, Jr. 2000. GD Schmidt and Larry S Roberts' *Foundations of Parasitology,* Sixth edition. Boston, New York, London: McGraw-Hill Higher Education.

Ukoli FMA. 1984. *Introduction to Parasitology in Tropical Africa.* Chichester, New York, Brisbane, Toronto, Singapore: John Wiley & Sons Limited.

4 Common Themes and Concepts in Helminthiasis

Some information about helminths, the diseases they induce, and approaches to their control have general application to each of the nine groups discussed in this handbook (Chapter 5 to Chapter 13). Each species of helminth considered here to be an agent of disease lives for part of its development or reproductive life inside the human body. The dynamic relationship between helminths and their human hosts exemplifies parasitism. Each is a partner of a pair of interacting species that, through evolutionary progression, have acquired integrated genomes so that the parasitic species is dependent on at least one gene or gene product of the host species (MacInnes, 1974). And at some stage in its life history, a parasite is more than likely to obtain nutrients and energy from the food, metabolites, or tissues of its host.

4.1 HISTORY AND DISCOVERY

A meticulous account of how our ancestors regarded the helminths with which they came into contact and how earlier generations of helminthologists set about studying helminth origins, transmission, and life histories has been compiled by Grove (1990). Contributors to a volume edited by Cox (1996) have given accounts of discoveries leading to the recognition of helminths as agents of disease while Kean et al. (1978) published a volume covering classic investigations into human helminthiasis. Contemporary health professionals are now able to embark on helminth control thanks to the body of work based on the efforts of the pioneers reviewed in these publications.

4.2 NOMENCLATURE

Helminthiasis is the term used to describe the disease that accompanies infection of humans with a species of endoparasitic worm. The suffix "-iasis," when attached to the name of a worm or group of worms such as ascariasis or cestodiasis, indicates disease due to *Ascaris lumbricoides* or to a group of cestodes. This system of disease nomenclature is convenient for communication and for the storage and retrieval of information; it does not hint at any natural classification (WHO, 1987). The system is not universally accepted. Infection with *Trichinella spiralis* is generally referred to as trichinosis or trichinellosis and veterinarians argue convincingly that the suffix "-osis" should replace "-iasis" and be adopted to indicate that the infection has given rise to disease (Kassai et al., 1988).

4.3 ZOONOTIC INFECTIONS

The diseases initiated by many of the helminths in the list (Chapter 1) are referred to as zoonoses. Zoonoses result from the natural transmission of helminths from vertebrates to humans (Hubbert et al., 1975; Soulsby, 1991; Holland, 1997). *Trichinella spiralis* is an example of a zoonotic helminth while *Ascaris lumbricoides* is not. Most of the burden of zoonotic helminthiasis arises through humans becoming intimately involved with helminths that depend for their existence on infection of nonhuman mammals. Association with companion, domesticated and wild vertebrate animals, methods of food production, and cultural influences on food habits form the basis of zoonotic helminthiasis (Macpherson, 2005).

4.4 DISEASE AND MORBIDITY

Disease and morbidity are terms that are used interchangeably. Disease is interference with normal bodily function while morbidity means that disease is present and detectable. Throughout this book, we refer to the use of anthelminthic treatment to reduce and control morbidity. The point is that by observing changes in morbidity, public health workers can record quantitative reductions in disease.

Disease may undoubtedly accompany a helminth infection, but many other factors including a host's genetic make-up and state of general health, which may be affected by the presence of other infections, can be involved. Evidence that helminth infections actually cause disease is difficult to obtain. Evidence that they are involved is rock solid. When we refer to a particular helminthiasis, the reader should remember that Koch's postulates will not have been satisfied (see Porter, 1997). For example, evidence supports the view that *Ascaris lumbricoides* adversely affects the nutritional status of children. Under similar circumstances, children who are worm free do better than those who are infected. Among those with worms, the heavily infected children fare worse than the lightly infected children. When infected children are treated to expel their worms, their growth rates improve. Children deserve to be freed from the burden of ascariasis. But perhaps *Ascaris* is the carrier of a pathogenic virus that is the real cause of the disease. Proof of causation is not straightforward in the field of helminthology and chronic helminthiasis.

4.5 IDENTIFICATION AND DIAGNOSIS

It seems unlikely that new helminth infections of lasting or large-scale public health significance will be discovered in human hosts. Health workers in areas where helminth infections are endemic (Table 1.1) know what to expect and how to confirm identification. Occasionally, local epidemics of zoonotic helminth infections occur as happened with *Capillaria philippinensis* in Northern Luzon in the Philippines (Whalen et al., 1969; Cross and Basaca-Seilla, 1991). If new or unexpected cases of infection emerge, perhaps as a result of the greater ease of the movement of people and their products around the world, consultation of the helminth list (Chapter 1) may offer a short cut to identification and reference to the treatise prepared by Yamaguti (1958a,b, 1959, 1961a,b) should help to confirm generic identification.

Until recently, examination of the anatomy of adult helminths has been the primary means of identification. The collection, fixation, and staining of specimens for identification by light microscopy is well described by Pritchard and Kruse (1982). Misidentification based on a helminth's anatomical features remains a hazard. For example, identification by light microscopy depends on comparing the unknown specimen with a description published by an experienced taxonomist. If the observer has used different fixatives or the same fixative for a different time than that used in preparing the reference specimen, different amounts of tissue shrinkage may occur thereby misleading the observer. Investigation of helminth anatomy by scanning electron microscopy can produce excellent images to aid in identification (Gibbons, 1986).

Parasitological diagnosis of infection relies on finding stages of a helminth's development, such as eggs and larvae, in the excretory products and tissues of the host. In most cases, parasitological diagnosis depends on the presence of a patent infection. The collection of specimens of stool, urine, and sputum is reasonably straightforward and is less intrusive than taking blood, skin, or other tissue samples. Knowledge of host and helminth activity enables the observer to collect samples at the most opportune time for obtaining evidence of the presence of helminths. For example, most eggs of *Schistosoma haematobium* are discharged in urine between 1100 and 1400 (Wilkins, 1987).

The Meade Readiview handheld microscope, manufactured by the Meade Instruments Corporation (Irvine, CA, U.S.) could prove to be an asset for parasitological surveys and diagnostic work. This light-weight instrument is portable, robust, and costs about a tenth of the price of a conventional compound microscope. Stothard et al., (2005) have undertaken a thorough evaluation of the instrument in an epidemiological survey of the distribution of *Schistosoma mansoni* infections in school children in Uganda. They concluded that the Meade Readiview handheld microscope could play an important role in the collection of prevalence data and strengthen the diagnostic capacity of health centers where compound microscopes are few or absent.

Failure to find helminth stages in a sample does not mean that the host is free from infection. An infection may not be patent, the release of stages sought in a parasitological diagnosis may be erratic, and the method may be insensitive when the worm burden is low or the worms are lodged in an ectopic site. Technology that exploits host immune responses offers another valuable and complementary means of diagnosing helminth infections (Table 4.1). A comprehensive review of the variety of methods appropriate for a range of helminth infections has been compiled by Ikeda and Akao (2001). The review cites 187 publications in which immunodiagnostic techniques are described. The application of immunodiagnostic tests requires resources including experienced technical staff, equipment, and laboratory facilities.

If technical resources permit, the use of DNA probes offers an extremely sensitive means of unequivocally establishing a helminth's identity and of distinguishing

TABLE 4.1

Combination of Conditions Selected for Immunodiagnosis of Helminth Infections. From Table 1 (Ikeda and Akao, 2001) with permission from Science Publishers Inc.

Detection of Specific Antibodies

Assay: a variety of assays

Parasite antigen

   Worm stage: adult, larva, egg

   Source: somatic extract, ES products, surface components, sections

   Preparation: crude antigen, fractionated antigen, purified antigen, recombinant antigen

Isotype: IgA, IgG (IgG1, IgG2, Ig3, IgG4), IgM, IgE

Specimen: Serum, pleural effusion, cerebrospinal fluid, vitreous fluid, urine

Detection of Parasite-Derived Antigens

Assay: DD, IHA, sandwich ELISA, competitive ELISA, dipstick assay

Parasite-specific antibody: polyclonal antibodies, monoclonal antibody, recombinant antibody

Specimen: serum, urine, cerebrospinal fluid, stool

DD, double diffusion test	ELISA, enzyme linked immunoabsorbent assay
ES, excretory antigens	IHA, indirect hemagglutination test

between subspecies (Hyde, 1990). Such sophistication may seem beyond the reach of public health in countries with minimal resources for health care, but DNA technology refines the work of epidemiologists and opens up new areas of research, the results of which will strengthen efforts to overcome helminthiasis.

In areas where helminthiasis is endemic, diagnosis may be managed using clinical signs and symptoms including information from the patient. Clinical diagnosis may not always be reliable unless access to radiography and ultrasound scanning is available. Infection with the lung fluke, *Paragonimus westermani,* is invariably accompanied by hemoptysis (rust-colored sputum), which is also a cardinal sign of tuberculosis (WHO, 2004a). Paragonimiasis is better confirmed by the detection of eggs in the host's sputum, but egg discharge is erratic and consecutive samples may be needed before eggs are found. Depending on the infection of concern, clinical diagnosis may have varying application in a public health setting where the extent of infections in large communities is best assessed by epidemiological sampling methods.

Planning, implementing, and sustaining programs for the control of helminthiasis require accurate information from trustworthy diagnostic techniques. There may be many techniques to choose from. Harnett (1991) lists three parasitological tests, eleven immunological tests, and two skin tests for the detection of infection with *Onchocerca volvulus.* What matters is that program managers choose methods that relate to sustainable resources including availability of trained staff, laboratory facilities, and funds for equipment and supplies (see Montresor et al., 2002).

4.6 IMMUNITY

The human immune system is endowed with two main components. One is the exquisitely sensitive mechanism that can detect and respond to molecules, particles, and organisms that are recognized as "not self." The other is a set of humoral and cellular effector mechanisms that are activated to neutralize or repel invaders. The system is under genetic control so individual variation occurs in the timing, strength, and efficiency of the responses. Innate and adaptive effector responses characterize the immune system. Innate responses occur rapidly while adaptive responses take longer to develop and are generated after repeated exposure to the foreign materials. An adaptive response depends on certain cells in the immune system acquiring a chemical memory of the invading or foreign molecules. The memory can be highly specific so that the immune system may respond precisely to a specific virus or bacterium. The discovery and characterization of these responses has led to the production of modern vaccines against such deadly pathogens as smallpox virus and yellow fever virus. Vaccination has become a most important intervention in the field of public health.

That summary of the workings of the human immune system becomes rather too simple once helminths challenge the surveillance component. Helminths release numerous molecules during the time they spend in the host and the array of these molecules tends to change as development proceeds. For example a schistosome cercaria, on passing into a human body, changes over several weeks to become a mature blood fluke (Chapter 10). As it changes and develops, so does its surface and the molecules it contains. The host's immune surveillance seems to be chasing an ever-changing shadow that has moved on by the time the effector response arrives. Many helminths have become adapted to avoid or counter host immune responses. The situation is reminiscent of a dynamic evolutionary arms race (Dawkins and Krebs, 1979). Furthermore, many of the molecules released by helminths do not elicit protective responses; rather they stimulate allergic responses causing the host to damage itself. If successful vaccination against helminth infection is to be accomplished, immunologists will have to be sure that the vaccines are totally protective and free from the possibility of causing an allergic response. That assurance will not come without sufficient support for basic research into human-helminth immune relationships. The need to find affordable vaccines is becoming urgent given the limited number of anthelminthic drugs that are available for public health use and the threat of drug resistance.

A concise explanation of how the human immune response may interact with an endoparasitic helminth infection has been published by Wakelin (2002). One feature of the host response during most helminth infections is a marked eosinophilia to the extent that eosinophils become the most numerous leukocytes in circulation. The role of eosinophils during helminth infections is still not completely understood (Ovington and Behm, 1997). Nor is the variety of protective and pathological responses that the host makes to helminth antigens (Moqbel, 1992). Immunology is a rapidly growing subject, particularly since it has been able to exploit the techniques of molecular genetics. Nevertheless, Moqbel's volume continues to illustrate the intricacy of host-helminth interactions that must be unravelled if vaccines are to become available.

4.7 EPIDEMIOLOGY AND POPULATION BIOLOGY

4.7.1 SURVEYS AND SAMPLING

The control of helminthiasis depends on as sound an epidemiological foundation as possible. Epidemiologists determine who gets which worms where and when and then identify those who are or will be most at risk of disease. The process depends on obtaining reliable data from community surveys. Ideally, surveys are best carried out through random sampling, but this is difficult where populations live in countries without accurate census data, address lists, and electoral rolls. Most information comes from people who respond to cross-sectional sampling of communities or from health center records. The results are then plotted to identify correlations between helminth prevalence or intensity of infection (see below) and host variables such as age, gender, socio-economic status, ethnicity, location, and climate. As an example, data relating the prevalence of *Ascaris lumbricoides* to host age, gender, environment, and socio-economic status have been published by Forrester et al. (1988) from Mexico, McCullough (1974) from Tanzania, Ratard et al. (1991) from Cameroon, and Holland et al. (1988) from Panama, respectively. Few longitudinal surveys have been carried out in which the infection status of the same subjects is followed through time. Survey data, gained at different times, under different conditions, by different observers, are often pooled to provide estimates of the distribution and abundance of helminth infections at local, national, and regional levels as displayed cartographically for schistosomiasis (Doumenge et al., 1987).

Geographical information systems (GIS) and remote sensing have the potential to make a significant contribution to the management of helminthiasis control programs (Brooker and Michael, 2000). GIS provides a means of collecting, storing, updating, and analyzing large amounts of information from epidemiological surveys and surveillance. The system offers a powerful tool for visualizing data so that program managers can decide where best to deploy resources in the interests of public health. How GIS may be used in the control of helminthiasis is illustrated with an example from dracunculiasis control in Mali (WHO, 1999a).

4.7.2 PREVALENCE AND INTENSITY

The prevalence of a helminth infection is the number or proportion (usually expressed as a percentage) of individuals in a population estimated to be infected at a given time. Prevalence is usually determined from data obtained by sampling a population of interest. The accuracy of the prevalence value depends on the quality of the sampling technique. The intensity of a helminth infection is the number of worms per infected person. Intensity may be measured directly by counting the worms expelled in stools for a set period after anthelminthic treatment or indirectly by counting helminth eggs and expressing the results as eggs per gram of stool or per ml of urine. Intensity gives a measure of an infected person's worm burden. Under normal circumstances, an infected person will be expected to gain new worms while losing old ones and we assume that the worm burden remains fairly constant in relation to the person's age, immune status, nutrition, and so on. As with prevalence, intensity gives information about the worm burden at the time the measurement was made. In some infections the number of adult or juvenile worms in a person may increase during the course of the infection. Knowledge of the growth of such worm populations is also important and must be obtained.

Prevalence and intensity values, even if difficult to measure accurately, are important because they help to (1) determine the public health importance of an infection, (2) identify risk groups, (3) provide quantitative targets for control interventions, and (4) enable resources such as numbers of anthelminthic tablets to be managed. Prevalence and intensity data give rough guidelines for assessing the likely health impact and level of endemicity of helminth infections (Table 4.2). When an infection is described as being endemic we mean that it is habitually present in a location due to circumstances that occur there.

Accepting the significance of the numbers of helminth eggs in stools does not always take account of the variability of daily fecal production (Hall, 1982). Measurements of average daily fecal production by individuals in a range of communities are shown in Table 16.4. Knowledge of the daily fecal production can change how the egg counts are interpreted regarding the intensity of infection. For example, 5000 epg is assumed to indicate that an individual has a heavy hookworm infection (Table 4.2). In theory if that individual were producing 200 g of feces daily then the hookworm burden would be releasing 1 million eggs daily. Another individual, however, with a count of 2000 epg would be assumed to have a moderate hookworm infection (Table 4.2), but if the daily fecal production was 500 g (not unreasonable, Table 16.4), then 1 million eggs would be being released daily, presumably by an equivalent worm burden. In a discussion about the response of helminths suspected to have become resistant to the therapeutic effects of anthelminthic drugs (chapter 14), attention was drawn to the need to make careful egg counts when dealing with intestinal worms. If the egg counts do not drop to the degree expected, given the known efficacy of the drug, the explanation may lie in a change in daily fecal production. Has the effect of anthelminthic drugs on human faecal production been investigated?

4.7.3 REPRODUCTIVE RATE

The theoretical basis of helminth population biology has established that in most cases a helminth's population in a community of hosts will decline and be destroyed if the helminth's basic reproductive rate (R_0) falls below 1 and remains there (Smith and Scott, 1994). Conversely, the helminth population will grow if R_0 is greater than 1. R_0 is defined as the average number of offspring (or female offspring if the helminth is of a dioecious species) produced by a mature helminth during its lifetime that then survive to maturity themselves in the absence of density-dependent constraints (Anderson and May, 1992). *Strongyloides stercoralis* is an exception to this

TABLE 4.2
Estimates of Degrees of Intensity (I) and Endemicity (E) of Various Helminth Infections

Helminth Infection (Chapter)	Intensity/Endemicity Status	References
Filarial nematodes		
Onchocerca volvulus (9)	E: MF (prevalence), < 10% (sporadic), 10–39% (hypo-), 40-69% (meso-), 70% (hyper-).	Bradley and Unnasch (1996)
Wuchereria bancrofti (8)	Antigenaemia ≥ 1%, universal (MDA) anthelminthic treatment should be applied.	Molyneux and Rio (personal communications)
Foodborne trematodes (7)		
Clonorchis sinensis	I: 1–999 (light), 1000–9999 (moderate), 10,000–29,999 (heavy), ≥ 30,000 (very heavy) epg, Kato Katz.	Rim (1986)
Fasciola hepatica	E: prevalence < 1% (hypoendemic), 1–10% (meso-), > 10% (hyper-). I: < 50, 51–300 and > 300 epg, respectively.	Mas-Coma (2004)
Heterophyid (intestinal) flukes	I: 1–100 (light), 101–1000 (moderate), > 1,000 (heavy) epg, Kato Katz.	Belizario et al. (2001)
Opisthorchis viverrini	I: 0 to > 6,000 epg gives increasing risk of liver cancer.	Haswell-Elkins et al. (1994)
Paragonimus uterobilateralis	I: 1–50 (low), 51–100 (moderate), > 100 (high) eggs/ml sputum.	Udonsi (1987)
Schistosomes (10)		
Schistosoma haematobium	I: < 50 eggs (light), ≥ 50 eggs (heavy), in 10 ml urine visible haematuria (heavy).	
S. mansoni	I: 1–99 (light), 100–399 (moderate), ≥ 400 (heavy) epg, Kato Katz.	WHO (2002)
Soil-transmitted nematodes (11)		
Ancylostoma duodenale *Necator americanus*	I: 1–1999 (light), 2000–3999 (moderate), ≥ 4000 (heavy) epg, Kato Katz.	
Ascaris lumbricoides	I: 1–4999 (light), 5000–49,999 (moderate), ≥ 50,000 (heavy) epg, Kato Katz.	WHO (2002)
Trichuris trichiura	I: 1–999 (light), 1,000–9,999 (moderate), ≥ 10,000 (heavy) epg, Kato Katz.	

TABLE 4.2 (continued)
Estimates of Degrees of Intensity (I) and Endemicity (E) of Various Helminth Infections

Helminth Infection (Chapter)	Intensity/Endemicity Status	References
Strongyloidiasis (12) *Strongyloides stercoralis*	Increasing stages of intensity: (1) < 100, (2) 100–349, (3) 350–499, (4) 500–1,000, (5) > 1,000 larvae/ml stool.	Grove (1989)

Note: These guidelines do not have any official status and should be expected to vary according to social and environmental conditions in the places where the infections prevail.

In the case of *Wuchereria bancrofti* (and *Brugia* spp.), community assessment is used to decide about the implementation of drug treatment.

concept of R_0 because of its property of autoinfection resulting from replication inside its human host (Muller, 2002). *Capillaria philippinensis* (Nematoda) and *Hymenolepis nana* (Cestoda) are other common helminths that replicate in human hosts (Grove, 1986).

In life, the real reproductive rate or effective reproduction ratio (R_e) is influenced by density-dependent constraints and by circumstances in the host's environment (Smith and Scott, 1994). Density-dependent constraints, directly influenced by the number of worms present in a host, result from competition between worms for resources such as nutrients and space, mating opportunities, and the level of stimulation of the host's immune system. Reproductive rates should be expected to differ from place to place because of differences in circumstances in the host's environment. The state of sanitation will directly affect exposure to infection and the availability of good food will affect the quality of the immune response. For example, reproductive rates of 1 to 2 and 4 to 5 have been determined for *Ascaris lumbricoides* in Bangladesh and Iran respectively (Anderson and May, 1992).

4.8 DEWORMING AND CONTROL STRATEGIES*

Deworming is treatment with anthelminthic drugs to kill or expel helminth infections. We encourage use of the drugs in WHO's list of essential drugs (WHO, 2000), which

---

* The use of WHO-recommended anthelminthic drugs brings many health benefits, some temporary side effects and extremely rare adverse events. Before rounds of treatment begin, public health professionals responsible for intervention with anthelminthic drugs in community programs must ensure that provision has been made to deal with adverse events. Advice about all aspects of anthelminthic use in the community is available from the Department of Control of Neglected Tropical Diseases, World Health Organization, 1211 Geneva 27, Switzerland (www.who.int/neglected_diseases/en/).

should be administered according to guidelines in the WHO Model Formulary (WHO, 2004b). We recognize the value of traditional deworming medicine when prescribed by those experienced in its use (see WHO, 1976; Ou Ming, 1989; WHO, 1999b). Deworming is now the major intervention for the control of helminthiasis in public health programs. Patents owned by the research-based pharmaceutical industry have expired enabling good-quality, cheap, generic anthelminthic drugs to become widely available. The control of morbidity due to soil-transmitted helminth infections can now be achieved at a cost as low as from USD 0.04 to USD 0.08 per school child depending on the epidemiological situation (Kabatereine et al., 2005). In addition, some of the major pharmaceutical companies have generously donated and continue to donate drugs to control programs.

Anthelminthic drugs for the control of helminthiasis in the community are described in Chapter 14. Three strategies, based on theoretical and practical experience of the control of soil-transmitted helminthiasis, but applicable to other forms of helminthiasis, are available for their use (Anderson, 1989):

- **Universal** (also called blanket, mass, or MDA [mass drug administration]) — is application of anthelminthic drug at the population level in the community irrespective of age, gender, infection status, or any social characteristic.
- **Targeted** — is application of anthelminthic drug at group level in the community where the group may be defined by age, gender, infection status, or social characteristic.
- **Selective** — is application of anthelminthic drug at individual level in the community where selection is based on the diagnosis of current infection.

Several ethical concerns have arisen over the use of anthelminthic drugs in the community. After evaluation of a risk/benefit analysis a strong case can be made for offering treatment to pregnant and lactating women to relieve them, their unborn children, and their infants from the burden of helminthiasis. Such a view, however, contradicts the manufacturers' instructions, probably because the manufacturers were prohibited from undertaking the necessary clinical trials. After consultation between toxicologists and clinicians and a review of the significant health benefits of treatment as compared with potential risks (WHO, 1996), the position has been reached whereby anthelminthic treatment may be offered under certain circumstances to pregnant and lactating women suffering from schistosomiasis (Chapter 10) and soil-transmitted helminthiasis (Chapter 11). Anthelminthic drugs may be given to children as young as 12 months to control soil-transmitted helminthiasis. Nevertheless, every effort should be made to avoid the use of anthelminthic drugs during the first trimester of pregnancy unless a medical practitioner intervenes for the health of the expectant mother.

Another concern connected with anthelminthic treatment and pregnancy could arise if expectant women were treated inadvertently in a control program that had decided to reserve treatment for them until after delivery on the grounds of the manufacturer's advice (Gyapong et al., 2003). If the decision to exclude pregnant women had been taken then program managers are faced with the need to find resources for identifying pregnant women in the community and also for returning to inform and offer them treatment after delivery at a later date. Those issues raise the question as to whether anthelminthic drugs should be offered in a public health setting without first carrying out diagnostic procedures throughout the community. Ultimately, decision makers must resolve the question, do the undoubted benefits of anthelminthic treatment for very many members of the community outweigh the adverse events that may overtake the few?

Efficacious anthelminthic treatment will fail as an agent for helminth control in the community unless a high degree of compliance is developed and extensive coverage is achieved and sustained. If an aim of drug use is to interrupt transmission, as is the case in efforts to eliminate lymphatic filariasis (Ottesen et al., 1997), adequate coverage becomes paramount and that cannot happen unless the community trusts the program and develops a sense of ownership of it. Drug distribution in a rural setting may be difficult if reliable knowledge of where and how people live is lacking. Nor should the needs of nomadic communities be ignored (Sheik-Mohamed and Velema, 1999). At least for people living in slums of high population density following unplanned urbanization, opportunities for good coverage with anthelminthic treatments may be easier to arrange (Ramaiah et al., 2005).

4.9 PREVENTION, ELIMINATION, AND ERADICATION

Experience from Europe and elsewhere shows that the prevention of communicable disease, including helminthiasis, depends on hygiene legislation, good housing, universal access to sanitation, safe drinking water, food inspection, immunization, waste collection and disposal, and comprehensive health and education services (Porter, 1997). Education has a vital part to play in the prevention of communicable disease because it helps people to understand and comply with health messages. Appropriate education should be offered in a way that avoids the imperialism of values. In many low-income countries where helminthiasis persists, there is an urgent need to increase scientific awareness in the population without denigrating confidence in traditional beliefs, undermining food security, or promoting measures that cannot be realized.

Elimination and eradication are technical terms in the public health arena (WHO, 2002). Elimination is the reduction of the number of new cases of a helminthiasis to zero in a defined territory as a result of control efforts. Having reached this situation, continued intervention and surveillance are needed. Eradication is the permanent reduction to zero of the global prevalence of a helminthiasis as a result of control efforts. No further control measures are needed. Formal certification is needed before eradication is declared and that stage cannot be achieved until pre-certification has taken place. The entire process involves the governments of member states of the United Nations and the Secretariat of the World Health Organization.

REFERENCES

Anderson RM. 1989. Transmission dynamics of *Ascaris lumbricoides* and the impact of che-motherapy. In: *Ascariasis and its Prevention and Control* (eds. DWT Crompton, MC Nesheim and ZS Pawlowski). London and Philadelphia: Taylor & Francis. pp. 253–273.

Anderson RM and May RM. 1992. *Infectious Diseases of Humans.* Oxford University Press (paperback edition).

Belizario VY, Bersabe MJ, de Leon WU et al. 2001. Intestinal heterophyidiasis: an emerging food-borne parasitic zoonosis in southern Philippines. In: *Proceedings of the 3rd Seminar on Food-borne Parasitic Zoonoses* (eds. J Waikagul, JH Cross and S Supajev). *Southeast Asian Journal of Tropical Medicine and Public Health* 32 (Supplement), 36–42.

Bradley JE and Unnasch TR. 1996. Molecular approaches to the diagnosis of onchocerciasis. *Advances in Parasitology* **37**, 57-106.

Brooker S and Michael E. 2000. The potential of geographical information systems and remote sensing in the epidemiology and control of human helminth infections. *Advances in Parasitology* **47**, 245–288.

Cox FEG (ed.) 1996. *The Wellcome Trust Illustrated History of Tropical Diseases.* London: The Wellcome Trust.

Cross JH and Basaca-Sevilla V. 1991. Capillariasis philippinensis: a fish-borne parasitic zoonosis. *Emerging Problems in Food-borne Parasitic Zoonosis: Impact on Agriculture and Public Health* (ed. JH Cross). Bangkok, Thailand: SEAMEO Regional Tropical Medicine and Public Health Project. pp. 153–157.

Dawkins R and Krebs JR. 1979. Arms races between and within species. *Proceedings of the Royal Society of London B,* **205**, 489–511.

Doumenge JP, Mott KE, Cheung C et al. 1987. *Atlas of the Global Distribution of Schistosomiasis.* Geneva and Bordeaux: World Health Organization and Universitaires de Bordeaux.

Forrester JE, Scott ME, Bundy DAP and Golden MHN. 1988. Clustering of *Ascaris lumbricoides* and *Trichuris trichiura* infections within households. *Transactions of the Royal Society of Tropical Medicine and Hygiene* **82**, 282–288.

Gibbons LM. 1986. *SEM Guide to the Morphology of Nematode Parasites of Vertebrates.* Farnham Royal, Buckinghamshire: CAB International.

Grove DI. 1986. Replicating helminth parasites of man. *Parasitology Today* **2**, 107–111.

Grove DI. 1989. Clinical manifestations. In: *Strongyloidiasis, a Major Roundworm Infection of Man* (ed. DI Grove). London, New York and Philadelphia: Taylor & Francis. pp. 155–175.

Grove DI. 1990. *A History of Human Helminthology.* Wallingford, Oxford: CAB International.

Gyapong JO, Chimbuah MA and Gyapong M. 2003. Inadvertent exposure of pregnant women to ivermectin and albendazole during mass drug administration for lymphatic filariasis. *Tropical Medicine and International Health* **8**, 1093–1101.

Hall A. 1982. Intestinal helminths of man: the interpretation of egg counts. *Parasitology* **85**, 605–613.

Harnett W. 1991. Molecular approaches to the diagnosis of *Onchocerca volvulus* in man and the insect vector. In: *Parasitic Nematodes — Antigens, Membranes and Genes* (ed. MW Kennedy). London, New York and Philadelphia: Taylor & Francis. pp. 195–218.

Haswell-Elkins, MR, Mairiang E, Mairiang P et al. 1994. Cross-sectional study of *Opisthorchis viverrini* infection and cholangiocarcinoma in communities within a high risk area in northest Thailand. *International Journal of Cancer* **59**, 505–509.

Holland C. 1997. *Modern Perspectives on Zoonoses.* Dublin: Royal Irish Academy.

Holland CV, Taren DL, Crompton DWT et al. 1988. Intestinal helminthiasis in relation to the socio-economic environment of Panamian children. *Social Science and Medicine* **26**, 209–213.

Hubbert WT, McCulloch WF and Schnurrenberger PR (eds). 1975. *Diseases Transmitted from Animals to Man,* 6th edition. Springfield, Illinois: Charles C Thomas, Publisher.

Hyde JE. 1990. *Molecular Parasitology.* Milton Keynes: The Open University Press.

Ikeda T and Akao N. 2001. Immunodiagnosis of helminthic diseases. In: *Perspectives on Helminthology* (eds. N Chowdhury and I Tada). Enfield (NH), U.S. and Plymouth, UK: Science Publishers Inc. pp. 397–418.

Kabatereine NB, Tukahebwa EM, Zazibwe F et al. 2005. Short communication: soil-transmitted helminthiasis in Uganda: epidemiology and cost of control. *Tropical Medicine and International Health* **10**, 1187–1189.

Kassai T, Cordero del Campillo M, Euzeby J et al. 1988. Standardized nomenclature of animal parasitic diseases (SNOAPAD). *Veterinary Parasitology* **29**, 299–326.

Kean BH, Mott KE and Russell AJ. 1978. *Tropical Medicine and Parasitology. Classic Investigations.* Volume I. Ithaca and London: Cornell University Press.

MacInnes AJ. 1974. A general theory of parasitism. Proceedings of the Third International Congress of Parasitology, Munich.

McCullough F. 1974. Observations on *Ascaris lumbricoides* infection in Mwanza, Tanzania. In: *Parasites in Man and Animals in Africa* (eds. C Anderson and WL Kilama). Nairobi: East African Literature Bureau. pp. 359–385.

Macpherson CN. 2005. Human behaviour and the epidemiology of parasitic zoonoses. *International Journal for Parasitology* **35**, 1319–1331.

Mas-Coma S. 2004. Human fascioliasis. In: *Waterborne Zoonoses* (eds. JA Cotruvo, A Cotruvo, A Dufour et al.). London: IWA Publishing on behalf of the World Health Organization. pp. 305–322.

Montresor A, Crompton DWT, Gyorkos TW and Savioli L. 2002. *Helminth Control in School-age Children. A Guide for Managers of Control Programmes.* Geneva: World Health Organization.

Moqbel R (ed.). 1992. *Allergy and Immunity to Helminths.* London and Washington DC: Taylor & Francis.

Muller R. 2002. *Worms and Human Disease,* 2nd edition. Wallingford, Oxford: CABI Publishing.

Ottesen EA, Duke BOL, Karam M and Behbehani K. 1997. Strategies and tools for the control/elimination of lymphatic filariasis. *Bulletin of the World Health Organization* **75**, 491–503.

Ou Ming. 1989. *Chinese-English Manual of Common-used Prescriptions in Traditional Chinese Medicine.* Hong Kong: Joint Publishing (HK) Co. Ltd and Guangdong Science and Technology Publishing House, China.

Ovington KS and Behm C. 1997. The enigmatic eosinophil: investigation of the biological role of eosinophils in parasitic helminth infection. In: *New Perspective in Eosinophils. Role in Inflammation Associated with Allergy, Asthma and Parasitic Disease* (eds. R Cordeiro, R Moqbel and PF Weller). *Memmorias do Instituto Oswaldo Cruz* **92** (Suppl. II) 93–104.

Porter R. 1997. *The Greatest Benefit to Mankind.* London: Harper Collins Publishers.

Pritchard MH and Kruse GOW. 1982. *The Collection and Preservation of Animal Parasites.* Lincoln and London: University of Nebraska Press.

Ramaiah KD, Vijay Kumar KN, Ravi R and Das PK. 2005. Situation analysis in a large urban area of India, prior to launching a programme of mass drug administration to eliminate lymphatic filariasis. *Annals of Tropical Medicine and Parasitology* **99**, 243–252.

Ratard RC, Kouemeni LE, Ekani Bessala et al. 1991. Ascariasis and trichuriasis in Cameroon. *Transactions of the Royal Society of Tropical Medicine and Hygiene* **85**, 84–88.

Rim H-J. 1986. The current pathobiology and chemotherapy of clonorchiasis. *Korean Journal of Parasitology* 24, Monograph series 3. pp. 1–141.

Sheik-Mohamed A and Velema JP. 1999. Where health care has no access: the nomadic populations of sub-Saharan Africa. *Tropical Medicine and International Health* **4**, 695–707.

Smith G and Scott ME. 1994. Model behaviour and the basic reproduction ratio. In: *Parasitic and Infectious Diseases* (eds. ME Scott and G Smith). New York and London: Academic Press. pp. 21–28.

Soulsby EJL. 1991. Parasitic zoonoses: new perspectives and emerging problems. *Health and Hygiene* **12**, 66–77.

Stothard JR, Kabatereine NB, Tukahebwa EM et al. 2005. Field evaluation of the Meade Readiview handheld microscope for diagnosis of intestinal schistosomiasis in Ugandan school children. *American Journal of Tropical Medicine and Hygiene* **73**, 949–955.

Udonsi JK. 1987. Endemic *Paragonimus* infection in Upper Igwun Basin, Nigeria: a preliminary report on a renewed outbreak. *Annals of Tropical Medicine and Parasitology* **81**, 57–62.

Wakelin D. 2002. Immunology of helminths. Chapter 7 in *Worms and Disease,* 2nd edition by R. Muller. Wallingford Oxon: CABI Publishing.

Whalen GE, Rosenberg EB, Strickland GT et al. 1969. A new disease in man. *Lancet* **1**, 13–16.

Wilkins HA. 1987. The epidemiology of schistosome infections in man. In: *The Biology of Schistosomes. From Genes to Latrines* (eds. D Rollinson and AJG Simpson). London and New York: Academic Press Limited. pp. 379–397.

WHO. 1976. The promotion and development of traditional medicine. Report of a meeting of experts in Africa. *Technical Report Series* **622**. Geneva: World Health Organization.

WHO. 1987. *International Nomenclature of Diseases.* Volume II *Infectious Diseases* Part 4: *Parasitic Diseases.* Geneva: World Health Organization.

WHO. 1996. Report the WHO Informal Consultation on hookworm infection and anaemia in girls and women. Geneva: World Health Organization. WHO/CTD/SIP/96.1.

WHO. 1999a. *WHO Recommended Surveillance Standards,* 2nd edition. Geneva: World Health Organization.

WHO. 1999b. *WHO Monographs on Selected Medicinal Plants* Volume 1. Geneva: World Health Organization.

WHO. 2000. The use of essential drugs. Ninth report of a WHO Expert Committee (including the revised Model List of Essential Drugs). *Technical Report Series* **895**. Geneva: World Health Organization.

WHO. 2002. Prevention and control of schistosomiasis and soil-transmitted helminthiasis. Report of a WHO Expert Committee. *Technical Report Series* **912**. Geneva: World Health Organization.

WHO. 2004a. *Food-borne Trematode Infections in Asia.* Report of a joint WHO/FAO workshop held in Ha Noi, Vietnam in November 2002. Manila, Philippines: WHO Regional Office for the Western Pacific.

WHO. 2004b. *WHO Model Formulary.* Geneva: World Health Organization.

Yamaguti S. 1958a. *Systema Helminthum.* Volume I. *The Digenetic Trematodes of Vertebrates.* Part I. New York and London: Interscience Publishers Inc.

Yamaguti S. 1958b. *Systema Helminthum.* Volume I. *The Digenetic Trematodes of Vertebrates.* Part II. New York and London: Interscience Publishers Inc.

Yamaguti S. 1959. *Systema Helminthum.* Volume II. *The Cestodes of Vertebrates.* New York and London: Interscience Publishers Inc.

Yamaguti S. 1961a. *Systema Helminthum.* Volume III. *The Nematodes of Vertebrates.* Part I. New York and London: Interscience Publishers Inc.
Yamaguti S. 1961b. *Systema Helminthum.* Volume III. *The Nematodes of Vertebrates.* Part II. New York and London: Interscience Publishers Inc.

Readers are advised that further information about helminthiasis is available in the collection of review articles to be found under the title "Control of human parasitic diseases," edited by DH Molyneux and published by Elsevier Ltd in 2006 in Advances in Parasitology, 61.

5 Cestodiasis

Six of the species of cestode or tapeworm responsible for zoonotic infections in humans (List, Chapter 1) are discussed in this chapter. Five species, *Echinococcus granulosus, E. multilocularis, Taenia saginata, T. asiatica,* and *T. solium,* are classified in the family Taeniidae. *Taenia asiatica* was not recognized as a valid species until relatively recently. Although *T. asiatica* resembles *T. saginata* much more closely than *T. solium,* humans become infected through eating undercooked pork rather than beef (Ito et al., 2003). Taeniid tapeworms are particularly interesting because species of mammal serve as both intermediate and definitive host. The sixth species, *Hymenolepis nana,* belongs to the family Hymenolepididae and is related to *H. diminuta,* a tapeworm that has been extensively studied in the laboratory because rats serve as its natural definitive host (Arai, 1980). The application of molecular techniques is demonstrating that systematics based on comparative anatomy alone do not reveal the variety of the genotypes and strains that exist within these species (Jenkins et al., 2005; McManus, 2006). In this chapter, strain differences will be largely ignored because public health interventions designed to control cestodiasis have general application.

The public health significance of disease caused by this selection of tapeworms has not featured as prominently as perhaps it should in estimates of DALYs or assessments of the burden of disease. Those people in whom infections are established may suffer from one of three terrible diseases: hydatid disease (CE) attributable to infection with *E. granulosus*; alveolar echinococcosis (AE) attributable to *E. multilocularis*; and neurocysticercosis (NCC) attributable to *T. solium.* These are serious, chronic, life-threatening, and invariably fatal illnesses. Perhaps as many as 50,000 people die annually from NCC and many others are left permanently disabled by neurological disorders (PAHO/WHO, 1997). Cestodiasis in the South has been neglected for too long and requires resources for intervention to increase prevention and establish control. Elimination or even eradication could be achievable in some of the world's endemic regions.

5.1 LIFE HISTORIES AND HOST RANGE

A flow chart of events in the indirect life history and developmental pattern of taeniid tapeworms is shown in Figure 5.1 and quantitative information is summarized in Table 5.1. Adult tapeworms live in the lumen of the mammalian small intestine, attached to the mucosa by means of their scolices (Figure 5.2). The worms are confined to the gastrointestinal tract because by not having an alimentary tract of their own they rely on the host's digestion to supply nutrient molecules for absorption across their surfaces. Sexual maturity is attained in the small intestine where the hermaphroditic adults cross fertilize or self fertilize if partners are unavailable. Eggs

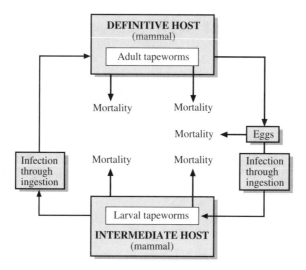

FIGURE 5.1 Flow chart of the life history and development of taeniid tapeworms. Based on Figure 4.2 (Keymer, 1982) with permission from Routledge/Taylor & Francis Group, LLC.

are carried out of the host contained in proglottides (segments) shed from the distal section of the strobila (main body) of the worm (Table 5.1). The eggs contaminate the environment and so become accessible to susceptible intermediate hosts (Table 5.1 and Table 5.2). Taeniid eggs are easily recognized as such because microscopic examination reveals the presence of three pairs of hooks in each oncosphere (see Appendix 3). Identification to generic and specific level, however, is rarely possible.

Although individual *Taenia saginata* and *T. solium* produce vast numbers of eggs compared with individual *Echinococcus granulosus* and *E. multilocularis* (Table 5.1) the contamination of the environment is not dissimilar because as many as 200,000 *Echinococcus* may be found in the gut of a dog or fox at any one time. The massive egg discharge compensates for the high mortality rates endured by taeniid tapeworms during development (Figure 5.1). The egg shells enclose an infective oncosphere that, after being ingested by the intermediate host, is released into the gut by digestive processes, penetrates the mucosa, and enters the blood and lymphatic systems. The timing of events during the development of these tapeworms is closely tied to the longevity of their intermediate hosts (Table 5.1).

5.1.1 DEVELOPMENT OF *ECHINOCOCCUS* AND CYST FORMATION

When a tapeworm's development in the intermediate host has been completed, metacestode is the name given to the stage that is infective to the definitive host. Cyst formation has been studied extensively by veterinarians because of concerns for the health of domesticated animals and the ensuing economic losses to the livestock industry (Georgi, 1985). After penetrating the intestinal wall by means of the six blade-like hooks and the action of proteolytic secretions, the oncospheres of *E. granulosus* reach the liver and other organs of the intermediate host where transformation takes place to produce a minute bladder containing an inner germinative

TABLE 5.1

Features of the Biology of Four Taeniid Tapeworms of Public Health Significance. Based on information in Keymer (1982), Pawlowski (1982), Schantz (1982), Kociecka (1987), and Muller (2002).

	Taenia saginata	*Taenia solium*	*Echinococcus granulosus*	*Echinococcus multilocularis*
Length of adult	4–12 m	1–8 m	3–6 mm	1.3–3.7 mm
No. of proglottides (segments)	1,000–2000	700–1000	3–4	3–5
Distinguishing features (see Figure 5.2)	Scolex with 4 suckers Many uterine branches	Scolex with 4 suckers and 22–32 hooks Few uterine branches	32 (28–40) scolex hooks 56 (45–65) testes	28 (20–36) scolex hooks 22 (16–26) testes
Life span of adult (years)	Up to 40 years	10–15 years	6 months	6 months
Stage infective to humans	Cysticercus (Cysticercus bovis)	1. Cysticercus (Cysticercus cellulosae) 2. Egg	Egg	Egg
Usual mode of infection	Ingestion of undercooked beef	Ingestion of undercooked pork — or of eggs	Ingestion	Ingestion
Prepatency — before eggs passed	6–8 weeks	5–12 weeks	48–61 days	30–35 days
Fecundity (eggs/worm)	480,000–720,000 in 6–9 proglottides daily	120,000–250,000 in 3–5 proglottides daily	200–400 in final proglottid released every 2 weeks	200–400 in final proglottid released every 2 weeks
Reservoir (definitive) hosts	None known	None known	Dog, coyote, dingo, wolf, jackal, hyena	Red fox, arctic fox, cat, dog
Intermediate hosts (mammals)	Cattle and bovines, reindeer, ruminants	Pig, dog, man, wild boar, bear, cat, rodents	Sheep and ruminants	Small rodents and insectivores
Infection of int. host	Ingestion of egg	Ingestion of egg	Ingestion of egg	Ingestion of egg
Development time in int. host to infectivity	12 weeks	8–10 weeks	Slow, up to 2 years	Rapid, 2 to 3 months
Human disease nomenclature	Taeniasis saginata	Taeniasis solium Neurocysticercosis (NCC)	Hydatid disease Cystic echinococcosis (CE)	Alveolar echinococcosis (AE)

layer from which the protoscolices are formed. The outer layer of the bladder consists mainly of collagen deposited by host cells. The bladders are metacestodes, better known as hydatid cysts, and their presence initiates the morbidity that defines hydatid

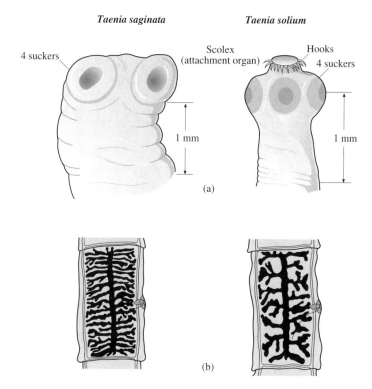

FIGURE 5.2 Distinguishing features of adult *Taenia saginata* and *T. solium* recovered from human hosts. (a) Scolices, reproduced from Figure 12.26 (Noble and Noble, 1982) for *T. saginata* with permission from Lea & Febiger, and from Figure 1, plate IX for *T. solium* from Ward (1907). (b) Mature proglottides (segments). From Figure 12.22 (Noble and Noble, 1982) with permission from Ward's Natural Science Establishment. The proglottides are shown as histological preparations to reveal prominent differences in uterine morphology.

disease. The structure of a hydatid cyst and the production of infective protoscolices are depicted diagrammatically in Figure 5.3 and details are provided by Schantz (1982). The cyst of *E. multilocularis* lacks a robust, outer layer of host tissue and, if dissected from the organ in which it is embedded, has the crude appearance of a bunch of grapes. The germinative layer increases in area by forming lobes or buds that grow exogenously while continuing to produce protoscolices.

The fluid-filled cyst of *E. granulosus* grows slowly, its diameter increasing at a rate of about 10 to 50 mm annually with protoscolices continuing to be formed. The protoscolices settle like sediment in the lumen of the bladder and are known as hydatid sand. Big cysts can contain liters of fluid and thousands of protoscolices. The multi-chambered cyst of *E. multilocularis* grows rapidly and infective protoscolices may be formed within a few weeks. In humans, the liver and then the lungs are the sites where most hydatid cysts develop; in Africa as many as 70% of cysts are found in the liver, usually the right lobe, and up to 20% in the lungs (Macpherson and Craig, 1991). In sheep this pattern is reversed. In about 80% of patients, one organ harbors a single cyst (Filice et al., 1997). The locations of cysts

TABLE 5.2
Echinococcus **Species, Strains, and Genotypes. From Table 1 (Jenkins et al., 2005) with permission from Elsevier.**

Species	Strain/ Genotype	Known Intermediate Hosts	Infective to Humans	Disease in Humans	Known Definitive Hosts
Echinococcus granulosus	Sheep/G1	Sheep (cattle, pigs, camels, goats, macropods)	Yes	Cystic (unilocular)	Dog, fox, dingo, jackal, and hyena
	Tasmanian sheep/G2	Sheep (cattle?)	Yes	Cystic (unilocular)	Dog, fox
	Buffalo/G3	Buffalo (cattle?)	?	?	Dog, fox?
	Camel/G6	Camels (sheep)	Yes	Cystic (unilocular)	Dog
	Pig/G7	Pigs	Yes	Cystic (unilocular)	Dog
	Cervid/G8 and G10	Cervids	Yes	Cystic (unilocular)	Wolf, dog
	?/G9		Yes	Cystic (unilocular)	
	Lion/?	Zebra, wildebeest, warthog, bushpig, buffalo, various antelope, giraffe?, hippopotamus?	?	?	Lion
Echinococcus equines	Horse/G4	Horses and other equines	No	—	Dog
Echinococcus ortleppi	Cattle/G5	Cattle	Yes	Cystic (unilocular)	Dog
Echinococcus multilocularis	Some isolate variation	Rodents, domestic and wild pig, dog, monkey	Yes	Alveolar (multivesicular)	Fox, dog, cat, wolf, raccoon-dog, coyote
Echinococcus shiquicus	?	Lagomorphs (pika)	?	?	Tibetan fox
Echinococcus vogeli	None reported	Rodents	Yes	Polycystic	Bush dog
Echinococcus oligarthrus	None reported	Rodents	Yes	Polycystic	Wild felids

depend on the migratory routes of the invading oncospheres, which may reach the spleen, kidney, brain, muscles, peritoneal cavity, and bones (Raether and Hanel, 2003). If a cyst is ruptured, protoscolices may escape and be carried round the body to give rise to more hydatid cysts. The bigger a cyst becomes, the greater the displacement and pressure imposed on surrounding organs. An impression of how humans are put at risk of contracting *Echinococcus* infections is shown in Figure 5.4.

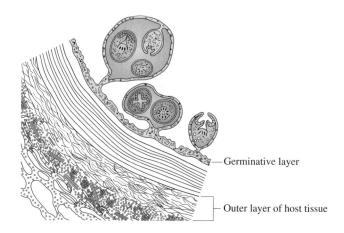

- Germinative layer

- Outer layer of host tissue

FIGURE 5.3 Diagrammatic representation of a section of an hydatid cyst of *Echinococcus granulosus*. From Figure 42 (Muller, 2002) with permission from CABI Publishing.

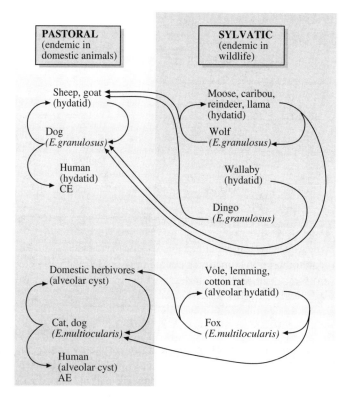

FIGURE 5.4 Examples of the host-*Echinococcus* interactions with which humans become involved. From Figure 5-20 (Georgi, 1985) with permission from Elsevier.

5.1.2 Development of *Taenia* and Cysticercus Formation

Eggs of *Taenia saginata* and *T. solium* leave the human host in stools and contaminate the food of cattle and pigs respectively, or the places where they are housed. On ingestion, the oncospheres are released from their egg shells and penetrate the intestinal wall. Those that reach the muscles of their intermediate hosts develop to form metacestodes or cysticerci, which are the stages infective to humans who are the sole definitive hosts of both species (Pawlowski, 1982). Cysticerci are often referred to as bladderworms. Before the life history of *T. solium* had been determined and the significance of the bladders found in pork was appreciated, the condition became known as cysticercus cellulosae; in beef the condition was called cysticercus bovis.

These terms are still widely used by veterinarians and meat inspectors (Table 5.1). Cysticerci of *T. asiatica* are more commonly found in the viscera of pigs than in the muscles to the extent that Eom and Rim (2001) refer to a cysticercus as cysticercus viscerotropica.

A cysticercus of *T. saginata* may be infective to humans within 10 weeks of the oncosphere's arrival in the muscle. A cysticercus resembles an ovoid bladder measuring about 10 × 5 mm. It contains an inverted scolex that lacks hooks and an opalescent fluid and is visible to the naked eye in raw beef. Cysticerci of *T. saginata* tend to be found in the lymphatic spaces in skeletal and heart muscle. They can survive for up to three years in their host's skeletal muscle.

A cysticercus of *T. solium* is bigger than that of *T. saginata* and can be identified if the hooks can be seen on the fully formed protoscolex (Figure 5.2.). The development of cysticerci of *T. solium* to infectivity takes about 8 to 10 weeks. Infective eggs of *T. solium* will hatch in the human small intestine and develop to form cysticerci in various parenteral sites (Flisser, 1998). Two forms of cysticerci exist; a simple bladder that was originally described as cysticercus cellulosae and a complex, multi-lobed bladder known as a racemose cysticercus. Racemose cysticerci do not appear to contain scolices and they are known only from the brain (Rabiela et al., 1989). The problem is most serious when cysticerci of either type develop in the brain giving rise to the disease known as neurocysticercosis (NCC, Table 5.1). NCC occurs in communities that eat undercooked pork and keep pigs under poor standards of animal husbandry. NCC appears to be increasing in parts of Africa, Asia, and Latin America (Engels et al., 2003).

A form of autoinfection sometimes occurs in people already infected with adult *T. solium*. Apparently gravid proglottides may be swept out of the proximal small intestine by reverse peristalsis into the stomach. Gastric digestion degrades the proglottides and the liberated eggs are carried back into the duodenum where they hatch and the oncospheres are then free to penetrate the intestinal wall and migrate through the body to produce cysticerci (Raether and Hanel, 2003).

5.1.3 Host Range

Echinococcus granulosus and *E. multilocularis* display relatively low host specificity and many species of mammal are available to ensure their survival and their persistence as a public health hazard (Table 5.1 and Table 5.2; Figure 5.4). *Taenia saginata* has relatively low intermediate host specificity (Table 5.1), but appears to be restricted to relying on humans as its definitive host. *Taenia solium* depends on pigs and humans to serve, respectively, as its intermediate and definitive host species. Although humans are capable of supporting the development of *T. solium* to its cysticercus stage, the process will have little impact on transmission unless cannibalism prevails. The host range of *T. asiatica* has been investigated for two island sites in Korea and for Taiwan (Eom and Rim, 2001). In addition to domestic pigs, wild boar, goats, cattle, muntjacs, and squirrels can serve as intermediate hosts.

5.1.4 *Hymenolepis nana*

Hymenolepis nana is a small tapeworm, the adults occasionally reaching a length of about 100 mm and having about 200 proglottides. The species interests helminthologists because it exploits both an indirect and a direct life cycle. Several insects serve as intermediate hosts including the flour beetle *Tribolium confusum,* which inhabits human food stores. Eggs, freed from disintegrating proglottides, pass out of the definitive host and after ingestion by susceptible insects the oncospheres develop to the cysticercoid stage in the insect's body. Development to form a cysticercoid takes from 2 to 4 weeks depending on the temperature in the insect's environment. If the insect is swallowed by a human the cysticercoid evaginates its scolex and attaches to the wall of the small intestine. After about a week, the tapeworms attain maturity and egg production begins. The direct life cycle of *H. nana* involves a process of autoinfection. Eggs may hatch in the small intestine where the oncospheres penetrate the villi and rapidly develop to the cysticercoid stage. On breaking out of the villi, the cysticercoids grow to become adult tapeworms *in situ* without leaving the host (Ash and Orihel, 1990; Chandler and Read, 1961).

5.2 TRANSMISSION TO HUMANS

Humans may unknowingly expose themselves to infection with *Echinococcus* and the risk of developing CE or AE primarily through a dependence on dogs for the management of sheep and cattle. The home environment becomes contaminated with eggs if the dogs are infected. The risk is increased wherever animals are slaughtered at home and the dogs are fed raw offal harboring either hydatid or alveolar cysts. The dogs will ingest numerous protoscolices and so perpetuate the infection. Scavenging wild animals also serve as sources of infection in the locality if they are able to feed on surplus offal or dead animals. The potential for retaining *Echinococcus* infections in an area is well illustrated in Figure 5.4 and the significance of human relationships with dogs is demonstrated by observations on pastoralists in Kenya (Macpherson and Craig, 1991). Ultrasonography showed that up to 9% of the Turkana in northwestern Kenya had hydatid disease with more women being infected than men, and parasitological surveys showed that from 39 to 70% of the dogs there were passing taeniid eggs. Macpherson (1983) describes the common presence of dogs in Turkana households and their intimate contact with women and children.

The dogs are encouraged to clean the babies after defecation and lick household goods; in this arid region water is so precious that it cannot always be spared for washing.

The predominant risk factors for the transmission of *Taenia saginata* to humans are the consumption of raw or undercooked beef and the consumption of raw meat from undercooked moose or other large herbivores that overlap with human activities and so serve as intermediate hosts (Figure 5.4). Exposure to cysticercus bovis has declined in communities of the north where meat inspection is enforced in licensed abattoirs. That public health measure permits individuals to choose to eat rare (undercooked) beef without a thought for personal health. The risk of infection is ever present where the local economy and food security depend on herds of semi-domesticated animals or on hunting. Poor health awareness, poor personal hygiene, and inadequate sanitation for the safe disposal of human excreta reinforce the risks.

The prominent risk factor for the transmission of *T. asiatica* to humans is the consumption of raw porcine viscera, particularly the tradition of eating warm liver as soon as possible after slaughter at ceremonial meals (Eom and Rim, 2001). When inadequately cooked porcine organs such as the stomach, spleen, and omentum are eaten they may contain cysticerci or have small portions of contaminated liver attached.

The predominant risk factors for the transmission of *Taenia solium* to humans are the consumption of raw or undercooked pork containing cysticercus cellulosae leading to the establishment of the adult tapeworm in the gut and the ingestion of eggs leading to the development of the cysticercus in the tissues and brain (NCC). People sharing their households with pigs under conditions of ignorance, poor sanitation, and inadequate animal husbandry will be at risk of taeniasis solium and NCC (Raether and Hanel, 2003).

5.3 DIAGNOSIS IN HUMANS

The complexity of human-taeniid interactions does not facilitate identification of the helminth species or clinical diagnosis of the form that the disease has taken. Infections in which humans harbor larval taeniids present a major challenge to health workers responsible for epidemiological surveys (see Craig et al., 1996).

5.3.1 *ECHINOCOCCUS* AND *TAENIA*

5.3.1.1 Parasitological Diagnosis

Diagnosis is important given the severity of the morbidity of cestodiasis and the prognosis for the different forms of disease (Section 5.5). Parasitological diagnosis of adult *Taenia* infections in humans can be made by examining the form of the uterine structures in proglottides passed in feces (Figure 5.2). Eggs are usually discharged inside mature proglottides and will not always be seen in isolation under the microscope. Eggs are often released from proglottides in the anal region and can be collected on perianal swabs or sticky tape. Up to 85% of people found to be infected with *T. saginata* have been identified by this method (Schantz and Sarti-Gutierrez, 1989). In any case, *T. saginata, T. asiatica,* and *T. solium* cannot be told apart on the basis of eggs passed in stools, and the intermittent nature of egg excretion may lead to the underestimation of prevalence values (Hall et al., 1981). If *T. solium* is suspected in a community, local pigs should be examined for the presence of cysticerci. The pig's tongue is a predilection site for cysticerci, which can be detected on the undersurface (Sarti et al., 1992).

The answers from people to questions about *Taenia* infections are valuable in diagnosis and surveys. Are individuals passing proglottides? Flisser et al. (2005) carried out an evaluation of a self-detection tool to aid in the identification of members of a community infected with either *T. saginata* or *T. solium.* An instructional guide was prepared to enable health workers to identify and then treat individuals considered to be infected. The use of the tool was strongly supported by a health awareness campaign. In the year before the evaluation seven tapeworm hosts were found during an epidemiological survey. The self-detection system identified 47 hosts and their treatment reduced the number of eggs that would have contaminated the environment. Parasitological diagnosis of cysticerci or hydatid and alveolar cysts in humans may not be reliable enough to justify the use of health care resources.

5.3.1.2 Clinical Diagnosis

In endemic areas, traditional healers and community leaders will be aware of prevailing illnesses and health professionals should expect the presence of CE and AE; such an approach is important when dealing with infectious agents that are not amenable to standard parasitological techniques. Long-established cysts in the liver may be detected by abdominal palpation. If modern hospital facilities are available, computed tomography (CT) and magnetic resonance imaging (MRI) can be used in diagnosis, but the differential diagnosis of cysts suspected of being due to *Echinococcus* infection is advised because other lesions may need to be considered and eliminated (Raether and Hanel, 2003). As an aid to diagnosis, WHO has published an ultrasound classification so that investigations can be standardized (WHO, 2003). The development of portable ultrasonography equipment has improved prospects for the detection of hydatid and alveolar cysts in people without access to hospital care (Macpherson et al., 1987; Macpherson and Milner, 2003). Immunodiagnostic methods are useful for resolving problems arising from scanning or imaging techniques (PAHO/WHO, 1997).

5.3.1.3 Immunodiagnosis and Molecular Diagnosis

The specificity and sensitivity of immunodiagnostic tests are steadily being refined and a variety of quality tests is available for both forms of taeniasis, for cysticercosis and NCC, and for CE and AE. The reader is referred to the reviews of Allan et al. (2003) and Raether and Hanel (2003) for evaluations of the different methods. The choice of test depends on the availability of resources and whether individual patients or communities are in need of investigation. The following methods could be considered for use in public health surveys:

- *Echinococcus granulosus* (CE) and *E. multilocularis* (AE). Specific serodiagnostic assays (Dot-ELISA and Em2-ELISA) described by Schantz et al. (2003).
- *Taenia saginata.* Enzyme-linked immunosorbent assay (ELISA) for the detection of coproantigens in stool samples described by Deplazes et al. (1991). In a similar procedure, Machnicka et al. (1996) found that freshly collected stool samples could be stored at 4°C for up to 72 h or at –20°C for up to 24 mo and still give reliable results when examined with their ELISA assay.
- *Taenia solium.* (1) Detection of intestinal infections based on coproantigen assay using ELISA techniques investigated by Allan et al. (1996a). (2) Detection of cysticercosis based on ELISA and EITB techniques compared by Diaz et al. (1992).

Polymerase chain reaction (PCR) technology offers precise results about the identity and biological status of an organism from which a minute amount of DNA was obtained. PCR is an extremely sensitive method for amplifying DNA sequences (Sullivan, 1994). Methods are discussed by McManus (2006). Singh (1997) has pointed out that while these methods are valuable for research, they are unlikely to replace routine methods of diagnosis in the countries of the south because of their high costs.

5.3.2 *Hymenolepis nana*

Diagnosis of *H. nana* continues to depend on the collection of stool samples and microscopic examination for eggs (Appendix 3). The eggs are small and ovoid in shape (30 × 47 μm). The egg shell is thin and the membrane around the oncosphere has structures called polar filaments at each end that are not seen in any of the other eggs produced by tapeworms infecting humans. Although the eggs may be seen in wet fecal smears or in Kato-Katz preparations, the chances of finding them in lightly infected individuals will be increased if the stools are first processed by a salt flotation technique (WHO, 1980).

5.4 EPIDEMIOLOGY

5.4.1 *Echinococcus granulosus* and *E. multilocularis*

Echinococcus granulosus has a worldwide distribution. As many as 3 million people are estimated to suffer from echinococcosis (Table 1.1) with many being afflicted with CE. Human infections (CE) occur wherever sheep and other species of domesticated herbivores are reared. The public health significance of the problem is exacerbated where meat inspection is not the rule and if the infection is entrenched in species of wild mammal that interact with domestic dogs (Raether and Hanel, 2003). A more detailed picture of the distribution will emerge once the distribution of the strains of the worm (Table 5.2) has been mapped. Jenkins et al. (2005) have reviewed recent reports of *E. granulosus* infections in a wide range of intermediate

TABLE 5.3

Recent Information about the Epidemiology of *Echinococcus granulosus* (complex). Abstracted from Jenkins et al. (2005).

Region	Observations	Key Reference
Australia	Provisional eradication of *E. granulosus* from Tasmania. Most active trasmission in area of Great Dividing Range. Coproantigen surveys of dog feces show 18% dogs infected in Victoria and 29% in New South Wales.	Jenkins (2005)
Western and Central Asia	Hydatid disease reemerged following disruption to veterinary services after break up of USSR. Sheep > 6 years old exhibit prevalence of 74-80%. By 1995, about 1000 cases of human infection annually in Kazakhstan.	Torgerson et al. (2000)
China	High endemic zone (human) from Western Xinjiang to Sichuan. Since 1950, about 35,000 cases treated surgically.	Craig (2004)
Africa	Highest infection rates in sheep-rearing areas. Focal distribution related to strain infectivity.	Macpherson and Wachira (1997)
Europe	Hydatid rare in Europe except Poland and further east. 4–8 cases of CE per 100,000 in parts of Spain, Sardinia, and Southern Italy.	Eckert et al. (2001)
North America	Human cases of CE rare, restricted to high-risk sheep farmers. Two strains exist; cervid strain in wildlife and sheep strain.	Schantz et al. (1995)
South America	Reemergence of transmission in Peru when control activities stopped.	Eckert et al. (2001)

and definitive host species in different regions and summarized evidence of the emergence and reemergence of infection (Table 5.3). Readers are referred to Table 4 in Muller (2002), which covers data from 28 countries representing all continents. Values in Muller's table are given for the percentage prevalences of hydatid cysts in humans, in sheep and cattle, and for the prevalence of adult infections in dogs. Hydatid cysts in humans are stated to vary from 0.0011% (Bavaria, Germany) to 10 to 16% (Kenya); the Kenyan figure may refer to pastoral community in part of the country where transmission is intense. Values in domesticated animals are stated to vary from 2 to 7% (Iran) to 89% (Xinjiang, China). Prevalence values in dogs range from 1.8 to 5.7% (Nepal) to 48 to 93% (wild dogs, Australia).

Echinococcus multilocularis is less common in humans than *E. granulosus* and is distributed in various species of canine mammals (definitive hosts) and rodents (intermediate hosts). Infections occur mainly in rural communities in some areas of North America, central Europe, the Middle East, Siberia and central Asia, Japan, and China (Craig, 2006). China accounts for most cases of CE and AE. Craig (2006) stated that 2% of China's 600,000 to 1.3 million cases of echinococcosis were due to *E. multilocularis*. Overall, 580 (2.4%) out of 24,017 people surveyed between

1992 and 2004 were found to be AE cases. Children comprise a third of the cases of CE and AE combined. Surveys based on ultrasound screening have found that most cases of AE occur in seminomadic pastoral communities from the Tibetan Plateau that extends into Western Sichuan. Foxes serve as definitive hosts in this region while voles appear to be the common intermediate hosts. Predation by domestic dogs exposes the infection to humans.

Schantz et al. (2003) carried out a survey of the prevalence of CE and AE among pastoral communities living on the Tibetan Plateau. CE was detected in 6.6% of the people and AE in 0.8%. The risk factors for both infections were livestock ownership, herding occupation, female gender, age over 25 years, limited occupation, water source, poor hygiene, offal disposal, and dog care.

5.4.2 Taenia saginata, Taenia asiatica, and Taenia solium

Authorities agree that *Taenia saginata* is much more common than *T. solium* (Table 1.1) but much less pathogenic. *Taenia saginata* may be found wherever people eat undercooked beef containing cysticerci, regardless of whether the beef is home produced or imported. If proof of that conclusion were needed, Wandra et al. (2006) demonstrated in Bali, with statistical significance, that undercooked beef in the form of a local dish (lawar), together with the source of the beef and the consumers' level of education were the main risk factors for *T. saginata* infection.

Adult taeniid worms expelled from people in China, parts of Indonesia, Korea, Taiwan, and Viet Nam were for many years assumed to be *T. saginata* until they were recognized as *T. asiatica,* an independent species closely related to *T. saginata*. The estimate of 70 million infections of *T. saginata* (Table 1.1) includes *T. asiatica*. Consumption of undercooked pig viscera and pork containing cysticerci of *T. asiatica* is the route of human infection wherever the infection is endemic (Fan and Chung, 1998; Eom and Rim, 2001; Ito et al., 2003). For readers interested in epidemiological detection, the investigation by Eom and Rim (2001) into how so many people in Korea were found to be infected with a *saginata*-type tapeworm while so few of them ate beef provides a most intriguing account.

Understandably, acceptance of neurocysticercosis (NCC) as a debilitating disease has raised concern over the distribution of *Taenia solium* culminating in the declaration of the World Health Assembly in 2003 that *T. solium* is of worldwide public health importance. NCC is diagnosed throughout the Americas, except perhaps for Argentina, Canada, the U.S., and Uruguay, in many countries of sub-Saharan Africa including Benin, Burundi, Cameroon, Dem. Republic of the Congo, Ghana, Ivory Coast, Madagascar, Republic of South Africa, Rwanda, Senegal, Togo, and Zimbabwe, and in parts of Asia including China, Korea, and to a lesser extent in Bali, Cambodia, Laos, Myanmar, Philippines, and Viet Nam (PAHO/WHO, 1997). The global figure for the number of *T. solium* infections in Table 1.1 may be an underestimate because many places where the infection is likely to occur may lack diagnostic resources. The current distribution of *T. solium* infections may tend to reflect the distribution of diagnostic services.

Raether and Hanel (2003) record the results of numerous surveys conducted to determine the prevalence of intestinal infections with *T. solium*. Values range from 0.13 to 37.5% with the infection being encountered more frequently in older people. Allan et al. (1996b) sampled 3,399 people in four villages in rural Guatemala. Of the 92 (2.7%) identified with the intestinal infection, females were significantly more likely to be infected than males, prevalence rose with age until the 30- to 39-year class, and cases were clustered in households. A single worm was passed by all but one of the infected subjects given anthelminthic drug. Nearly 10% of people interviewed stated that they worked away from their home villages, some in the U.S. Economic migration and labor mobility, facilitated by easier and cheaper travel, might contribute to the dispersal of long-lived helminths such as *Taenia* spp. (Pawlowski et al., 2005).

Accurate quantitative information about NCC in humans depends on the availability of scanning and imaging equipment in hospitals and on efforts to correlate neurological disorders with the contacts that patients have with pigs and pork. An example of how to investigate the problem without the use of sophisticated equipment is the survey conducted in Mexico by Sarti et al. (1992). Results from an immunoblot assay identified 10.8% of those sampled (n = 1,552) as seropositive for cysticercosis and seropositivity increased with age reaching a peak in the 46- to 55-year group. History of seizures (strong circumstantial evidence for NCC) was significantly associated with seropositivity, which in turn was significantly associated with households in which one member reported a history of passing proglottides. In regions where *T. solium* infection is endemic, surveys designed to assess proglottid expulsion will be relatively inexpensive and can be informative.

5.4.3 *Hymenolepis nana*

Hymenolepis nana has a worldwide distribution and occurs where poor standards of personal and community hygiene prevail. *Hymenolepis nana* var. *fraterna* is a form that occurs in rats and mice and is morphologically indistinguishable from that in humans (Kociecka, 1987). Whether these are one and the same species needs to be resolved as does the possibility of cross infectivity. Many surveys such as one carried out by Sahbra et al. (1967) have established that *H. nana* is mainly an infection of childhood (Figure 5.5). The decline in prevalence with age (Figure 5.5) may result from better personal hygiene acquired by adults or by the development of some degree of adaptive immunity.

5.5 MORBIDITY AND PATHOLOGY

Intestinal infections with *T. saginata*, *T. asiatica*, *T. solium*, and *H. nana* do not pose a major threat to human health and well being. Parenteral infections with *Echinococcus* spp. and *T. solium* cause serious disease and the associated morbidity is described below. Community interventions to reduce the morbidity of parenteral cestodiasis require resources beyond most public health services in the countries of the South. The morbidity must not be ignored, however, because it highlights the case for interventions that can protect communities from infection (Section 5.8).

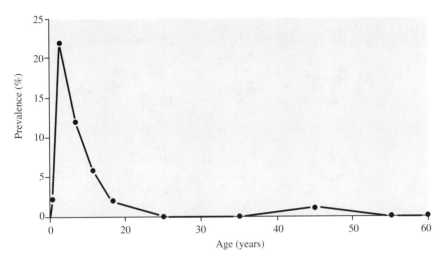

FIGURE 5.5 Plot of the relationship between host age and the prevalence of *Hymenolepis nana* in southwest Iran. From Figure 7 (Crompton, 1987) with permission from Elsevier. The original figure is from Sahbra et al. (1967), *Annals of Tropical Medicine and Parasitology* 61, page 354 with the permission of the Liverpool School of Tropical Medicine.

5.5.1 ECHINOCOCCOSIS (CE AND AE)

Hydatid disease (CE) is not usually confirmed until patients have reached adulthood unless the cysts have developed in the brain or eye. Infected people may be asymptomatic for years because hydatid cysts take years to reach a size that induces morbidity. CE involving the liver may include hepatic enlargement, pain, nausea, vomiting, abscesses, calcified lesions, portal hypertension, compression of the inferior vena cava, cirrhosis, biliary complications, and peritonitis (Filice et al., 1997; Flisser, 1998). CE involving the lungs may include pain reminiscent of pulmonary tumors, chronic cough, fever, and abscesses. If a cyst should rupture as a result of trauma or surgical intervention, the release of cyst fluid may cause a range of allergic reactions and even fatal anaphalaxis. Equivalently serious complications occur when hydatid cysts develop in the brain, bones, or other organs.

In humans, cysts of *E. multilocularis* are most frequently found in the liver. The course of infection (AE) resembles the developmental pattern of a malignant tumor (Kern et al., 2006). There can be infiltration of the growing cyst throughout the liver and neighboring organs with the formation of metastases in more distant organs. Mortality rates are known to reach 100% in untreated patients (Flisser, 1998). *Echinococcus multilocularis* is the most deadly helminth infection of humans. Prognosis is improved through early diagnosis leading to early treatment.

5.5.2 TAENIASIS SAGINATA AND TAENIASIS ASIATICA

Infection with *Taenia saginata* is not usually severe and morbidity relates to digestive disturbance (Figure 5.6). Once diagnosis has been made, the infection responds well to anthelminthic treatment (Section 5.7). Anecdotes to the effect that the presence

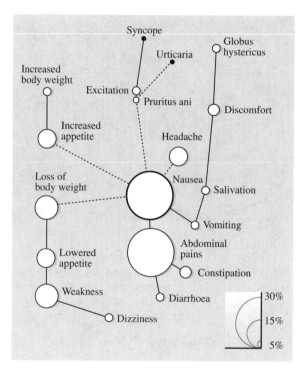

FIGURE 5.6 A cluster analysis of symptoms that accompany or are reported to accompany infection with *Taenia saginata*. From Figure 2 (Kociecka, 1987) with permission from Elsevier and the approval of ZS Pawlowski.

of the tapeworm in the gut stimulates food intake still await quantitative verification. Perhaps the most distressing symptom for people living where good hygiene exists is finding a live proglottid that has left the anus and is wriggling around in under-clothes or crawling on the bed linen. Proglottides can be observed to crawl out of freshly passed stools; this behavior is indicative of infection with *T. saginata*. Morbidity due to *T. asiatica* is similar to that observed for *T. saginata* (Figure 5.6).

Morbidity has an economic impact. Fan (1997) estimated the losses due to taeniasis asiatica in three endemic regions in Asia. The infection was estimated to deprive communities in the mountainous part of Taiwan about USD 11.3 million annually, the community on Cheju Island, Korea, about USD 13.6 million annually, and the community on Samosir Island, Indonesia, about USD 2.4 million annually. The major source of economic loss in Fan's calculations was the loss of seven days wages for each infected person followed by the overall costs of medical care.

5.5.3 Taeniasis solium (NCC)

Morbidity due to intestinal *T. solium* is not generally serious apart from the risk of autoinfection (Section 5.1.2) that can then lead to cysticercosis. The symptoms are much the same as those described for *T. saginata* (Figure 5.6) by Botero et al. (1993). In the body, cysticerci may develop beneath the skin, in skeletal and heart muscle, in the eyes, and in the brain. Subcutaneous cysticerci are more common in Africa than elsewhere and are readily detected as prominent protrusions in the skin. NCC is often present in individuals with subcutaneous cysticerci. Acute muscular inflammation accompanies cysticerci in muscles and abnormal heart rhythms are detected when cysticerci develop in the heart (Botero et al., 1993).

Our knowledge of the extent and understanding of the pathology of NCC increased once scanning and imaging equipment became available. The form the pathology takes depends on whether the cysticerci are simple or racemose, alive or dead, and the degree of inflammation, immune response, and calcification that these states stimulate (Botero et al., 1993). Patients are likely to suffer a range of neurological and psychiatric disorders, which begin about 3.5 years after the formation of the cysticerci. NCC has been studied extensively in Mexico and is recognized there as the main cause of late-onset epilepsy as it now is elsewhere in Latin America, Asia, and Africa (Del Brutto, 2005). Botero et al. (1993) have summarized procedures for the treatment and management of individual cases of NCC and ocular cysticercosis. The costs of medical supervision and hospital facilities for individual cases are beyond the resources of a public health program designed to control morbidity.

5.5.4 HYMENOLEPIASIS DUE TO *HYMENOLEPIS NANA*

Light infections appear to be asymptomatic, but morbidity is observed when the worm burden increases to perhaps as many as 2,000 worms in a child. The symptoms include enteritis, abdominal pain, diarrhea, nausea, vomiting, loss of appetite, and dizziness (Muller, 2002).

5.5.5 BURDEN OF CESTODIASIS

Torgerson (2003), Budke et al. (2004), and Carabin et al. (2005) have developed methods for estimating the economic and social burdens of echinococcosis and neurocysticercosis. Such information is required if public health priorities are to be set (Chapter 3). There are two main elements in the assessment: first, the monetary and nonmonetary costs to human health; second, the costs to livestock production. Costs to livestock production involve infected sheep giving birth to fewer lambs, producing less milk, providing wool of poor quality, and having lower rates of feed conversion (Torgerson, 2003). This reduced productivity has a direct impact on the income and spending power of those involved in livestock production at whatever level.

Reliable estimates of medical care costs and lost income are available from some countries (see Carabin et al., noting the years to which the figures apply). For example, the annual cost of hospitalization for a case of NCC in the U.S. was about USD 6,500 and the loss to an individual's income was about USD 1,400. With 1,100 cases of NCC treated in the U.S. at the time, the combined national cost of these aspects of the disease would be about USD 8.75 million annually. In Mexico during 1982, the annual loss in wages associated with NCC was calculated to be USD 225 million. Figures like these are accurate insofar as they apply to the cases that were recorded. How many cases go unrecorded? How efficient is the financial management for determining the costs of medical care? In some places, where NCC, CE, and AE are endemic, resources for diagnosis, data collection, and financial management may be inadequate for use in economic models.

Budke et al. and Carabin et al. managed to calculate the total DALYs lost to echinococcosis (CE and AE) in Shiqu County, China. Overall, the DALYs lost to CE and AE amounted to 50,933 (41,995 to 61,026) with an average of about 0.81 DALYs lost per person in Shiqu County. The Global Burden of Disease study carried out under the auspices of the World Health Organization had already concluded that the average DALYs lost per person throughout China because of infectious and non-infectious disease was 0.18. The studies are important because they highlight the impact of specific helminthiasis in a region of high endemicity. If the average DALYs lost per person due to echinococcosis were to be calculated for the entire population of China, echinococcosis might seem to be a minor problem.

Torgerson (2003), in an evaluation of the costs of *E. granulosus* infections to the state of Jordan, applied methods that allowed the variables needed for the analysis to be randomly shuffled simultaneously along likely frequency distributions. In other words, uncertainty caused by lack of precision in the data was able to be reduced. The main results concerning human health costs and animal health costs are summarized in Table 5.4. Two general points emerge. First is the matter of purchasing power parity. Losses presented in Table 5.4 may seem small when compared with costs in G7 countries, but one USD in many countries of the south will purchase much more than one USD in the U.S. Second, with control in mind, Torgerson's work demonstrates that zoonotic infections affect various sectors in a nation's economy and so the sharing of resources for control should be discussed between government ministries and affected institutions.

5.6 IMMUNITY

Much information is available about mammalian resistance to larval taeniid worms, with more being known about humoral than cellular responses. The spur for research into this branch of immunobiology has been concern to develop reliable diagnostic tests and to produce vaccines to protect livestock from infection. If the vaccination of livestock becomes widely available, humans will also be protected to a significant extent. Flisser (1998) has summarized the immune response as occurring in two phases. The first phase is directed against oncospheres during their attempts to penetrate the intestinal wall. The second phase is directed at the established metacestodes in their parenteral sites. Evidence indicates that the immune response against oncospheres is more successful than that against metacestodes, which seem to be endowed with adaptations for avoiding immune effectors.

5.7 ANTHELMINTHIC TREATMENT

Health authorities in areas where cestodiasis is endemic are urged to ensure that access to essential drugs is available for the treatment of those in need. The anthelminthic drugs recommended for the treatment of intestinal infections with *Taenia* spp. and *Hymenolepis nana* are niclosamide and praziquantel (WHO, 2004; Table 5.5).

TABLE 5.4
Summary of the Economic Evaluation of the Costs (USD) of Echinococcosis due to *Echinococcus granulosus* Infection in Jordan. Abstracted from Table 2 and Table 3 (Torgerson, 2003) with permission from Elsevier.

Human Health Costs in Jordan		Animal Health Costs in Jordan	
Cost Item	Median Value (USD)	Cost Item	Median Value (USD)
Hospital costs	70,500	Loss of edible liver	850,588
Reduced quality of life following treatment	36,402	Loss of lamb production	1,494,355
Morbidity due to undiagnosed CE	96,525	Loss of kid production	157,422
Loss of income in fatal cases	34,220	Loss of sheep meat production	298,722
Overall human health costs	185,517	Loss of goat meat production	28,839
		Loss of sheep's milk production	266,286
		Loss of goat's milk production	92,794
		Loss of value of fleece from infected sheep	227,267
		Overall animal health losses	3,576,753

Note: USD, U.S. Dollars

Clearly niclosamide is not suitable for universal application in large-scale programs intended to reduce morbidity given the range of doses required according to age and the need for purgatives. Praziquantel has been used in communities in single oral doses of 5 mg/kg body weight (Cruz et al., 1989; Sarti et al., 2000). In such cases, the drug has an effectiveness of about 67% (Sarti and Rajshekhar, 2003). A system should always be in place to deal with any adverse events. A person infected with intestinal *T. solium* might also have early or asymptomatic neurocysticercosis. Praziquantel at doses above 5 mg/kg may provoke an intense inflammatory reaction against cysticerci and so initiate neurological symptoms (Pawlowski et al., 2005). Although not mentioned in the *WHO Model Formulary* (WHO, 2004), presumably the drugs recommended for the treatment of *T. saginata* and *T. solium* will be appropriate for the treatment of *T. asiatica.*

Echinococcosis (CE and AE) may under certain circumstances be treated with either albendazole or mebendazole, but the use of these drugs for this purpose is complicated and requires medical supervision for individual patients. Similarly, neurocysticercosis may respond to treatment with praziquantel, but close supervision of patients should be provided. Guidance is provided in the *WHO Model Formulary* (WHO, 2004).

5.8 PREVENTION AND CONTROL

The transformation from an agrarian to an industrial society in the North during the 19th century released most people from a subsistence economy; food production

TABLE 5.5
WHO-Recommended Anthelminthic Drugs for the Treatment of the Intestinal Stages of Cestode Infections in Humans, Abstracted from information in WHO (2004).

Anthelminthic Drug (Drug Form)	Species of Cestode (Disease)	Dosage (By Mouth)
Niclosamide (500mg, chewable tablet)[a]	*Hymenolepis nana* (hymenolepiasis)	Adults and children over 6 years, 2 g as a single dose, then 1g daily for 6 days. Children 2–6 years, 1g as a single dose, then 500 mg daily for 6 days. Children under 2 years, 500 mg as a single dose, then 250 mg daily for 6 days.
Niclosamide (500mg, chewable tablet)	*Taenia saginata*[b,c] (taeniasis saginata) *Taenia solium*[d] (taeniasis solium)	Adults and children over 6 years, 2 g as a single dose after a light breakfast followed by a purgative after 2 hours. Children 2–6 years, 1 g as a single dose; children under 2 years, 500 mg as a single dose; followed by a purgative after 2 hours in both cases.
Praziquantel (150 mg and 600 mg tablets)	*Hymenolepis nana* (hymenolepiasis)	Adults and children over 4 years, 15–25 mg/kg body weight as a single dose.
Praziquantel (150 mg and 600 mg tablets)	*Taenia saginata* (taeniasis saginata) *Taenia solium*[d] (taeniasis solium)	Adults and children over 4 years, 5–10 mg/kg body weight.

[a] Niclosamide tablets must be chewed thoroughly (or crushed) before swallowing with water.

[b] In the case of *T. saginata*, half the appropriate dose of niclosamide may be taken after breakfast, followed by the other half an hour later and then the purgative after a further 2 hours.

[c] Niclosamide may be used to treat infections of *Diphyllobothrium latum* by following the same doses and procedure as described for *T. saginata*.

[d] Botero et al. (1993) advise caution when treating infections of *T. solium* in case the drugs induce retroperistalsis (see section 5.1.2).

including the supply of meat became a specialized industry. Hygiene laws introduced the public health intervention of meat inspection and the population has enjoyed protection from infection with *Echinococcus, Taenia,* and *Trichinella.*

Several interventions are available for the prevention and control of cestodiasis. (1) The introduction or strengthening of meat inspection if the process is lacking or under resourced. (2) The introduction or strengthening of diagnostic services for patient care and community surveillance in endemic countries, including statutory reporting of cases of CE, AE, and NCC. (3) The promotion of measures to improve standards of animal husbandry in the livestock industry and to help people who rely on their animals for food and household needs. (4) Health awareness and education campaigns for people who consume beef and pork and for people who work in branches of food supply including livestock production, slaughter, butchery, processing, distribution, and the restaurant sector. (5) Quality control for the safety of meat and meat products for home consumption and for export. (6) Regular anthelminthic distribution to treat dogs for *Echinococcosis* and pigs for cysticercosis. (7) Health checks on arrival for migrants from endemic countries. (8) Vaccination of domestic animals in endemic countries. Aspects of these interventions also apply to the exclusion of trematodiasis and trichinellosis from the human food chain are discussed in some detail in Chapter 7 and Chapter 13, respectively. Attention here is paid to the animal vaccination strategy and the prospect it offers for protecting the public from infection and disease (Lightowlers et al., 2000; Lightowlers, 2003; Heath et al., 2003). The basic idea is to vaccinate sheep, goats, pigs, and cattle (intermediate hosts) so that cestode metacestodes would be unable to develop, thereby protecting humans and dogs (definitive hosts) from infection and steadily disrupting transmission. If humans were exposed to less risk of consuming infected pork and subsequently harboring mature *T. solium*, they would be much less likely to ingest infective eggs and so develop NCC. Nor would they serve to perpetuate the infection. Exploitation of the host response to oncosphere antigens has led to the development of protective vaccines using recombinant DNA technology. With such vaccines, high levels of protection have been demonstrated against *E. granulosus* infection in sheep (96 to 100%), against *T. ovis* infection in sheep (92 to 98%), and against *T. saginata* infection in cattle (94 to 99%) (Lightowlers et al., 2000). How to use this form of intervention most effectively is discussed by Heath et al. (2003).

REFERENCES

Allan JC, Velasquez-Tohom M, Torres-Alvarez R et al., 1996a. Field trial of the coproantigen-based diagnosis of *Taenia solium* taeniasis by enzyme-linked immunosorbent assay. *American Journal of Tropical Medicine and Hygiene* **54**, 352–356.

Allan JC, Velasquez-Tohom M, Garcia-Noval J et al. 1996b. Epidemiology of intestinal taeniasis in four, rural, Guatemalan communities. *Annals of Tropical Medicine and Parasitology* **90**, 157–165.

Allan JC, Wilkins PP, Victor CW et al. 2003. Immunodiagnostic tools for taeniasis. *Acta Tropica* **87**, 87–93.

Arai H (ed.). 1980. *Biology of the Tapeworm* Hymenolepis diminuta. New York and London: Academic Press.

Ash LR and Orihel TC. 1990. *Atlas of Human Parasitology,* 3rd edition. Chicago: American Society of Clinical Pathologists Press.

Botero D, Tanowitz HB, Weiss L and Wittner M. 1993. Taeniasis and cysticercosis. *Infectious Disease Clinics of North America* **7**, 683–697.

Budke CM, Jiamin Q, Zinststag J et al. 2004. Use of disability adjusted life years in the estimation of the disease burden of echinococcosis for a high endemic region on the Tibetan Plateau. *American Journal of Tropical Medicine and Hygiene* **71**, 56–64.

Carabin H, Budke CM, Cowan LD et al. 2005. Methods for assessing the burden of parasitic zoonoses: echinococcosis and cysticercosis. *Trends in Parasitology* **21**, 327–333.

Chandler AC and Read CP. 1961. *Introduction to Parasitology* 10th edition. New York and London: John Wiley & Sons Inc.

Craig PS. 2004. Epidemiology of echinococcosis in western China. In: *Echinococcosis in Central Asia: Problems and Solutions* (eds. P Torgerson and B Shaikenov). Almaty, Duir: INTAS Network Project 01-0505. pp. 43–58.

Craig PS. 2006. Epidemiology of human alveolar echinoccosis in China. *Parasitology International* **55**, S221–S225.

Craig PS, Rogan MT and Allan JC. 1996. Detection, screening and community epidemiology of taeniid cestode zoonoses: cystic echinococcosis, alveolar echinoccosis and neurocysticercosis. *Advances in Parasitology* **38**, 169–250.

Crompton DWT. 1987. Human helminthic populations. In: *Balliere's Clinical Tropical Medicine and Communicable Diseases. Intestinal Helminthic Infections* (ed. ZS Pawlowski). London and Philadelphia: Balliere Tindall. pp. 489–510.

Cruz M, Davis A, Dixon H et al. 1989. Operational studies on the control of *Taenia solium* taeniasis/cysticercosis in Ecuador. *Bulletin of the World Health Organization* **67**, 401–407.

Del Brutto OH. 2005. Neurocysticercosis. *Seminars in Neurology* **25**, 243–251.

Deplazes P, Eckert J, Pawlowski ZS et al. 1991. An enzyme-linked immunosorbent assay for diagnostic detection of *Taenia saginata* copro-antigens in humans. *Transactions of the Royal Society of Tropical Medicine and Hygiene* **85**, 391–396.

Diaz JF, Verastegui M, Gilman RH et al. 1992. Immunodiagnosis of human cysticercosis (*Taenia solium*): a field comparison of an antibody-enzyme-linked immunosorbent assay (ELISA), an antigen-ELISA, and an enzyme-linked immunoelectrotransfer blot (EITB) assay in Peru. The cysticercosis working group in Peru (CWG). *American Journal of Tropical Medicine and Hygiene* **46**, 610–615.

Eckert J, Schantz PM, Gasser RB et al. 2001. Geographic distribution and prevalence. In: *WHO/OIE Manual on Echinococcosis in Animals and Humans: a Public Health Problem of Global Concern* (eds. J Eckert, MA Gemmel, FX Meslin and ZS Pawlowski). Paris: Office International des Epizooities. pp. 100–142.

Engels D, Urbani C, Belotto A et al. 2003. The control of (neuro)cysticercosis: which way forward? *Acta Tropica* **87**, 177–182.

Eom KS and Rim H-J. 2001. Epidemiological understanding of *Taenia* tapeworm infections with special reference to *Taenia asiatica* in Korea. *Korean Journal of Parasitology* **39**, 267–283.

Fan PC. 1997. Annual economic loss caused by *Taenia saginata asiatica* taeniasis in three endemic areas of East Asia. *Southeast Asian Journal of Tropical Medicine and Public Health* **28**, (Suppl.) 217–222.

Fan PC and Chung WC. 1998. *Taenia saginata asiatica*: epidemiology, infection, immunological and molecular studies. *Journal of Microbiology, Immunology and Infection* **31**, 84–89.

Filice C, Brunetti E, D'Andrea F and Filice G. 1997. Minimal invasive treatment for hydatid abdominal cysts: PAIR (Puncture, Aspiration, Injection, Reaspiration) — state of the art. Geneva: World Health Organization. WHO/CTD/SIP/97.3.

Flisser A. 1998. Larval cestodes. In: *Topley and Wilson's Microbiology and Microbial Infections,* 9th edition. Volume 5 *Parasitology* (eds. FEG Cox, JP Kreier and D Wakelin). London, Sydney, Auckland: Arnold.

Flisser A, Vazquez-Mendoza A, Martinez-Ocana J et al., 2005. Short report: evaluation of a self-detection tool for tapeworm carriers for use in public health. *American Journal of Tropical Medicine and Hygiene* **72**, 510–512.

Georgi, JR. 1985. *Parasitology for Veterinarians,* 4th edition. Philadelphia and London: WB Saunders Company.

Hall A, Latham MC, Crompton DWT and Stephenson LS. 1981. *Taenia saginata* (Cestoda) in Western Kenya: the reliability of faecal examinations in diagnosis. *Parasitology* **83,** 91–101.

Heath DD, Jensen O and Lightowlers MW. 2003. Progress in the control of hydatidosis using vaccination — a review of formulation and delivery of the vaccine and recommendations for practical use in control programmes. *Acta Tropica* **85,** 133–143.

Ito A, Nakao M and Wandra T. 2003. Human taeniasis and cysticercosis in Asia. *Lancet* **362,** 1918–1920.

Jenkins DJ. 2005. Hydatid control in Australia: where it began, what we have achieved and where to from here. *International Journal of Parasitology* **35,** 733–740.

Jenkins DJ, Romig T and Thompson RCA. 2005. Emergence/reemergence of *Echinococcus* spp. — a global update. *International Journal for Parasitology* **35,** 1205–1219.

Kern P, Wen H, Sato N et al. 2006. WHO classification of alveolar echinococcosis: principles and application. *Parasitology International* **55,** S283–S287.

Keymer A. 1982. Tapeworm infections. In: *Population Dynamics of Infectious Diseases* (ed. RM Anderson). London and New York: Chapman and Hall. pp. 109–138.

Kociecka W. 1987. Intestinal cestodiases. In: *Balliere's Clinical tropical medicine and Communicable Diseases. Intestinal Helminthic Infections* (ed. ZS Pawlowski). London and Philadelphia: Balliere Tindall. pp. 677–694.

Lightowlers MW. 2003. Vaccines for the prevention of cysticercosis. *Acta Tropica* **87,** 129-135.

Lightowlers MW, Flisser A, Gauci CG et al. 2000. Vaccination against cysticercosis and hydatid disease. *Parasitology Today* **16,** 191–196.

Machnicka B, Dziemian E and Zwierz C. 1996. Detection of *Taenia saginata* antigens in faeces by ELISA. *Applied Parasitology* **37,** 106–110.

McManus DP. 2006. Molecular discrimination of taeniid cestodes. *Parasitology International,* **55,** S31–S37.

Macpherson CNL. 1983. An active intermediate host role for man in the life cycle of *Echinococcus granulosus* in Turkana, Kenya. *American Journal of Tropical Medicine and Hygiene* **32,** 397–404.

Macpherson CNL and Craig PS. 1991. Echinococcosis — a plague on pastoralists. In: *Parasitic Helminths and Zoonoses in Africa* (eds. CNL Macpherson and PS Craig). London: Unwin Hyman. pp. 25–53.

Macpherson CNL and Milner R. 2003. Performance characteristics and quality control of community based ultrasound surveys for cystic and alveolar echinococcosis. *Acta Tropica* **85,** 203–209.

Macpherson CNL and Wachira TWM. 1997. Cystic echinococcosis in Africa south of the Sahara. In: *Compendium of Cystic Echinococcosis in Africa and in Middle Eastern Countries with Special Reference to Morocco* (eds. FL Anderson, H Ouhelli and M Kachani). Brigham Young University. pp. 245–277.

Macpherson CNL, Zeyle E, Romig T et al. 1987. Portable ultrasound scanner vs. serology in screening for hydatid cysts in a nomadic population. *Lancet* ii, pp. 259–262.

Muller R. 2002. *Worms and Human Disease,* 2nd edition. Wallingford, Oxon: CABI Publishing.

Noble ER and Noble GA. 1982. *Parasitology. The Biology of Animal Parasites.* Philadelphia: Lea & Febiger.

PAHO/WHO (Pan American Health Organization/World Health Organization).1997. PAHO/WHO Informal Consultation on the Taeniosis/Cysticercosis Complex. Geneva: World Health Organization. Series HCT/AIEPI-5.

Pawlowski ZS. 1982. Taeniasis and cysticercosis. In: *Handbook Series in Zoonoses*. Section C. *Parasitic Zoonoses* Volume I (eds. L Jacobs and P Arambulo). Boca Raton, Florida: CRC Press Inc. pp. 313–348.

Pawlowski ZS, Allan J and Sarti E. 2005. Control of *Taenia solium* taeniasis/cysticercosis: from research towards implementation. *International Journal for Parasitology* **35**, 1221–1232.

Rabiela MT, Rivas A and Flisser A. 1989. Morphological types of *Taenia solium* cysticerci. *Parasitology Today* **5**, 357–359.

Raether W and Hanel H. 2003. Epidemiology, clinical manifestations and diagnosis of zoonotic cestode infections: an update. *Parasitology Research* **91**, 412–438.

Sahbra GH, Arfaa F and Bijan H. 1967. Intestinal helminthiasis in the rural area of Khuzestan, South-west Iran. *Annals of Tropical Medicine and Parasitology* **61**, 352–357.

Sarti E and Rajshekhar V. 2003. Measures for the prevention and control of *Taenia solium* taeniosis and cysticercosis. *Acta Tropica* **87**, 137–143.

Sarti E, Schantz PM, Plancarte A et al. 1992. Prevalence and risk factors for *Taenia solium* taeniasis and cysticercosis in humans and pigs in a village in Morelos, Mexico. *American Journal of Tropical Medicine and Hygiene* **46**, 677–685.

Sarti E, Schantz PM, Avila G et al. 2000. Mass treatment against human taeniasis for the control of cysticercosis: a population-based intervention study. *Transactions of the Royal Society of Tropical Medicine and Hygiene* **94**, 85–89.

Schantz PM. 1982. Echinococcosis. In: *Handbook Series in Zoonoses*. Section C. *Parasitic Zoonoses* Volume I (eds. L Jacobs and P Arambulo). Boca Raton, Florida: CRC Press Inc. pp. 231–277.

Schantz PM and Sarti-Gutierrez E. 1989. Diagnostic methods and epidemiologic surveillance of *Taenia solium* infection. *Acta Leidensia* **57**, 153–163.

Schantz PM, Chai J, Craig PS et al. 1995. Epidemiology and control of hydatid disease. In: *Echinococcus and Hydatid Disease* (eds. RCA Thompson and AJ Lymbery). Wallingford, Oxon: CABI Publishing.

Schantz PM, Wang H, Qiu J et al. 2003. Echinococcosis on the Tibetan Plateau: prevalence and risk factors for cystic and alveolar echinococcosis in Tibetan populations in Qinghai Province, China. *Parasitology* **127**, S109–S120.

Singh B. 1997. Molecular methods for diagnosis and epidemiological studies of parasitic infections. *International Journal for Parasitology* **27**, 1135–1145.

Sullivan KM. 1994. Polymerase chain reaction. In: *The Encyclopaedia of Molecular Biology* (ed. J Kendrew). Oxford: Blackwell Science. pp. 864-866.

Torgerson PR. 2003. Economic effects of echinococcosis. *Acta Tropica* **85**, 113–118.

Torgerson PR, Shaikenov BS and Kuttubaev O. 2000. Cystic echinococcosis in Central Asia: new epidemic in Kazakhstan and Kyrgystan. In: *Cestode Zoonoses: Echinococcosis and Cysticercosis* (eds. P Craig and Z Pawlowski). Nato SCIENCE Series 341. Oxford: IOS Press. pp. 99–105.

Wandra T, Sutisna P, Dharmawan NS et al. 2006. High prevalence of *Taenia saginata* taeniasis and status of *Taenia solium* cysticercosis in Bali, Indonesia, 2002 to 2004. *Transactions of the Royal Society of Tropical Medicine and Hygiene* **100**, 346–353.

Ward HB. 1907. *Iconographia Parasitorum Hominis*. Lincoln, Nebraska: The University of Nebraska.

WHO. 1980. *Manual of Basic Techniques for a Health Laboratory*. Geneva: World Health Organization.

WHO. 2003. (Informal Working Group). International classification of ultrasound images in cystic echinococcosis for application in clinical and field settings. *Acta Tropica* **85**, 253–261.

WHO. 2004. *WHO Model Formulary 2004*. Geneva: World Health Organization.

6 Dracunculiasis

Infection with the nematode *Dracunculus medinensis* causes dracunculiasis, also known as dracontiasis or guinea worm disease (Ukoli, 1984; Edungbola, 1985; Edungbola and Kale, 1991). The burden of this disease is borne by poor subsistence farming communities; human health is impaired and agricultural productivity is diminished. At the start of the 1980s as many as 15 million cases of this disabling disease were estimated to exist, but by 2003 the number of cases had been reduced to about 32,000 (WHO, 2004a). India, Pakistan, and seven other countries where dracunculiasis was endemic have been certified free from transmission. This significant progress followed the establishment of a major campaign, launched in 1991 by the World Health Assembly, to eradicate dracunculiasis based on partnerships between the governments of endemic countries and institutions including the Carter Center, UNICEF, U.S. Centers for Disease Control and Prevention, the Japanese Government, and various corporate and private donors.

Remarkable success in the control of dracunculiasis in Nigeria, where 25% of the world's dracunculiasis cases were located when the global effort began, has been achieved through a national eradication campaign (Edungbola et al., 1992). The original date set for the global eradication of dracunculiasis was 1995, but this date has had to be extended to 2005. The decline in the extent of dracunculiasis has been managed without recourse to anthelminthic treatment programs or vaccines (Section 6.6). Although much of the content of this book places emphasis on the importance of deworming in the control of morbidity due to helminthiasis, the dracunculiasis experience merits inclusion because it should inspire and encourage public health workers dealing with other forms of helminthiasis. Helminthiasis can be controlled in countries where the terrain is challenging, the infrastructure is minimal, resources are scarce, and the population endures the deprivation that exudes from poverty.

6.1 LIFE HISTORY AND HOST RANGE

Dracunculus medinensis has an indirect life history with humans serving as definitive hosts and freshwater cyclopoid crustaceans as intermediate hosts (Kale, 1982; Muller, 2002). Sexual reproduction takes place in the human host where male and female worms mate, probably in the subcutaneous tissues about 4 months after establishment of the infection. Mature female worms are long and thin with a mean length of 600 mm (range 500 to 1200 mm) and a breadth of 1 to 2 mm. Male worms are much smaller and are rarely found except in experimental infections in dogs. In humans, male worms are assumed to die after mating and become calcified; radiography occasionally reveals images that may be calcified worms.

Female *Dracunculus* are essentially viviparous and they release first-stage larvae (L1) into water. The female body alters to become little more than a sac containing embryos and L1, other organs having atrophied. The mature female migrates to position the anterior end of her body just below the surface of the skin. The worms have a predilection for sites in the lower limbs and ankles. A papule develops at this site followed by a blister. The host's reaction to the escape of a few larvae from the parent worm initiates blister formation. After the blister has burst, thousands of L1 are discharged when the exposed end of the worm makes contact with cold water. The larvae may escape from the female's vulva or ruptured anterior end. After either event, part of the worm is likely to continue to protrude from the site of the blister. This permits accurate diagnosis and enables the worm to be trapped and gradually wound out of the limb during several weeks. Discharges of L1 occur until the female worm's brood is exhausted, most being released during the first contact with water (Kale, 1982). About a year passes from the ingestion of infective larvae to the release of L1 by a mature female *D. medinensis.*

The L1 are ingested by cyclopoid copepods belonging to the genera *Mesocyclops, Metacyclops,* and *Thermocyclops.* After ingestion by the copepod the L1 penetrates the gut wall to reach the body cavity (Muller, 2002). Two molts occur in the copepod's body cavity to produce the third-stage larva (L3), which is infective to humans. Development to this stage takes about two weeks at the water temperature in places where dracunculiasis still occurs. When water containing such copepods is swallowed, the L3 are freed from the copepods' tissues during digestion and they pass through the intestinal wall and migrate to the subcutaneous tissues where two more molts occur and growth takes place before mating.

There are reports of *Dracunculus* infections in nonhuman primates and other wild and domesticated mammals in areas where dracunculiasis is endemic (Kale, 1982; Muller, 1991). Emerging female worms have been observed from nonhuman mammals on many occasions. Whether these animals pose any threat to prevention and control programs in the capacity of reservoir hosts is unresolved, but seems doubtful assuming the worms have been identified accurately as *D. medinensis.* Since experimental infections have been established in dogs and several species of monkeys, awareness of the possibility of animal reservoirs of infection should not be lost if eradication is to be achieved.

6.2 TRANSMISSION TO HUMANS

Poor rural people generally depend on ponds and open wells for drinking water in areas where dracunculiasis is endemic. This lack of safe drinking water puts the population at risk of infection because swallowing copepods carrying infective larvae of *D. medinensis* is the only known route of infection of any significance (WHO, 2003), although there is the possibility that the vagina may offer an occasional route of infection (St George, 1975). Infection tends to be seasonal (Belcher et al., 1975). During the dry seasons, water levels fall so that the copepod populations become greater per unit volume of water and the chance of their being included in a household's supply increases. By the time the rains arrive and the ponds are filling, mature worms are ready to release larvae, a process facilitated by the opportunity for easier human contact with water. The blisters are extremely painful and infected people tend to seek relief by immersing their ankles and legs in water (Muller, 2002).

6.3 DIAGNOSIS

In endemic areas diagnosis can be made with confidence from clinical signs and symptoms and from information from the infected person once the female *D. medinensis* has reached a site for the discharge of larvae. The formation of a blister and the accompanying itching and pain confirm the presence of the infection. Larvae may be retrieved from the blister after it has burst. The shape of a female worm may be seen under the skin and the anterior end of a worm may be protruding from a limb (Ukoli, 1984). Immunological tests have been developed to enable prepatent infections to be detected. Such tests have little use in the public health setting because infected individuals rarely experience any morbidity until worms are fully mature by which time diagnosis is cheap, simple, and accurate (Muller, 2002).

6.4 DISTRIBUTION, PREVALENCE, AND INTENSITY

Dracunculiasis is now restricted to Africa. Between 1989 and 2003, the number of cases fell from 892,005 to 32,193 and were to be found in Benin, Burkina Faso, Cote d'Ivoire, Ethiopia, Ghana, Mali, Mauritania, Niger, Nigeria, Sudan, Togo, and Uganda (Figure 6.1; WHO, 2003, 2004a). At the time 20,299 of the known cases were to be found in Sudan where political dispute had probably disrupted the implementation of control efforts. Governments are obliged not to announce that the eradication of dracunculiasis has been achieved in their territory until a precertification process has been carried out. Precertification is now in progress in Cameroon, Central African Republic, Chad, and Kenya, where disease transmission has been successfully interrupted. Despite the progress made toward the eradication of dracunculiasis, the possibility will remain that villages in remote areas may have been missed. Vigilance will be needed to ensure that this disease does not reemerge, perhaps through infectivity changes in the populations of dracunculid worms in nonhuman mammals.

Before the success of measures for prevention and control (Section 6.6), prevalences as high as 83% were recorded in Nigerian village communities with peak values being observed in the 15 to 40 age group. Some individuals frequently became reinfected indicating that protective immunity had not developed in them while others did not succumb to infection despite repeated exposure to infected cyclopoids. There is not always any obvious gender difference in rates of infection (Ukoli, 1984). Variations in infection rates with age or gender can usually be explained by reference to behavior, drinking habits, and mobility (WHO, 2004a). The need for quantitative information about the prevalence of *D. medinensis* in communities is rapidly becoming obsolete when entire countries such as Benin and Cote d'Ivoire now have less than 100 cases each (WHO, 2004a). Usually the intensity of infection is restricted to 1 to 2 worms per person (Kale, 1977), indicating that many infective larvae perish in the body after contaminated water has been swallowed.

Percentage of cases reported by endemic countries in 2003

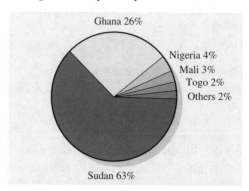

Comparison of guinea-worm cases reported by Sudan and other endemic countries, 1989-2003

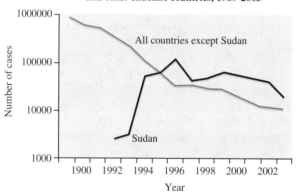

FIGURE 6.1 The distribution of dracunculiasis cases in Africa. From WHO (2004a) with permission from the World Health Organization (www.who.int).

6.5 MORBIDITY

The recorded mortality rate associated with dracunculiasis has always been low, of the order of 0.1% (Ukoli, 1984). Acute morbidity involving blister formation, intense itching, and severe pain is experienced by every person harboring mature female *D. medinensis* that have reached the stage of larval discharge. Burst blisters regularly become sites of secondary bacterial infection and develop into ulcers or abscesses if bacteria invade the track made as the worm migrates down the limb. Infected people may be seriously incapacitated following bacterial invasion. Many cases have been observed of impairment of joints leading to permanent reduced mobility.

When dracunculiasis had free reign, many of the people suffering from it would be confined to bed for about a month at the time of peak agricultural activity (Belcher et al., 1975; WHO, 2004a). Combinations of pain, incapacitation, and reduced mobility invariably coincided with the planting and harvesting routines of subsistence farmers. The contribution of infection with *D. medinensis* to the persistence of rural poverty can be appreciated by reference to literature cited by Kale (1982) and Ukoli (1984), both authors having lived and worked in Nigeria where dracunculiasis once had a stranglehold on the health and prosperity of millions of Nigerians for many years. Smith et al. (1989) found that in a community in Imo State, Nigeria, the duration of dracunculiasis symptoms lasted for an average of 12 weeks (range 3 to 29 weeks) with more than half the adults being bedfast for some of the time. The observed immobility coincided with the peak time for harvesting yams and rice. Hours and Cairncross (1994) selected at random 195 sufferers from two sites in Ghana and tested their ability to perform six activities. They were able to show that seemingly mild dracunculiasis affected the ability of people to perform physical tasks.

Dracunculiasis has had a devastating effect on the economy of endemic regions. In an area of southeastern Nigeria, with a population of about 1.6 million people, dracunculiasis was implicated in reducing the gross profit from rice production in the region by about USD 20 million annually. Estimates indicate that the economic rate of return on eradication investment will be about 29% once the task has been completed (WHO, 2003). Of equal importance in terms of development and the future of the region, dracunculiasis was judged to be responsible for about 33% of school absenteeism, low rates of school enrollment, and high rates of school drop-out rates.

6.6 PREVENTION AND CONTROL

The fact that drinking water contaminated with infected cyclopoid crustaceans is the key factor in the transmission of *D. medinensis* focused attention on the need to reduce significantly the risk of transmission for those living in endemic areas by improving the safety of drinking water. Dracunculiasis would be controlled and eradicated if people could be protected from the source of infection in water, leading to sustainable prevention of infection. Measures have included treating water sources with temephos (Abate) to kill crustaceans and encouraging people to boil drinking water. Both measures have environmental consequences; chemicals pollute the environment and boiling requires firewood, an energy resource of increasing scarcity across Africa. Large cisterns have been constructed for the collection of rain water (Edungbola et al., 1992). Efforts have been made to enable people to draw water from wells and ponds without getting wet by building steps and parapets, but immersion of an affected limb in water helps to relieve pain. Dressing blisters with clean bandages reduces the discharge of L1 into the water and reduces the risk of bacterial infection. Oral doses of metronidazole (25 mg/kg body weight daily for 10 days, with a daily maximum of 750 mg for children) relieve symptoms and weaken the anchorage of mature female worms so that they are more easily withdrawn by the traditional winding technique (WHO, 2004b).

The national eradication campaign in Nigeria placed great emphasis on health education and raising public awareness about dracunculiasis in order to develop good compliance and participation with control measures. Health messages were promoted through traditional town criers, posters, newspapers, magazines, cartoons, commemorative postage stamps, radio jingles, documentary videos, and a special

TABLE 6.1
Perceived Socioeconomic Benefits of Hand Pump on Guinea Worm Reduction among 368 Households Interviewed. From Table 3 of Edungbola et al. (1988) with permission from the American Society of Tropical Medicine and Hygiene.

	Yes		No	
Benefit	**Households**	**%**	**Households**	**%**
Larger farm size and greater harvest	308	83.7	60	16.3
Better school enrollment and attendance	342	92.9	26	7.1
Higher annual income	260	70.6	108	29.4
Reduced problem of water scarcity	348	94.6	20	5.4
Greater social respect and more visitors coming to the community	304	82.6	64	17.4
Better health	356	96.7	12	3.3

film (Edungbola et al., 1992). Health workers embarking on the control of other forms of helminthiasis can learn much from this aspect of the Nigerian campaign against dracunculiasis; there is no need to reinvent the wheel.

If resources are available, the provision of protected water supplies provides benefits in addition to relief from dracunculiasis. There are other obvious health benefits from safe water supplies such as a reduction in the risk of contracting schistosomiasis or having episodes of diarrhea. In a large-scale study in rural Nigeria, Edungbola et al. (1988) demonstrated that providing villages with boreholes dedicated for drinking water not only reduced the prevalence of the disease from about 50% to nearly 0% in three years, but also brought economic and social benefits to the communities. A series of interviews revealed that in 368 households most people with access to the boreholes could identify improvements in their quality of life in addition to being freed from the pain and disability of dracunculiasis (Table 6.1). Most of those interviewed attributed some of these benefits to increased income because of increased productivity from farming. Villagers could afford school fees, buy radios, and even feel that making a pilgrimage to Mecca was no longer just a dream.

Much of the success of the eradication program has been achieved by a cheaper and simpler process than installing boreholes. All that is needed is to provide households with nylon filters to remove the crustaceans from the water before drinking. This cheap and robust technique has met with excellent compliance and support from the communities where dracunculiasis was entrenched. The prevention and control of the infection and disease represents a triumph for community involvement, the dedication of health workers, the response of governments, the contribution of partners, and the efforts of scientists who investigated the ecology, life history, and transmission of *D. medinensis*.

REFERENCES

Belcher DW, Wurapa FK, Ward WB and Lourie IM. 1975. Guinea worm in Southern Ghana: its epidemiology and impact on agricultural productivity. *American Journal of Tropical Medicine and Hygiene* **24**, 243–249.

Edungbola LD. 1985. A general appraisal of dracunculiasis and its implications. In: *Dracunculiasis in Nigeria.* Proceedings of the first national conference on dracunculiasis in Nigeria (eds. LD Edungbola, OO Kale and SJ Watts). Emene, Enugu, Nigeria: Federal Social Development Directorate Unit. pp. 11–21.

Edungbola LD and Kale OO. 1991. Guinea-worm disease. *Surgery* **98**, 2351–2354.

Edungbola LD, Watts SJ, Alabi TO et al. 1988. The impact of a UNICEF-assisted rural water project on the prevalence of guinea worm disease in Asa, Kwara State, Nigeria. *American Journal of Tropical Medicine and Hygiene* **39**, 79–85.

Edungbola LD, Withers PC, Braide EI et al. 1992. Mobilization strategy for guinea worm eradication in Nigeria. *American Journal of Tropical Medicine and Hygiene* **47**, 529–538.

Hours H. and Cairncross S. 1994. Long-term disability due to guinea worm disease. *Transactions of the Royal Society of Tropical Medicine and Hygiene* **88**, 559–560.

Kale OO. 1977. The clinico-epidemiological profile of guinea worm in the Ibadan District of Nigeria. *American Journal of Tropical Medicine and Hygiene* **26**, 208–214.

Kale OO. 1982. Dracontiasis. In: *Handbook Series in Zoonoses, Section C: Parasitic Zoonoses,* Volume II (ed. MG Schultz). Boca Raton, Florida: CRC Press Inc. pp. 111–122.

Muller R. 1991. Dracunculiasis in Africa. In: *Parasitic Helminths and Zoonoses in Africa* (eds. CNL Macpherson and PS Craig). London: Unwin Hyman. pp. 204–223.

Muller R. 2002. *Worms and Human Disease,* 2nd edition. Wallingford, Oxford: CABI Publishing.

St George J. 1975. Bleeding in pregnancy due to retroplacental situation of guinea worms. *Annals of Tropical Medicine and Parasitology* **69**, 383–386.

Smith GS, Blum D, Huttly SRA et al. 1989. Disability from dracunculiasis: effect on mobility. *Annals of Tropical Medicine and Parasitology* **83**, 151–158.

Ukoli FMA. 1984. *Introduction to Parasitology in Tropical Africa.* Chichester, New York, Brisbane, Toronto, Singapore: John Wiley and Sons Limited.

WHO. 2003. *Global Defence against the Infectious Disease Threat* (ed. MK Kindhauser). Geneva: World Health Organization.

WHO. 2004a. *Eradicating Guinea-worm Disease.* Geneva: World Health Organization (WHO/CDS/CPE/CEE/2004.47).

WHO. 2004b. *WHO Model Formulary 2004.* Geneva: World Health Organization.

7 Food-Borne Trematodiasis

From 600 million to 750 million people are reckoned to be at risk from infection with food-borne trematodes (FBTs) that have a worldwide distribution (WHO, 1995, 2000). Although many species of digenean fluke are involved in zoonotic FBT infections (see Coombs and Crompton, 1991) we have concentrated mainly on the public health significance of seven species (Table 1.1 and Table 7.1). This approach allows the principles that underpin the prevention and control of any FBT infection to be identified. Whether an infection has public health significance or not is not always easy to establish. We consider that public health status applies when health professionals have concluded that measures for the protection and treatment of a community should be implemented. Provision to treat and support individuals suffering from one of the rarer forms of food-borne trematodiasis is equally important, but may not justify a program of public health intervention.

7.1 FOOD SECURITY AND FOOD SAFETY

For most people in places where FBTs are endemic, infections arise from the consumption of infective metacercariae contaminating raw, pickled, smoked, or undercooked fish, crabs, crayfish, and vegetables that have been harvested from fresh or brackish water (Anantaphruti, 2001). The health professional is faced with a dilemma. How can food be made safe without putting food security at risk? A strong educational campaign aimed at reducing the consumption of food in ways that kill metacercariae would seem to be the approach to adopt. But such measures should be introduced with care and sensitivity to cultural attitudes and prepared customs (see Table 9.2 in Murrell and Crompton, 2006) and should not endanger food security. Fish is a most important source of protein for millions of people living where FBT infections are endemic. Also, aquaculture is now a major industry in Asia where most of the world's finfish and shellfish are produced. Acceptable quality and safety standards must be agreed and practiced for aquaculture products intended for consumption locally, nationally, and internationally (see Section 7.8).

7.2 LIFE HISTORIES AND HOST RANGES

All FBTs considered here undergo asexual reproduction in a snail host and sexual reproduction in a mammalian host (Yamaguti, 1975; Table 7.2). Although the adult flukes are hermaphrodites, cross fertilization rather than self fertilization appears to be the basis of sexual reproduction. The eggs pass out of the definitive mammalian host into fresh or brackish water inhabited by susceptible snails. Depending on the

TABLE 7.1
Food-Borne Trematode Infections of Public Health Significance

Family	Genus and Species	Human Infections (Millions)[1]	Principal Morbidity
Fasciolidae	*Fasciola gigantica* (PB, LiF)[2] *F. hepatica* (PB, FW, LiF)	2.4	Biliary abnormalities and fibrosis, pain, weight loss
Heterophyidae[3]	*Haplorchis taichui* (AB, IF) *Metagonimus yokogawai* (AB, IF)	1.3	Abdominal pain, maldigestion, diarrhea
Opisthorchidae	*Clonorchis sinensis* (AB, LiF)	7.0	Liver malfunction, weight loss,
	Opisthorchis viverrini (AB, LiF)	10.3	cholecystitis, cholethiasis, cholangiocarcinoma
Paragonimidae	*Paragonimus westermani*[4] (AB, LuF)	20.7[5]	Chest pain, fever, pneumonia and lung malfunction

[1]The estimates, taken from data provided by endemic countries to a WHO Study Group (WHO, 1995) were not markedly different from estimates discussed at a more recent WHO/FAO Workshop (WHO, 2004a) with the exception of *Paragonimus* infections (see below).

[2]AB animal-borne metacercariae carried in food of animal origin; FW, metacercariae sometimes free in fresh water; PB plant-borne metacercariae carried in food of plant origin; LiF, liver fluke; LuF, lung fluke; IF, intestinal fluke.

[3]These species have been chosen arbitrarily to represent the assemblage of species from the 11 families of digenean fluke (see Coombs and Crompton, 1991) known to infect humans in addition to the Heterophyidae. An example of the diversity of intestinal fluke infections has been published by Chai and Lee (2002) with reference to the Republic of Korea.

[4]*Paragonimus* taxonomy is in a process of revision (see Blair et al., 1999). In this chapter we assume that information relevant to *P. westermani* applies to all species of *Paragonimus* infecting humans irrespective of the reliability of the identification (see Yokogawa, 1982).

[5]This figure may now be an overestimate. Since the publication of the Study Group's report (WHO, 1995), a major reduction in the number of cases of *Paragonimus* spp. has been reported in China (WHO, 2004a).

FBT species, the eggs either release miracidia that penetrate the snail's surface or are eaten by the snail in which the miracidia escape from the eggshells. There then follows the typical digenean stages of sporocyst and redia culminating in the release of cercariae (Muller, 2002). Usually the cercariae leave the snail to encyst and develop into metacercariae on or in other aquatic animals (crabs, crayfish, and fish) and plants. Cercariae have a short lifespan, rarely surviving for more than 48 h under favorable conditions. The rate of development in an intermediate host is dependent on the ambient temperatures in its habitat. The cycle is completed when humans, or the many other species of susceptible definitive hosts, become infected after swallowing metacercariae (Table 7.3).

TABLE 7.2
Life Cycles of Selected Food-Borne Trematodes Infecting Humans[1]

Species	First Intermediate Hosts (Snails)[2]	Second Intermediate Hosts[2]	Nonhuman Definitive Hosts[3]
Fasciolidae			
Fasciola gigantica (LiF)	*Lymnaea* spp.	(Metacercaria adhere to water plants)	Buffalo, cattle, goat, sheep, and other mammals
F. hepatica (LiF)	*Lymnaea* spp.	(Metacercariae adhere to water plants)	Cattle, goats, sheep, and other mammals
Fasciolopsis buski (IF)	*Gyraulus chinensis*	(Metacercariae adhere to water plants)	Dogs, pigs
Heterophyidae			
Haplorchis taichui (IF)	*Melania obliquegranosa*	*Puntius* spp.	Cats, cattle, dogs
Metagonimus yokogawi (IF)	*Semisulcospira libertina*	*Plecoglassus ativelis* and other species of fish	Cats, dogs, rats
Opisthorchidae			
Clonorchis sinensis (LiF)	*Parafossarulus manchouricus*	113 species of freshwater fish	Cats, dogs, pigs, rats, and other mammals
Opisthorchis viverrini (LiF)[4]	*Bithynia* spp.	*Cyclocheilichthys siaja* and other fish	Cats, dogs, and other mammals
Paragonimidae			
Paragonimus westermani (LuF)[5]	*Semisulcospira* spp.	51 species of freshwater crab and crayfish	Cats, dogs, pigs, and other mammals

[1] Information for liver flukes (LiF) taken from Mas-Coma and Bargues (1997), for the lung fluke (LuF) from Blair et al. (1999), and WHO (1995), and for the intestinal flukes (IF) from Chai and Lee (2002) and Coombs and Crompton (1991).

[2] Detailed information about the range of first and second intermediate host species utilized by FBTs in different locations is to be found in Muller (2002) and WHO (1995).

[3] The hosts listed are those serving as important "reservoir" hosts in human communities. Wild birds and mammals are frequently found to be or are suspected of being infected.

[4] In some places, including the former USSR, *O. felineus* has significant public health importance (IARC, 1994).

[5] The mostly widely distributed species of *Paragonimus* known to infect humans (Blair et al., 1999) but see Yokogawa (1982) for details of other species.

The low host specificity that has evolved for many species of FBT and their mammalian definitive hosts, many of which are domesticated animals, presents a formidable challenge to attempts to control morbidity due to FBT infections (WHO, 1995). *Opisthorchis viverrini* is an exception, having a high specificity for humans. In addition, paratenic hosts are known in which the ingested metacercariae develop

TABLE 7.3
Biological Features of Selected Species of Food-Borne Trematodes Infecting Humans. Compiled from information in Blair et al. (1999), Chai and Lee (2002), Mas-Coma and Bargues (1997), Muller (2002), and Rim (1986).

	Fasciola hepatica	*Metagonimus yokogawai*	*Clonorchis sinensis*	*Opisthorchis viverrini*	*Paragonimus westermani*
Life span (years)	9–13	—	15–25	10	6–20
Prepatency (days)	112	—	28	56	65[1]
Fecundity (egg production/ fluke/day)	—	1500	600/g feces[2]	3000	—
Generation time (months)	6	—	3	4.5	—
Location of adults	Biliary system and gall bladder	Jejunum	Bile ducts and gall bladder	Distal bile ducts	Lungs
Size of adults (mm)	$20–50 \times 6–13$	$1–2 \times 0.4–0.6$	$11–20 \times 3–5$	$5–10 \times 0.8–1.7$	$8–16 \times 4–8$

[1] In experimental nonhuman hosts.

[2] Such an egg count suggests the presence of a heavy infection (Muller, 2002, p. 38). In nonhuman definitive hosts infected experimentally, egg production has been found to vary from about 1,000 to 4,000 per fluke per day according to the host species (Rim, 1986).

partially but do not mature. For example, wild boar (*Sus scrofa*) can serve as a paratenic host for *P. westermani* and so provide another route for human infection with this fluke (Blair et al., 1999). There are also reports of frogs, snakes, and birds being able to contribute to the persistence of FBT infections.

7.3 RISK FACTORS INVOLVED IN TRANSMISSION AND ESTABLISHMENT OF INFECTION

The morbidity experienced by humans from infection with FBTs results from the integration of human behavior and activity with the flukes' natural life cycles. Each category of risk factor (Maclean et al., 1999; Mas-Coma, 2004; Mas-Coma and Bargues, 1997; Murrell, 2002; Fried et al., 2004; Murrell and Crompton, 2006; WHO, 1995) has developed unwittingly through human ignorance.

Transmission of FBTs to humans occurs through daily nutrition when metacercariae are consumed in raw, undercooked, or inadequately preserved food. Transmission occurs on social occasions when special dishes are eaten at parties, weddings, and so on. Alcohol consumption may increase the risk and as a result men may be more vulnerable than women. Asian men regularly gather to drink and enjoy fish dishes with their friends, oblivious to health hazards. In Viet Nam, De et al.

(2003) recorded higher infection rates with *C. sinensis* in males than in females. Out of samples of 3725 males and 3075 females, the prevalence rates were 32 and 14.6%, respectively (p < 0.0001). Similar observations have been made in Korea and China. In Japan a form of "sashimi" made from raw wild boar meat is now popular. Although eating freshwater crabs is known to be the risk factor for infection with *Paragonimus,* over 70% of paragonimiasis cases in a reemergence of the disease had arisen through eating contaminated wild boar "sashimi" (Nawa, 2000). Unsanitary practices in the kitchens of restaurants easily lead to chopping boards and utensils acquiring infective metacercariae that may then contaminate other foods. Metacercariae are roughly spherical and small (diameter 100 to 200 µm); thousands may be harbored by one fish. Information about the retention and destruction of infectivity of FBT metacercariae is summarized in Table 7.4.

Famine, whether caused by natural disasters, climatic change, or political upheaval, is a risk factor making poor people vulnerable to infection with FBTs. Famine and displacement threaten food supplies and food security. After the Nigerian civil war (1967 to 1970) there was an outbreak of paragonimiasis because starvation forced people to change their food habits and ingest infective metacercariae

---

TABLE 7.4
Conditions Reported to Inactivate the Metacercariae of Selected Species of Food-Borne Trematodes

Genus and Species

Clonorchis sinensis[1]
- 50% viable after 60 days between 3–6°C
- Die after 3 min at 65°C
- Die after 2.5 h at 39°C

Fasciola hepatica[1]
- Resistant, remain viable for a long period but killed by excessive heat and dryness

Fasciolopsis buski[2]
- Killed by direct exposure to sunlight after 20–30 min
- Killed by 1% HCl in 18 days
- Killed by 3% acetic acid in 6 days
- Killed by Japanese soy sauce in 30 min

Opisthorchis viverrini[3]
- Inhibited after 1 h in 4% commercial vinegar
- Inhibited after 10 days in 0.9% aqueous NaCl
- Inhibited after 1 h in 30% aqueous NaCl

Paragonimus westermani[4]
- Survive heating in crab muscle for 37 min at 45°C
- Survive 3 h in 1% salt solution
- Remain viable for 43 h at 22°C in millet wine (10% alcohol)

[1] Mas-Coma and Bargues (1997).
[2] Suzuki quoted by Rim (1982).
[3] Waigakul quoted in WHO (1995).
[4] Yokogawa et al. (1960).

(Nwokolo, 1972). An increase in infection with *P. uterobilateralis* occurred in Nigeria more recently when people living in the region of the Igwun Basin increased their consumption of freshwater crabs. The overall prevalence rate reached 16.8%; the value was higher in children who ate more raw crabs than adults (Udonsi, 1987).

Another major risk factor that consolidates the persistence of human exposure to FBT infection is the use of fresh human and animal feces to enrich the nutrient levels in household fish ponds where susceptible snails reside. This practice is common in Asia (Naegel, 1990). For example, latrines are often positioned directly above fish ponds in rural Viet Nam. Contamination of water with pig feces is a most important risk factor for infection with *Fasciolopsis buski* (De et al., 2003). That household fish ponds require nutrient enrichment is beyond dispute if food is to be produced, but alternative fertilizers to fresh human and animal excreta will need to be available if the practice is to decline.

The risk of becoming infected depends on water and responds to seasonal changes in rainfall. In parts of Thailand, where the dry season extends from November to May, contaminated feces make contact with dry ground rather than water and so fewer snails are exposed to eggs of *O. viverrini*. Meanwhile the dry conditions also affect snail and fish populations. Studies have shown that humans are most at risk of infection during the last third of the rainy season, when infected fish populations have built up, and at the start of the dry season, when water levels fall making fish easier to catch (Wykoff et al., 1965). After investigating the ecology of *Bithynia siamensis goniomphalos,* an important snail in the life cycle of *O. viverrini,* Brockelman et al. (1986) concluded that the snail's ability to survive a drought, its lack of a specific habitat, its capacity for dispersal in surface waters, and its generalized feeding habits would allow it to overcome antisnail control measures.

7.4 DIAGNOSIS

7.4.1 PARASITOLOGICAL DIAGNOSIS

For epidemiological surveys and community control programs the microscopic detection of fluke eggs in stool and sputum samples remains the best means for the diagnosis of human infection by liver and intestinal flukes (Table 7.1; Cheesborough, 1998). The recommended procedure is the Kato Katz technique (WHO, 1994; Appendix 3). Eggs of *Clonorchis, Opisthorchis,* and various intestinal flukes are extremely difficult to distinguish from each other as are eggs of the two species of *Fasciola.* Eggs of *O. felineus* tend to be longer and narrower than those of *O. viverrini.* Comprehensive morphological descriptions of FBT eggs are to be found in WHO (1980). Information about known FBT infections in the locality should be sought at the time when the survey is being planned. Eggs of *Paragonimus* spp. are also detectable in stool samples, but the presence of an infection is better confirmed by finding the fluke's eggs in sputum samples (Blair et al., 1999). Technicians should be aware of *spurious* infections such as the detection of *Fasciola* eggs in stool samples from an individual who has recently eaten sheep's liver harboring a large burden of patent flukes.

The Kato Katz procedure is unlikely to detect infections consisting of a few worms; there will be too few eggs in a 50 mg sample of stool. In a clinic or hospital setting stool samples may be examined after processing with a sedimentation or concentration technique. Thienpont et al. (1979) give details of different techniques for processing and examining stool samples and for collecting stool samples directly from cats, cattle, dogs, pigs, and rats that are known to be important reservoir hosts for FBTs. In a hospital, biliary drainage, duodenal aspirates, and biopsy specimens may also be examined microscopically for FBT eggs. Obtaining egg counts, expressed in a standard manner such as eggs per gram feces (epg), is of paramount importance. This quantitative measurement provides indirect information about the intensity of infection or worm burden. Generally, the greater the intensity, the greater the severity of morbidity provided the helminth is one that does not replicate in its host. Following changes in intensity allows the effectiveness of anthelminthic treatment to be assessed.

Diagnosis of FBT infections in snails, crustaceans, and fish or free cercariae may be beyond the resources of a control program. Accurate identification of the stages requires expert knowledge applied without time constraints. Scholz et al. (1991) described a method for distinguishing between the metacercariae of opisthorchiid and heterophyid metacercariae from cyprinid fish in Laos. Sukontason et al. (2001) tested the method to their satisfaction in a survey of infections of *Haplorchis taichui* and *O. viverrini* in cyprinid fish in Thailand.

7.4.2 Immunodiagnosis

Numerous methods based on host responses have been tried for the diagnosis of FBT infections. The following tests have been used effectively in community surveys and in support of individual case diagnosis. The possibility of cross-reactivity caused by the presence of concurrent infections should not be overlooked:

- *Fasciola* — FAST-ELISA (Falcon assay screening test-enzyme-linked immunosorbent assay) and the EITB (enzyme-linked immunoelectrotransfer blot) procedure (see Hillyer et al., 1992).
- *Clonorchis* — ELISA and other immunodiagnostic tests are reviewed by Rim (2005). Immunoblot method based on *Clonorchis*-specific antigens gives results that correlate with the intensity of infection (Choi et al., 2003).
- *Opisthorchis* — MAb-ELISA (monoclonal antibody-based ELISA) for the detection of coproantigen evaluated by Sirisinha et al. (1995). Immunoblot method based on *Opisthorchis*-specific antigens gives results that correlate with the intensity of infection (Choi et al., 2003).
- Intestinal flukes — Chai and Lee (2002), in an extensive review of the identification, based largely on egg morphology, of 19 species of intestinal fluke infections in Korea comment that serological tests such as ELISAs are helpful in false negative cases and cite two references.
- *Paragonimus* — ELISA using cysteine proteinase antigens (Ikeda et al., 1996) and ELISA using ES antigens (Narain et al., 2005).

7.4.3 CLINICAL DIAGNOSIS

Clinical diagnosis depends on attributing signs and symptoms to the presence of particular worm infections (Table 7.1). As the following information shows (abstracted from Mas-Coma and Bargues, 1997; WHO, 1995, 2000, 2004a) there is considerable overlap between the infections of concern, so confirmation with parasitological evidence is desirable. Individuals with light patent infections in the liver or intestine (passing too few eggs to be detected by the Kato Katz procedure) may be asymptomatic. Secure clinical diagnosis is likely to require health facilities beyond the reach of many poor people suffering from food-borne trematodiasis.

7.4.3.1 Fascioliasis

Patient lives where livestock is infected with *Fasciola* and has a history of eating raw freshwater water plants. Patients may present with fever, sweating, abdominal pain, dizziness, cough, bronchial asthma, urticaria, and eosinophilia. The variety and complexity of the signs and symptoms are related to the intensity of infection and its duration. The use of noninvasive techniques, including computed tomography, radiology, and ultrasound (Richter et al., 2002) will provide further information about the course of the disease and will reduce the chances of confusion with other diseases.

7.4.3.2 Fasciolopsiasis

Apparently most infections with *Fasciolopsis buski* are of light intensity and asymptomatic. Patients may present with nausea, vomiting, epigastric pain, and diarrhea. Sometimes edema of the face, abdominal wall, and lower limbs may be observed. Suspected cases require parasitological confirmation that might involved the microscopic examination of stool samples collected at regular intervals.

7.4.3.3 Clonorchiasis

Patient lives where *Clonorchis* is endemic and has a history of eating freshwater fish that has not been fully cooked. Abdominal discomfort, distension and pain, anorexia, indigestion, and irregular bowel movements are experienced. Heavily infected patients are weak, lose weight, and may be jaundiced. Bleeding from the upper gastrointestinal tract may occur and be accompanied by anemia (Rim, 2005).

7.4.3.4 Opisthorchiasis

Patient lives where *Opisthorchis* is endemic and has a history of eating freshwater fish that has not been fully cooked. Fever, abdominal pain, dizziness, flatulence, intolerance of fatty food, and jaundice are some of the more common signs and symptoms. Opisthorchiasis in its early stage may be difficult to distinguish from clonorchiasis. If facilities are available, ultrasonography offers a reliable means of diagnosing the contribution of opisthorchiasis to hepatobiliary disease including cholangiocarcinoma (Elkins et al., 1990).

7.4.3.5 Metagonimiasis (Representative of Intestinal Fluke Infections)

Patients present with abdominal pain, intermittent mucous diarrhea, colicky pain, and lethargy. Interestingly, individuals with heavy infections may experience little more than minor discomfort. A patient from whom over 60,000 flukes were recovered complained of mild indigestion and epigastric pain (WHO, 1995).

7.4.3.6 Paragonimiasis

Patient lives where *Paragonimus* is endemic and has a history of eating freshwater crabs or crayfish that have not been fully cooked. The early stage of infection may not stimulate any obvious signs and symptoms. As the flukes settle in the lungs chronic coughing, thoracic pain, fever, fatigue, and dyspnea are detected. Blood-stained viscous sputum (hemoptysis) is characteristic of paragonimiasis and also of tuberculosis. Paragonimiasis is not infrequently assumed to be tuberculosis and *vice versa.* Health professionals should be made aware of this diagnostic problem. *Paragonimus* sometimes migrates to the brain. This form of ectopic paragonimiasis should be suspected in patients with severe headaches, blurred vision, mental confusion, and other cerebral symptoms.

7.5 EPIDEMIOLOGY

Food-borne trematode infections have a wide distribution, which have the potential to expand further as a consequence of globalization. The movement of people through economic migration, displacement, tourism, or illegal trafficking has the potential to facilitate the spread of infection. When transmission depends on food and food habits, opportunities for spread seems to be boundless or at least "boundary-less." European colonialism, with its export of infected livestock, probably dispersed *F. hepatica* to pastures new. Extra vigilance must be introduced and sustained to protect people from exposure to the metacercariae of FBTs.

Public health programs designed to (1) prevent infection, (2) control morbidity, (3) deploy resources for individual case management, and (4) protect communities from the spread of infections depend on the best available epidemiological information. Such information is rarely complete and requires constant revision and refinement. This information provides the base line against which the impact of health care and preventative actions can be measured and sustained. The distribution of FBTs in endemic countries is determined by the distribution of host and vector snail species. That factor explains why food-borne trematodiasis often has an uneven distribution in a country. A hypothetical national prevalence of 2% may well conceal considerable pockets of disease where the prevalence is 50%.

7.5.1 FASCIOLIASIS

Cases of human fascioliasis have been reported from 61 countries in Africa, the Americas, Asia, Europe, and Oceania (WHO, 1995) and numbers of cases have been increasing recently in these countries (Mas-Coma, 2003, 2004). *Fasciola hepatica* is found in Europe, Africa, the Americas, Asia, and Oceania, and *F. gigantica* in Africa and Asia. Fascioliasis is the vector-borne disease with the widest known latitudinal, longitudinal, and altitudinal distribution. Curiously high prevalence of *Fasciola* infection does not always correlate with high prevalence in livestock (Mas-Coma, 2003).

Human fascioliasis can be classified into three epidemiological situations: hypoendemic, mesoendemic, and hyperendemic (Mas-Coma et al., 1999; Mas-Coma, 2003, 2004). This scheme offers a useful framework for assessing the significance of fascioliasis in an area and planning a response to its impact on public health (Table 4.2).

Hypoendemic fascioliasis prevails when the prevalence of infection is > 1%, the mean intensity is > 50 epg (eggs per gram feces), and the role of human participation in transmission through passing *Fasciola* eggs is judged to be negligible because adequate sanitation and waste disposal facilities are available and are used properly.

Mesoendemic fascioliasis prevails when the prevalence of infection ranges from 1 to 10%. The prevalence in children aged between 5 and 15 years may be higher. The mean intensity ranges from 50 to 300 epg; individual egg counts above 1000 epg would rarely be encountered. Humans may participate in transmission through passing *Fasciola* eggs into the community because sanitation and waste disposal would be unsatisfactory and outdoor defecation might be practiced.

Hyperendemic fascioliasis prevails when the prevalence of infection is > 10%. The prevalence in children aged between 5 and 15 years is usually higher than the average for the community. The mean intensity will be > 300 epg and counts > 1000 epg will not be unusual. Humans will be contributing to transmission through passing *Fasciola* eggs into the community. Sanitation and waste disposal will be inadequate and indiscriminate outdoor defecation will be a common practice.

A striking example of hyperendemic fascioliasis is to be found in northern Bolivia where the prevalence of infection has been found to reach 38% in some communities. Prevalence and intensity reached a peak in children aged between 9 and 11 years; some had egg counts > 5000 epg (Hillyer et al., 1992; Esteban et al., 1997). Children are not always the group most at risk. Over 500 cases of infection with *F. gigantica* have been diagnosed recently in people living mainly in the central and southern parts of Viet Nam (De et al., 2003). Most of the patients were aged between 21 and 50 years. There is no satisfactory explanation for this outbreak, particularly since *F. gigantica* is well known from cattle in northern Viet Nam.

7.5.2 CLONORCHIASIS

Clonorchis sinensis is habitually found in Cambodia, China, Japan, Korea, Laos, Taiwan, Viet Nam, and the East Asian region of the former USSR. The transport of aquaculture products and the movements of people from endemic areas explains how *C. sinensis* infections have been detected in Australia, Brazil, Canada, France, Hong Kong, Malaysia, Panama, Philippines, Saudi Arabia, Singapore, Surinam, Thailand, and U.S. (WHO, 1995; Mas-Coma and Bargues, 1997; Rim, 2005).

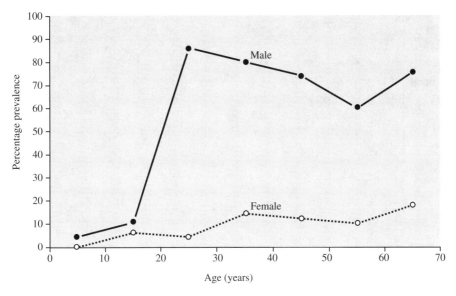

FIGURE 7.1 Relationship between host age and gender and prevalence of *Clonorchis sinensis* in an area of low intensity in Korea. From Figure 4a (Rim, 1986; IARC, 1994) with permission from the International Agency for Research on Cancer and the World Health Organization (www.who.int).

The epidemiology of *C. sinensis* in Korea has been investigated in detail (Rim, 1986) and the results have wider application. Peak prevalence is reached in middle age, with more men being infected than women (Figure 7.1, Figure 7.2, and Figure 7.3). The intensity of infection, measured indirectly as egg counts, follows the prevalence trend (Figure 7.4 and Figure 7.5). *Clonorchis* has also been shown to be aggregated by families in its distribution. It is to be expected that the frequency distribution of numbers of flukes per host will be overdispersed with most flukes in the community being carried by relatively few people. Prevalence rates are extremely variable, with values collected from the literature by Mas-Coma and Bargues (1997) ranging from about 1 to 83%. Variations in prevalence and intensity rates reported from different locations are determined by social customs, food habits, and the distribution and ecology of intermediate hosts. It is clear, however, that clonorchiasis declines in response to community anthelminthic treatment applied in the context of improving economic prosperity (Mas-Coma and Bargues, 1997).

7.5.3 INTESTINAL TREMATODIASIS

Infection with *Fasciolopsis buski* is known from Bangladesh, Cambodia, China, India, Indonesia, Laos, Myanmar, Taiwan, and Vietnam (Rim, 1982). Pigs are the important reservoirs of infection in human communities. Children are usually more heavily infected than adults. The metacercariae are ingested when contaminated water plants are eaten. Some evidence suggests that the prevalence rate is higher in children living close to where the plants grow compared with the rate in those who

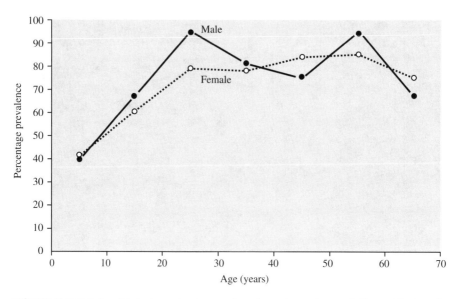

FIGURE 7.2 Relationship between host age and gender and prevalence of *Clonorchis sinensis* in an area of high intensity in Korea. From Figure 4c (Rim, 1986; IARC, 1994) with permission from the International Agency for Research on Cancer and the World Health Organization (www.who.int).

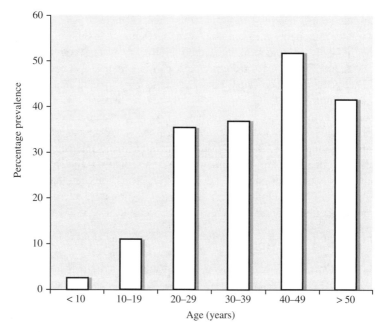

FIGURE 7.3 Relationship between host age and prevalence of *Clonorchis sinensis* in Vietnam. From Figure 2 (De et al., 2003) with permission from SEAMEO, Bangkok.

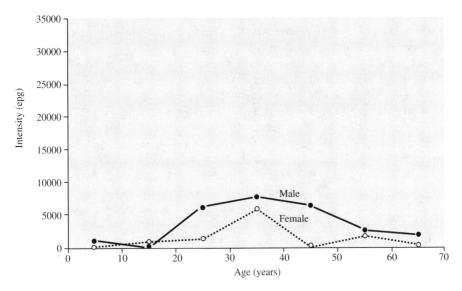

FIGURE 7.4 Relationship between host age and gender and fecal egg counts of *Clonorchis sinensis* infection in an area of low intensity in Korea. From Figure 4b (Rim, 1986; IARC, 1994) with permission from the International Agency for Research on Cancer and the World Health Organization (www.who.int).

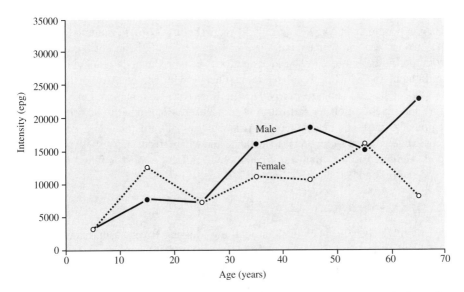

FIGURE 7.5 Relationship between host age and gender and fecal egg counts of *Clonorchis sinensis* infection in an area of high intensity in Korea. From Figure 4d (Rim, 1986; IARC, 1994) with permission from the International Agency for Research on Cancer and the World Health Organization (www.who.int).

eat the same plants but live further away from the water. How contaminated water plants are handled before being eaten is crucial to the persistence of fasciolopsiasis. In some endemic areas, people in cities are more likely to boil water plants before consumption. In some rural areas, water plants tend to dry out before consumption if people live some distance from the source; *Fasciolopsis* metacercariae lose infectivity in dry conditions. In some areas women maintain the freshness of the plants by regular dipping in cold water thereby preserving infectivity (see Rim, 1982).

The many other intestinal trematodes, sometimes called minute flukes or heterophyid flukes, even though flukes from other families are included, have a global distribution in tropical and subtropical countries (Fried et al., 2004) especially in Asia (Harinasuta et al., 1987). Human infections with these small flukes have been diagnosed in Australia, Brazil, China, Egypt, Greenland, Hawaii (U.S.), India, Indonesia, Iran, Israel, Japan, Korea, Malaysia, Palestine, Philippines, Spain, Taiwan, Thailand, Tunisia, Turkey, and states of the former USSR (Velasquez, 1982; WHO, 1995). Information about the epidemiology of intestinal trematodes and the complexity of their life cycles can be inferred from knowledge gained in Korea where 17 indigenous species have been identified to date (Chai and Lee, 2002). Cats, dogs, and rats serve as the common reservoir hosts and human infections arise from swallowing metacercariae carried by freshwater and brackish water fish, snakes, snails, bivalves, and insects.

Intestinal infection with heterophyid flukes has recently been considered to be an emerging public health problem in a region of the Philippines, where 36% of the community surveyed was found to be infected (Belizario et al., 2001). Whether the problem is emerging or one that has only been recognized recently is not clear. A plot of the relationship between host age and prevalence reveals the highest prevalence in people aged between 15 and 30 years (Figure 7.6). The same infection was detected in men and women, but more men were found to be infected than women. With morbidity in mind, Belizario et al. (2001) proposed that heterophyid infections could be classified as light, moderate, and heavy with egg counts of 1 to 100, 101 to 1000, and > 1000 epg, respectively, after Kato Katz examination. One conclusion from the investigation by Beliziario et al. (2001) is that health authorities in this region of Asia should be aware of the risk of intestinal trematodes infections to poor people living where sanitation is inadequate and where freshwater fish is an important food. The intensity of numbers of heterophyid flukes per host is found to be overdispersed (Giboda et al., 1991; Figure 7.7).

7.5.4 Opisthorchiasis

Opisthorchis viverrini is endemic around the Mekong River basin, being found in Cambodia, Laos, and Thailand (WHO, 1995). The fluke may also be endemic in Malaysia. Infected human emigrants from these endemic countries have been diagnosed in the Czech Republic, France, Germany, Japan, Kuwait, and the U.S. (Mas-Coma and Bargues, 1997; Rim, 2005). *Opisthorchis felineus* is endemic in territories of the former USSR including Belorussia, Kazakstan, Russia, Siberia, and Ukraine. The risk of becoming infected and contracting opisthorchiasis is considerable for rural people living in areas of river basins (Mas-Coma and Bargues, 1997).

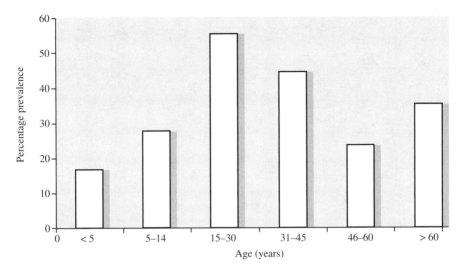

FIGURE 7.6 Relationship between host age and the prevalence of *Heterophyes heterophyes* infection in the southern Philippines. From data in Belizario et al. (2001).

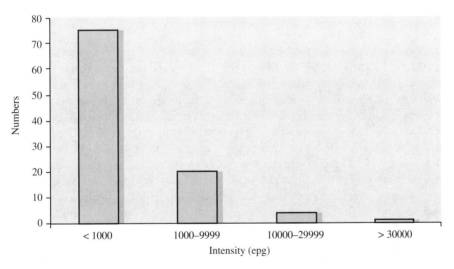

FIGURE 7.7 Overdispersed frequency distribution of numbers of heterophyid fluke infections in people from Lao PDR. From Figure 1 (Giboda et al., 1991) with permission from SEAMEO, Bangkok.

The public health impact of *O. felineus* in Russia is demonstrated by the fact that between 1986 and 1992 health authorities detected from 40,000 to 95,000 cases annually in river basin areas (WHO, 1995). Infection with *Opisthorchis felineus* occurs in fish-eating mammals in parts of Europe, but information about the infection in humans is not available (WHO, 1995).

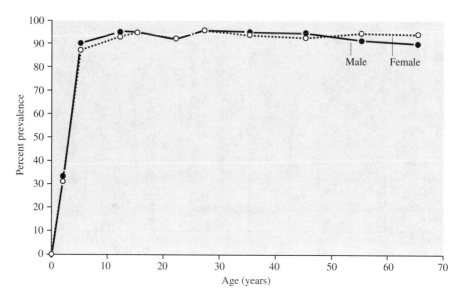

FIGURE 7.8 (Thailand). Relationship between host age and gender and prevalence of *Opisthorchis viverrini* in an area of high intensity in Thailand. From Figure 4e (Upatham et al., 1984; IARC, 1994) with permission from the International Agency for Research on Cancer and the World Health Organization (www.who.int).

The relationship between host age and prevalence of *O. viverrini* infection in part of Thailand is shown in Figure 7.8 and between host age and intensity in Figure 7.9. Gender effects are not obvious in this data set. Intensity is seen to increase steadily with age, an observation that correlates with the incidence of cholangiocarcinoma (Section 7.6.1.2). Numbers of flukes are highly aggregated in an infected community with most being found in relatively few people (Sithithaworn et al., 1991; Haswell-Elkins and Elkins, 1998). For example, in a survey in northeast Thailand, 8,991 flukes of the 11,027 flukes recovered from 241 people treated with praziquantel were passed in the stools of 27 of the people (Haswell-Elkins et al., 1991). A related study by the same team demonstrated that egg counts represent a reasonably secure indication of the intensity of infection, there being a positive correlation between the number of worms passed in the stools after anthelminthic treatment and the number of eggs per gram of stool before treatment (Elkins et al., 1991).

7.5.5 PARAGONIMIASIS

Species of *Paragonimus* have a global distribution. *Paragonimus westermani* and *P. heterotremus* are the common species in Asia (Blair et al., 1999), *P. uterobilateralis* and *P. africanus* in West Africa (Ukoli, 1984), and *P. mexicanus* in the Americas (WHO, 1995). In endemic countries the distribution of lung fluke infections is focused on locations where people traditionally eat raw crabs or use crab juice for therapeutic purposes. Comprehensive lists of intermediate and reservoir hosts for each species of *Paragonimus* found in human hosts have been provided by Yokogawa

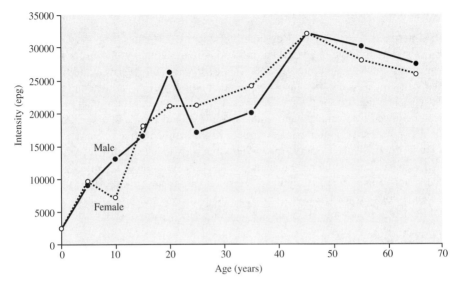

FIGURE 7.9 Relationship between host age and gender and fecal egg counts of *Opisthorchis viverrini* in an area of high intensity in Thailand. From Figure 4f (Upatham et al., 1984; IARC, 1994) with permission from the International Agency for Research on Cancer and the World Health Organization (www.who.int).

et al. (1960), Yokogawa (1982), Blair et al. (1999), and (WHO, 1995). The extraordinarily low specificity of these lung flukes for nonhuman mammalian hosts presents a formidable obstacle to control programs designed to relieve people from the burden of paragonimiasis.

7.6 MORBIDITY

The signs and symptoms noted during a clinical diagnosis of infection indicate that overt morbidity is occurring (Section 7.4, WHO, 2004a). Little is known about subtle morbidity. The extent and severity of disease due to food-borne trematodiasis and its impact on the health of individuals, households, and communities is assumed to depend directly on the intensity of infection, but few studies have been undertaken to demonstrate that increasing morbidity correlates with increasing intensity. Breaking transmission must seem to be an intractable problem for health authorities because of the network of host species that sustains the flukes. Transmission rates are also likely to be determined by the intensity of infection, and morbidity would decline if transmission rates were to be reduced. Tentative guidance about levels of infection intensities that determine an individual's likelihood of being ill are shown in Table 4.2. These values should always be interpreted with caution because so many other factors, including nutrition, concurrent infections, and immune competence, may be involved. In long-standing FBT infections involving the biliary system, the number of worms may reach the point at which bile ducts are blocked so that few eggs reach the intestine to be discharged in the feces.

7.6.1 Liver Flukes and Disease

The International Agency for Research on Cancer (IARC) has been most concerned to investigate the role of liver fluke infections in the etiology of two forms of liver cancer in humans. Attention is focussed on *C. sinensis, O. viverrini,* and *O. felineus.* The cancers are classified as cholangiocarcinoma (CCA) and hepatocarcinoma (HCC) and evidence for the role of liver fluke infections in these malignancies has been reviewed by Watanapa and Watanapa (2002) and Okuda et al. (2002). CCA is the second most common primary cancer in the world with a very poor prognosis (Okuda et al., 2002). Sadly, many individuals who eventually receive care for cholangiocarcinoma will have been unaware of the duration of infection before their malignant cancers were recognized (Haswell-Elkins and Elkins, 1998). Wherever the infections are endemic, *Clonorchis* and *Opisthorchis* deserve high priority in public health programs (Haswell-Elkins et al., 1994; Rim, 2005).

7.6.1.1 Fascioliasis

An incubation, an acute, a latent, and a chronic phase can be recognized during the course of chronic fascioliasis (Chen and Mott, 1990; Mas-Coma and Bargues, 1997). The phases tend to overlap in areas where fascioliasis is endemic and infection is repetitive. Disease begins with the incubation phase, which runs from ingestion of metacercariae to the detection of the first symptoms. Next comes the acute phase in which the mechanical destruction of the liver tissue occurs as the invading flukes feed on the parenchyma following migration across the body cavity after penetrating the intestinal wall. Body temperature may rise to 42°C, abdominal pain may be severe, and physical examination may reveal hepatomegaly, jaundice, and anemia.

During the latent phase, which may last for years, egg release begins and infected individuals may appear asymptomatic although eosinophilia is usually detectable. The chronic or obstructive phase is initiated by the adult flukes in the biliary passages and gall bladder. Epithelial hyperplasia follows and the accumulation of flukes leads to mechanical obstruction and the disruption of bile flow. This phase of fascioliasis involves cholangitis, cholecystitis, and cholelithiasis that cannot be attributed directly to *Fasciola* unless the patient is known to be infected. During any of the phases flukes may die releasing toxic and allergenic molecules before calcification occurs. Eggs may become trapped in granulomatous nodules and adhesions may form.

The proportion of infected individuals that suffers such severe disease as has been outlined above is unknown. There is no evidence that *Fasciola* is carcinogenic. There are numerous reports of immature flukes having been found in locations other than the liver and biliary system. Flukes that die in ectopic sites become calcified or encapsulated in granulomatous tissue. The death rate appears to be low (Mas-Coma, 2003).

7.6.1.2 Clonorchiasis

If clinical diagnosis has been made, a patient must be ill and in need of treatment (Section 7.4.4.3). In heavy infections cholangitis, cholecystitis, and cholelithiasis develop. Inflammation and hyperplasia of the biliary epithelium occurs leading to the deposition of fibrous tissue. The pancreatic duct may also be invaded (WHO, 2004a). Clonorchiasis has been associated with liver cancer for many years and cases have been reported from infected people living in China, Hong Kong, Japan, and Korea, and from Asian people who had migrated to the U.S. (Rim, 2005). In a region of Korea where the prevalence of *C. sinensis* was found to be high, infection was reckoned to increase significantly the risks of CCA in particular, but also of HCC (Chung and Lee, 1976). In 1996 in Guangzhou, China, out of 10,480 cases of clonorchiasis 37 also presented with primary liver cancer, while out of 87,639 patients without clonorchiasis only 44 presented with liver cancer (WHO, 2004a).

In the human liver, *C. sinensis* has been found to induce adenomatous hyperplasia through which a tumor might develop (Kim, 1984). In experiments with hamsters, mice, rabbits, and rats infected experimentally with *C. sinensis* somewhat differing histological pictures emerged from the liver lesions (Lee and Rim, 1996) making difficult the extrapolation of results to infected humans. Hamsters served as the best vehicle for studying tumor development; the presence of *C. sinensis* seemed to predispose the animals to acquire liver tumors. Since the causal mechanism of carcinogenesis during clonorchiasis is not yet fully understood, the IARC (1994) concluded that infection with *C. sinensis* is probably carcinogenic to humans rather than definitely so.

7.6.1.3 Opisthorchiasis

Since 1971 in Thailand, epidemiological surveys, clinical studies, and postmortem examinations have demonstrated a secure relationship between infection and both CCA and HHC. There are marked geographical variations in CCA and HHC rates in the country that relate to variations in the prevalence of the infection. Most convincing is the comparison between the incidences of cholangiocarcinoma outside Thailand with that in the country at the time of the IARC review. The liver cancer rate in men outside Thailand was 0.2 to 2.8 per 100,000 annually compared with 84.6 per 100,000 annually in the North of the country where the prevalence of *O. viverrini* was about 35% (WHO, 1995). The age group for most cases of cholangio-carcinoma associated with opisthorchiasis is 25 to 44 years, a peak that correlates well with peak prevalence of infection and includes peak intensity of infection (Figure 7.8 and Figure 7.9). In the former USSR, an association has been detected between infection with *O. felineus* and cholangiocarcinoma. In the part of Tyumen where the prevalence of opisthorchiasis is high, liver cancer was revealed in 119 of 146 autopsies in which the fluke was also found compared with 43 in 125 autopsies in which the fluke was not found (WHO, 1995).

The histopathology of liver carcinogenesis has been studied extensively in hamsters infected with *O. viverrini* and *in vitro* using a cell line established from a CCA in patients seropositive to antibody to *O. viverrini* (see Watanapa and Watanapa, 2002). The hamster studies have shown compelling evidence of a synergism between mutagenic nitrogenous compounds and *O. viverrini* infection. Hamsters developed liver cancers when given 100 metacercariae of *O. viverrini* and 0.0025% dimethyl-nitrosamine (DMN) in their drinking water. DMN is produced by bacteria in salted

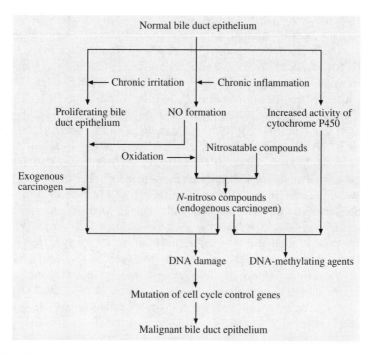

FIGURE 7.10 Possible mechanisms of carcinogenesis during opisthorchiasis and clonorchiasis. From Figure 2 (Watanapa andWatanapa, 2002). Copyright British Journal of Surgery Society Ltd; with permission granted by John Wiley & Sons Ltd on behalf of the BJSS Ltd.

fish while other popular foods contain other compounds with carcinogenic potential such as nitrates, nitrites, and N-nitropyrolidine (WHO, 2004a). In addition to deleterious effects on the liver and biliary system, *O. viverrini* is deemed to be carcinogenic (IARC, 1994). A scheme outlining possible mechanisms of carcinogenesis during liver fluke disease is depicted in Figure 7.10. There is insufficient evidence at present to convince authorities unequivocally that infection with *O. felineus* causes CCA or HHC.

Does anthelminthic treatment to remove liver flukes have any effect on the development or fate of CCA and HHC? A positive answer would further strengthen the case for implementing measures for the control of morbidity due to food-borne trematodiasis. Treatment of 913 people infected with *O. viverrini* with praziquantel was followed by improved clinical symptoms and reduced hepatobiliary pathology as examined by ultrasound (Pungpak et al., 1997), a finding that suggests that the risk of CCA and HHC should be reduced. The outcome of anthelminthic treatment of infected hamsters is equivocal. Removal of the worm burden by use of praziquantel caused a reduction in proliferative cholangitis, a form of precancerous inflammation (Thamavit et al., 1992). By contrast removal of the worm burden appeared to have little effect on the emergence of cancer in the hamsters if permanent changes to the system had already occurred (Thamavit et al., 1993).

7.6.2 Intestinal Flukes and Disease

Morbidity during infection with any of the species of intestinal fluke other than *Fasciolopsis buski* is poorly understood. In some cases, fasciolopsiasis involves allergic symptoms, nausea, vomiting, and edema. Deaths of children have been attributed to fasciolopsiasis in China, India, and Thailand. Modification of the intestinal mucosal architecture appears to be the lesion common to intestinal fluke infections, but the effects seem to be relatively mild (Carney, 1991). Infection with heterophyid flukes gives symptoms of acid peptic disease or peptic ulcer disease (Belizario et al., 2001). The general rule that intensity of infection correlates positively with severity of morbidity may not apply. An individual who passed over 60,000 adult *Metagonimus yokogawai* experienced only mild indigestion and epigastric pain. Diarrhea, colicky pain, indigestion, and lethargy characterize morbidity during intestinal fluke infections (WHO, 1995). Morbidity associated with infections with 17 species of intestinal fluke in Korea has been summarized by Chai and Lee (2002).

Small intestinal flukes have been found in ectopic sites. There are reports of brain invasion by adult *Heterophyes heterophyes* and encapsulated eggs of this species have been found in brain tissue. Eggs identified as those of *Haplorchis pumilio* have been found in brain tissue. Infection with small intestinal flukes poses a problem over the parasitological diagnosis of infection with *Clonorchis* and *Opisthorchis* infection because of the morphological similarity of the eggs as seen with light microscopy (WHO, 1995). More research is needed to investigate subtle morbidity during intestinal fluke infections. The recent review by Fried et al. (2004) provides a contemporary summary of knowledge of these tiny trematodes.

7.6.3 Lung Flukes and Disease

The taxonomy of *Paragonimus westermani* is in a state of flux (Blair et al., 1999). In this description of morbidity due to the infection with paragonimid flukes we have largely ignored the research that has led to recognition of many species of *Paragonimus* because the course of paragonimiasis and the ill health endured by infected people are essentially the same (Yokogawa, 1982). *Paragonimus westermani* is the most widely distributed lung fluke (Section 7.5.3) and it exists in both a diploid and triploid form and very rarely in a tetraploid form. Triploid adults are larger than diploid adults and their eggs are larger and of different shape. Also, triploid adults can reproduce by parthenogenesis. Triploid *P. westermani* are more pathogenic than the diploid form (Blair et al., 1999). These comments about diploid and triploid *P. westermani* apply to that species and not to all the paragonimids known to infect humans.

Two types of paragonimiasis may be recognized; pleuropulmonary paragonimiasis and ectopic paragonimiasis. Pleuropulmonary disease begins about three weeks after the ingestion of metacercariae. Flukes will have then entered the lungs and after a further nine weeks their presence will have provoked fibrotic encapsulation by host tissue. In the case of the diploid form, these cysts develop around pairs of flukes, whereas an individual triploid fluke may occupy a cyst. Light infections are often asymptomatic apart from eosinophilia. When heavier infections are established, fever, chest pain, coughing, dyspnea, and pneumothorax and pleural effusion occur. Coughing is associated with hemoptysis and the passing of eggs in sputum. Pleuro-pulmonary paragonimiasis may last for years and may be misdiagnosed as chronic bronchitis, bronchial asthma, or tuberculosis. In regions where paragonimiasis is endemic, patients diagnosed as having tuberculosis, but failing to respond to therapy, should be reexamined with paragonimiasis in mind (WHO, 1995).

Ectopic paragonimiasis usually occurs in individuals harboring flukes in their lungs or having a history of lung involvement. Immature flukes en route to the lungs, adult flukes seeking partners, or flukes from species of *Paragonimus* not well suited to the human host may cause ectopic morbidity (Blair et al., 1999). Cerebral com-plications occur when flukes enter the brain causing hemorrhage, edema, or meningitis. The death rate is high, particularly when intercranial hypertension occurs. Other ectopic sites occupied by paragonimids include the reproductive and urinary tracts, thoracic and cardiac muscles, the pericardial cavity, and the spleen. Immature *Para-gonimus skrjabini,* an example of a paragonimid for which the human is not an ideal host, become trapped in subcutaneous nodules.

7.7   ANTHELMINTHIC TREATMENT

Praziquantel has transformed the treatment of patients with clonorchiasis, fasciol-opsiasis, opisthorchiasis, metagonimiasis and the gamut of infections with heterophyid and other minute intestinal flukes, and paragonimiasis. The drug is recommended by WHO (2004b) and its use is summarized in Table 7.5 and Table 14.1. Praziquantel is not effective for the treatment of fascioliasis. Triclabendazole is now recommended for the treatment of fascioliasis and is also effective for the treatment of paragon-imiasis (WHO, 2004b; Table 7.5). A wide range of anthelminthic drugs is available for the treatment of trematode infections in domesticated animals and livestock (WHO, 1995), but these drugs must not be used for the treatment of food-borne trematodiasis in humans.

Praziquantel is exceptionally well tolerated although side effects from its use have been reported occasionally. Some discomfort may arise when trapped flukes that have been killed by the drug decompose *in situ.* There is no evidence to suggest that praziquantel has any adverse effects on the developing fetus, but the drug should not be given to women known to be or suspected of being pregnant until expert evaluation has taken place and a positive recommendation made. Currently, treatment should be delayed if possible until after delivery (WHO, 2004b). A recent Informal Consultation convened by the World Health Organization concluded that prazi-quantel may be offered to both pregnant and lactating women suffering from schis-tosomiasis because the health benefits of treatment substantially outweigh the risks (WHO, 2002). This conclusion regarding schistosomiasis provides some reassurance if pregnant women were found to have received praziquantel during a public health program aimed at controlling morbidity due to food-borne trematodiasis.

Paragonimus flukes have a tendency to migrate to ectopic sites in the body (Section 7.6.3). This property may put infected individuals at risk of serious side effects when anthelminthic treatment is forming part of a community program in which medical supervision will be minimal. Cerebral paragonimiasis carries the

TABLE 7.5
Anthelminthic Treatment for Food-Borne Trematode Infections with Praziquantel (PZQ)[1] and Triclabendazole (TCZ)[2]

Helminthiasis	Recommended Drug	Recommended Dose
Clonorchiasis (*Clonorchis sinensis*)	(a) PZQ, by mouth, 3 times daily for 2 consecutive days (b) PZQ, by mouth, single dose	(a) 25 mg/kg body weight (b) 40 mg/kg body weight
Fascioliasis (*Fasciola gigantica* and *F. hepatica*)	TCZ by mouth, single dose	10 mg/kg body weight
Fasciolopsiasis (*Fasciolopsis buski*)	PZQ, by mouth, single dose	25 mg/kg body weight
Megagonimiasis and minute intestinal fluke infections	PZQ, by mouth, single dose	25 mg/kg body weight
Opisthorchiasis (*Opisthorchis viverrini* and *O. felineus*)	(a) PZQ, by mouth, 3 times daily for 2 consecutive days (b) PZQ, by mouth, single dose	(a) 25 mg/kg body weight (b) 40 mg/kg body weight
Paragonimiasis (*Paragonimus* spp.)	(a) PZQ, by mouth, 3 times daily for 2 consecutive days (b) PZQ, by mouth, single dose	(a) 25 mg/kg body weight (b) 40 mg/kg body weight

Note: These anthelminthic drugs should be used as recommended in the *WHO Model Formulary 2004* (WHO, 2004b) noting guidance about contraindications, precautions, and side effects.

[1]PZQ is supplied in 150mg and 600 mg tablets and [2]TCZ in 250 mg tablets.

greatest risks and patients with this form of the disease should not be treated until they have been admitted to hospital.

The fact that praziquantel is such an effective drug for the control of morbidity due to FBT infections is reassuring. The greater its use, however, the greater becomes the threat of emerging drug resistance. Perhaps triclabendazole may gain greater use for the treatment of FBT infections and so relieve the selection pressure generated by the widespread use of praziquantel. A single dose of 10 mg/kg body weight was well tolerated by 19 children infected with *P. uterobilateralis* in a pilot study in Cameroon (Ripert et al.,1992).

7.8 PREVENTION AND CONTROL OF FOOD-BORNE TREMATODIASIS

A simplified conceptual framework showing how FBT infections have become integrated into the human food chain is set out in Figure 7.11. The food chain consists of all involved with the production, processing, distribution, trading, and consumption of food, including governments at home and abroad that provide the necessary structure (WHO, 2004a). The framework (Figure 7.11) glosses over the complexities of human behavior and social interactions in the diverse places where FBT infections are endemic. Attention is drawn to three important points. First, morbidity control in infected people is probably the most realistic public health intervention at present in countries where FBT infections prevail. Second, the food security of poor rural

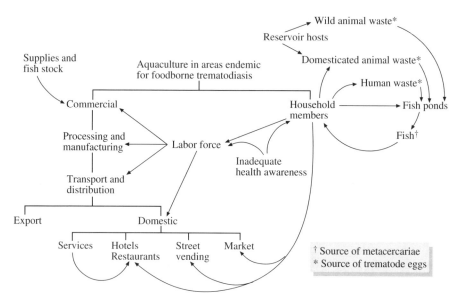

FIGURE 7.11 A conceptual framework to interpret contamination of the human food chain by foodborne trematode infections transmitted through the consumption of fish.

people who rely heavily on fish protein from household ponds should not be put at risk by the rigid implementation of measures intended for prevention and control. Third, the commercial aquaculture industry must continue to protect national and international consumers through its procedures for ensuring food safety and quality control. The Asian aquaculture sector produces over 90% of the world's finfish and shellfish, products that contribute significantly to nutrition and economic development (Earth Policy Institute, 2000; Lima dos Santos, 2002; WHO, 2004a).

7.8.1 Prevention and Control in the Community

There is no master plan for dealing with FBT infections wherever they exist. Knowledge should be obtained about local infection variables in the community, fluke and snail populations, wild and domesticated reservoir hosts, climate, food and food habits, traditions, integrated animal husbandry and aquaculture practices, hygiene standards, residents' knowledge of transmission and infection, and health care provision. Information acquired over several years would be a bonus because seasonal trends could be revealed to identify opportunities for the most effective interventions.

Once the public health significance of food-borne trematodiasis has been established in a community, decisions must be taken to plan, implement, monitor, and sustain an appropriate control program. One approach to consider is the adoption, and modification as required, of Green and Kreuter's PRECEDE-PROCEED model, which has been tested by Jimba and Joshi (2001) in Nepal for the control of food-borne parasitic zoonoses (FBPZ) involving meat in the food chain. The main elements of the model are shown in Figure 7.12. Phase 1 to Phase 5 analyze risk factors and identify optimum points for feasible actions. Phase 6 to Phase 9 deal with

PRECEDE

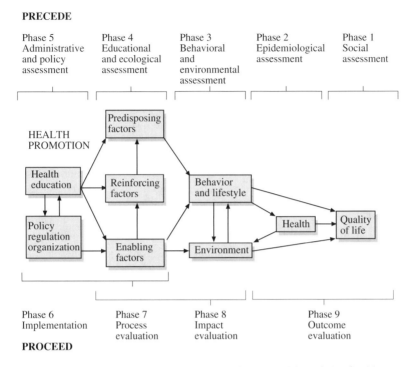

FIGURE 7.12 Green and Kreuter's PRECEDE-PROCEED model applied to food-borne trem-atodiasis. From Figure 1 (Jimba and Joshi, 2001) with permission from SEAMEO, Bangkok.

operational aspects. After application of the model, Jimba and Joshi (2001) confi-dently recommended to Nepalese authorities that the control of FBPZ in the food chain required (1) meat inspection, (2) training programs on safe meat production and safe selling practices, and (3) improved slaughter houses and slaughtering methods. There is a strong educational component in Jimba and Joshi's report based on the evidence they obtained about ignorance of FBPZ among relevant health professionals, meat producers, and meat handlers. Promoting health education throughout the community is of major importance in every aspect of attempts to prevent and control FBT infections in the community (WHO, 2000).

Progress can be made in the control of FBT infections despite their intricate fusion with the food chain (Figure 7.11). In 1984, the Korean government launched its national program for the control of clonorchiasis in endemic areas. Between 1984 and 1990, stool samples were collected and examined from over three million people and those found to be egg-positive were treated with a single oral dose of praziquantel (40 mg/kg body weight) at health centers under medical supervision (Rim, 1997). During this period the prevalence of *C. sinensis* fell from 13.3 to 0.9% and the proportion of people judged to have heavy infections also fell. The progress made by the control program coincided with economic growth in Korea. Cycles of re-infection with *C. sinensis* were reduced by mechanization on farms, urbanization, and other developments. Health education helped to reduce the consumption of raw fish, but Rim stressed that changing food habits can be exceedingly difficult in the face of centuries-old traditions.

Operational research into the control of opisthorchiasis in Thailand has demonstrated that annual anthelminthic treatment with praziquantel (40 mg/kg body weight) combined with regular health education and improved sanitation is accompanied by a decrease in prevalence and intensity of infection (Saowakontha et al., 1993). Three villages (I, II, and III) in an area of Thailand where *O. viverrini* is endemic were selected for study. Villagers underwent stool examinations every six months during a three-year period. Those in village I were treated every six months and efforts were made to increase health education and improve sanitation. The same regimen was followed in village II except that the anthelminthic treatment was given annually. No interventions were provided for village III, which served as the control community. The prevalence of infection fell in village I from 64.4 to 13.2% and in village II from 56.8 to 14.1% (Figure 7.13). The intensity of infection followed the same pattern in villages I and II (Figure 7.14). Increases in knowledge about *O. viverrini* infection, in latrine construction and use, and decrease in the consumption of raw fish are shown in Figure 7.15, Figure 7.16, and Figure 7.17. Beneficial changes were also detected in village III despite the absence of interventions from the study team; news travels quickly when health matters are in focus.

Two important points should be noted from this investigation. First, a single annual dose of praziquantel was as effective as two 6-monthly doses in reducing prevalence and intensity, a finding of significance when the costs and logistics of

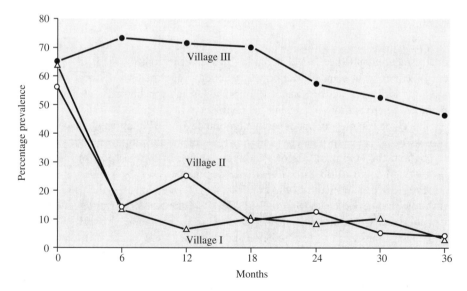

FIGURE 7.13 Impact of control intervention on the prevalence of *Opisthorchis viverrini* infection. Village I 6-monthly praziquantel, village II annual praziquantel, and village III no praziquantel. From Figure 1 (Saowakontha et al., 1993) with permission from Cambridge University Press.

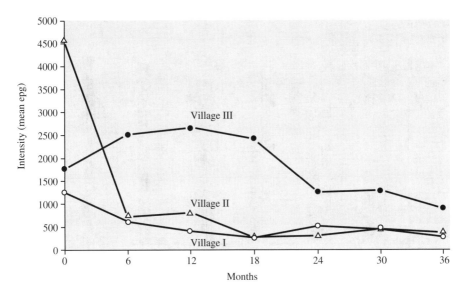

FIGURE 7.14 Impact of control intervention on the intensity (epg) of *Opisthorchis viverrini* infection. Village I 6-monthly praziquantel, village II annual praziquantel, and village III no praziquantel. From Figure 2 (Saowakontha et al., 1993) with permission from Cambridge University Press.

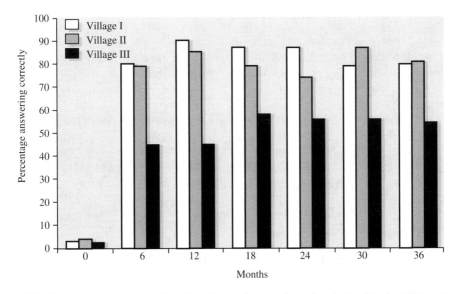

FIGURE 7.15 Impact of health education in knowledge of opisthorchiasis. Village I 6-monthly praziquantel, village II annual praziquantel, and village III no praziquantel or health education. From Figure 5 (Saowakontha et al., 1993) with permission from Cambridge University Press.

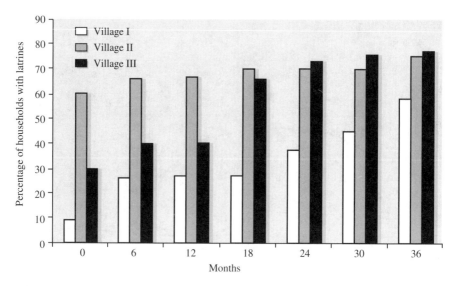

FIGURE 7.16 Changes in the numbers of latrines of the villages in the opisthorchiasis control program. From Figure 6 (Saowakontha et al., 1993) with permission from Cambridge University Press.

drug delivery are considered. Also the reinfection rate was not found to be significantly different between villages I and II despite the difference in the number of doses given. Secondly, the villagers were clearly eager to comply with and support this pilot control program. There must have been a strong "felt need" for such a marked change in food habits (Figure 7.17) to occur. Control programs will succeed if health workers gain the community's trust and address a community's own health agenda in a culturally acceptable manner (Sormani, 1987).

Liver fluke control is a component of the National Public Health Development Plan in Thailand. The components of the program are shown as a flow diagram in Figure 7.18 (WHO, 2004a). Self reliance — local ownership and empowerment — are recognized as essential for sustaining progress made in the control of opisthorchiasis. Shortly before his death from SARS in March 2003 as he struggled to understand and curtail the spread of that infection, Dr. Carlo Urbani proposed a protocol to guide health workers with responsibility for the control of FBT infections in the community. His guidelines are reproduced as Appendix 4.

7.8.2 PREVENTION AND CONTROL IN COMMERCIAL AQUACULTURE

The experts attending the joint WHO/FAO workshop on food-borne trematodes infections in Asia drew attention to the need for education for all stakeholders involved in the production, processing, distribution, trade, and quality control of aquaculture products (WHO, 2004a). The importance of education cannot be stressed too strongly if consumers are to be protected from the risks of infection. Reliance on changes in the food habits of consumers regarding raw fish is not a secure strategy without action by the aquaculture industry to keep metacercariae out of the food

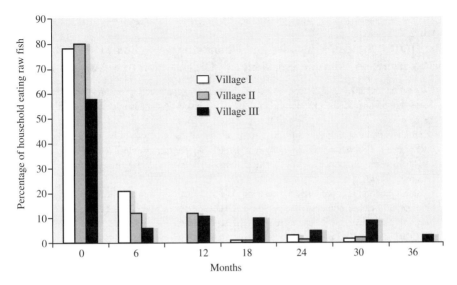

FIGURE 7.17 Changes in the habit of eating raw fish by the villagers in the opisthorchiasis control program. From Figure 7 (Saowakontha et al., 1993) with permission from Cambridge University Press.

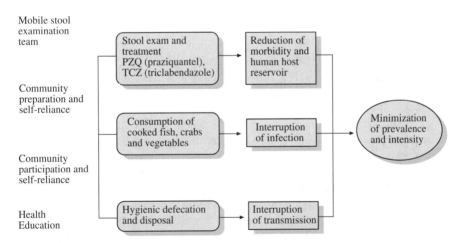

FIGURE 7.18 A conceptual framework for the control of foodborne trematodiasis. From Annex 5 (WHO, 2004a) with permission from the World Health Organization (www.who.int).

chain. Successful action depends on educated managers and staff for whom they have responsibility.

There are reasons to believe that inspection for the presence of pathogens such as metacercariae in aquaculture end-products may not be as effective as best practice in food safety demands because it does not deal with the management of risks (Lima dos Santos, 2002). Inspection cannot be comprehensive because it relies on sampling.

TABLE 7.6
The HACCP System: Principles and Definitions. From Box 11 (WHO, 2000)
with permission from the World Health Organization (www.who.int).

1. Conduct hazard analysis (i.e., identify hazard, evaluate risk and specify measures for risk control). **Hazard**: a biological, chemical, or physical agent or factor with the potential to cause an adverse health effect.
2. Determine critical control points (CCPs). **Critical control point**: A step at which control can be applied that is essential to prevent or eliminate a food safety hazard or reduce it to an acceptable level.
3. Establish critical limits at each CCP. **Critical limit**: A criterion which separates acceptability from unacceptability.
4. Establish monitoring procedures. **Monitoring**: The act of conducting a planned sequence of observations or measurements of control of parameters to assess whether a CCP is under control.
5. Establish corrective actions. **Corrective action**: Actions to be taken when the results of monitoring the CCP indicate a loss of control.
6. Establish verification procedures. **Verification**: The application of methods, procedures or tests, in addition to those used in monitoring, to determine compliance with the HACCP plan and whether the HACCP plan needs modification.
7. Establish documentation procedures.

The Hazard Analysis Critical Control Point (HACCP) system has proved to be more effective than inspection in the aquaculture sector in several countries including Australia, Brazil, Canada, Chile, Cuba, Ecuador, Indonesia, Scotland, Thailand, and the U.S. (Table 7.6; WHO, 2000).

Application of the HACCP approach may seem to require the resources of large-scale producers rather than those of small-scale fish farmers. A project in Chiang Mai, Thailand, however, has tested HACCP in a scheme designed to reduce opisthor-chiasis in people consuming infected cultured carp (Khamboonruang et al., 1997). The project involved two ponds, one to be experimental and one control, in a place where traditional fish farming occurred and where opisthorchiasis was endemic (13.9% prevalence). At the start of the project the experimental pond was drained, treated with molluscicide, flushed with fresh water, and then allowed to refill with water during the rainy season. Fish fry were introduced and monthly monitoring was carried out to assess water quality, visits by dogs, fecal contamination, and so on. Two hazards were detected that the study team could not remedy. First, snail hosts had returned to the experimental pond within a month despite the precautions taken to eliminate them. Secondly, although simple fencing prevented dogs from visiting the pond, rats were able to burrow under the fence and enter the ponds. Interestingly, the systematic monitoring required by HACCP detected the presence of a species of *Haplorchis* in the area and so warned health officials of a further risk to residents.

The project in Chiang Mai was carried out by a multidisciplinary research team. Whether small-scale fish farmers will be able to implement HACCP is uncertain and poor rural people, who maintain their own ponds, cannot be expected to under-take extensive monitoring. Regular anthelminthic treatment and access to anthel-minthic drugs when required is the realistic public health approach to be adopted at present to reduce and control morbidity due to food-borne trematodiasis.

REFERENCES

Anantaphruti MY. 2001. Parasitic contaminants in food. In: Proceedings of the 3rd Seminar on Food-borne Parasitic Zoonoses (eds. J Waikagul, JH Cross and S Supavej). *Southeast Asian Journal of Tropical Medicine and Public Health* **32** (Supplement), 218–228.

Belizario VY, Bersabe MJ, de Leon WU et al. 2001. Intestinal heterophyidiasis: an emerging food-borne parasitic zoonosis in southern Philippines. In: Proceedings of the 3rd Seminar on Food-borne Parasitic Zoonoses (eds. J Waikagul, JH Cross and S Supajev). *Southeast Asian Journal of Tropical Medicine and Public Health* **32** (Supplement), 36–42.

Blair D, Xu Z-B and Agatsuma T. 1999. Paragonimiasis and the genus *Paragonimus*. *Advances in Parasitology* **42**, 113–222.

Brockelman WY, Upatham ES, Viyanant V et al. 1986. Field studies on the transmission of the human liver fluke, *Opisthorchis viverrini*, in Northeast Thailand: population changes of the snail intermediate host. *International Journal for Parasitology* **16**, 545–552.

Carney WP. 1991. Echinostomiasis — a snail-borne intestinal trematodes zoonosis. In: Food-borne Parasitic Zoonosis: Impact on Agriculture and Public Health (ed. JH Cross) *Southeast Asian Journal of Tropical Medicine and Public Health* **22** (Supplement), 206–211.

Chai J-Y and Lee S-H. 2002. Food-borne intestinal trematodes infections in the Republic of Korea. *Parasitology International* **51**, 129–154.

Cheesbrough M. 1998. *District Laboratory Practice in Tropical Countries Part 1*. March, Cambridgeshire: Tropical Health Technology.

Chen MG and Mott KE. 1990. Progress in assessment of morbidity due to *Fasciola hepatica* infection: a review of recent literature. *Tropical Diseases Bulletin* **87**, R1–R38.

Choi MH, Ryu JS, Lee M et al. 2003. Specific and common antigens of *Clonorchis sinensis* and *Opisthorchis viverrini* (Opisthorchidae, Trematoda). *Korean Journal of Parasitology* **41**, 155–163.

Chung CS and Lee SK. 1976. An epidemiological study of primary liver carcinomas in Busan area with special reference to clonorchiasis. *Korean Journal of Pathology* **10**, 33–46. (In Korean, quoted by IARC, 1994).

Coombs I and Crompton DWT. 1991. *A Guide to Human Helminths*. London, New York, Philadelphia: Taylor & Francis Ltd.

De NV, Murrell KD and Cong LD et al. 2003. The food-borne zoonoses of Vietnam. *Southeast Asian Journal of Tropical Medicine and Public Health* **34** (Supplement 1), 12–34.

Earth Policy Institute. 2000. World aquacultural production by principal producers in 1998. Sourced from UN FAO, Yearbook of Fisheries Statistics: Aquaculture Production (Rome: 2000). www.earth-policy.org/Alerts/Alert9.

Elkins DB, Haswell-Elkins MR, Mairiang E et al. 1990. A high frequency of hepatobiliary disease and suspected cholangiocarcinoma associated with heavy *Opisthorchis viverrini* infection in a small community in north-east Thailand. *Transactions of the Royal Society of Tropical Medicine and Hygiene* **84**, 715–719.

Elkins DB, Sithithaworn P, Haswell-Elkins MR et al. 1991. *Opisthorchis viverrini*: relationships between egg counts, worms recovered and antibody levels within an endemic community in Northeast Thailand. *Parasitology* **102**, 283–288.

Esteban JG, Flores A, Aguirre C ct al. 1997. Presence of very high prevalence and intensity of infection with *Fasciola hepatica* among Aymara children from the Northern Bolivian Altiplano. *Acta Tropica* **66**, 1–14.

Fried B, Graczyk TK and Tamang L. 2004. Food-borne intestinal trematodiases in humans. *Parasitology Research* **93**, 159–170.

Giboda M, Ditrich O, Scholz T et al. 1991. Current status of food-borne parasitic zoonoses in Laos. In: Food-borne Parasitic Zoonoses: Impact on Agriculture and Public Health (ed. JH Cross). *Southeast Asian Journal of Tropical Medicine and Public Health* **22** (Supplement), 56–61.

Harinasuta T, Bunnag D and Radomyos P. 1987. Intestinal fluke infections. In: *Balliere's Clinical Tropical Medicine and Communicable Diseases. Intestinal Helminthic Infections* (ed. ZS Pawlowski). London and Philadelphia: Balliere Tindall. pp. 695–721.

Haswell-Elkins MR, Elkins DB, Sithithaworn P et al. 1991. Distribution patterns of *Opisthorchis viverrini* within a human community. *Parasitology* **103**, 97–101.

Haswell-Elkins MR, Mairiang E, Mairiang P et al. 1994. Cross-sectional study of *Opisthorchis viverrini* infection and choliangiocarcinoma in communities within a high-risk area in northeast Thailand. *International Journal of Cancer* **59**, 505–509.

Haswell-Elkins MR and Elkins DB. 1998. Lung and liver flukes. In: *Topley and Wilson's Microbiology and Microbial Infections*, 9th edition. Volume 5. Parasitology (eds. FEG Cox, JP Kreier and D Wakelin). London, Sydney, Auckland: Arnold. pp. 507–520.

Hillyer GV, Soler de Galanes M, Rodriguez-Perez J et al.1992. Use of the Falcon assay screening test — enzyme-linked immunosorbent assay (FAST-ELISA) and the enzyme-linked immunoelectrotransfer blot (EITB) to determine the prevalence of human fascioliasis in the Bolivian Altiplano. *American Journal of Tropical Medicine and Hygiene* **46**, 603–609.

IARC. 1994. *IARC Monographs on the Evaluation of Carcinogenic Risks to Humans: Liver Flukes*. Volume 61. Geneva: World Health Organization, International Agency for Research on Cancer.

Ikeda T, Oikawa Y and Nishiyama T. 1996. Enzyme-linked immunosorbent assay using cysteine protinase antigens for immunodiagnosis of human paragonimiasis. *American Journal of Tropical Medicine and Hygiene* **55**, 435–437.

Jimba M and Joshi DD. 2001. Health promotion approach for the control of food-borne parasitic zoonoses in Nepal: emphasis on environmental assessment. In: Proceedings of the 3rd Seminar on Food-borne Parasitic Zoonoses (eds. J Waikagul, JH Cross and S Supavej). *Southeast Asian Journal of Tropical Medicine and Public Health* **32** (Supplement), 229–235.

Khamboonruang C, Keawvichit R, Wongworapat K et al. 1997. Application of Hazard Analysis Critical Control Point (HACCP) as a possible control measure for *Opisthorchis viverrini* infection in cultured carp (*Puntius gonionotus*). *Southeast Asian Journal of Tropical Medicine and Public Health* **28** (Supplement), 65–72.

Kim Y-I. 1984. Liver carcinoma and liver fluke infection. *Arzneim-Forsch./Drug Research* **34**, 1121–1126.

Lee J-H and Rim H-J. 1996. Cholangiocarcinoma caused by *Clonorchis sinensis* infection in laboratory animals. Proceedings of the 2nd Japan-Korea Parasitologists' Seminar. pp. 43–49.

Lima dos Santos CA. 2002. Hazard analysis critical control point and aquaculture. In: *Public, Animal, and Environmental Aquaculture Health Issues* (eds. ML Jahncke, ES Garrett, A Reilly et al.). Chichester: Wiley-Intersciences, Inc. pp. 103–119.

Maclean JD, Cross J and Mahanty S. 1999. Liver, lung and intestinal fluke infections. In: *Tropical Infectious Diseases* (eds. RL Guerrant, DH Walker and PF Weller). Philadelphia: Churchill Livingstone. pp. 1349–1369.

Mas-Coma S. 2003. Human fascioliasis: epidemiological patterns in human endemic areas of South America, Africa and Asia. 4th Seminar on Food- and Water-borne Parasitic Zoonoses (ed. J Waikagul). Bangkok: Faculty of Tropical medicine, Mahidol University. pp. 44–60.

Mas-Coma S. 2004. Human fascioliasis. In: *Waterborne Zoonoses* (eds. JA Cotruvo, A Dufour, G Rees et al.). Geneva: World Health Organization. pp. 305–322.

Mas-Coma S and Bargues MD. 1997. Human liver flukes: a review. *Research and Reviews in Parasitology* **57**, 145–218.

Mas-Coma S, Esteban JG and Bargues MD. 1999. Epidemiology of human fascioliasis: a review and proposed new classification. *Bulletin of the World health Organization* **77**, 340–346.

Muller R. 2002. *Worms and Human Disease*, 2nd edition. Wallingford, Oxfordshire: CABI Publishing.

Murrell KD. 2002. Fishborne zoonotic parasites: epidemiology, detection and elimination. In: *Safety and Quality Issues in Fish Processing* (ed. HA Bremner). Cambridge: Woodhead Publishing Ltd.

Murrell KD and Crompton DWT. 2006. Foodborne trematodes and helminths. In: *Emerging Foodborne Pathogens* (eds. Y Motarjemi and M Adams). Cambridge: Woodhead Publishing Ltd.

Naegel LCA. 1990. A review of public health problems associated with the integration of animal husbandry and aquaculture, with emphasis on Southeast Asia. *Biological Wastes* **31**, 69–83.

Narain K, Devi KR and Mahanta J. 2005. Development of enzyme-linked immunosorbent assay for serodiagnosis of human paragonimiasis. *Indian Journal of Medical Research* **121**, 739–746.

Nawa Y. 2000. Reemergence of paragonimiasis. *International Medicine* **39**, 353–354.

Nwokolo, C. 1972. Endemic paragonimiasis in Eastern Nigeria. Clinical features and epidemiology of the recent outbreak following the Nigerian civil war. *Tropical and Geographical Medicine* **24**, 138–147.

Okuda K, Nakanuma Y and Miazaki M. 2002. Cholangiocarcinoma: Recent progress. Part 1: Epidemiology and etiology. *Journal of Gastroenterology and Hepatology* **17**, 1049–1055.

Pungpak S, Viravan C, Radomyos B et al. 1997. *Opisthorchis viverrini* infection in Thailand: studies on the morbidity of the infection and resolution following praziquantel treatment. *American Journal of Tropical Medicine and Hygiene* **56**, 311–314.

Richter J, Knipper M, Gobels K and Haussinger D. 2002. Fascioliasis. *Current Treatment Options in Infectious Diseases* **4**, 313–317.

Rim H-J. 1982. Fasciolopsiasis. In: *Handbook Series in Zoonoses. Section C: Parasitic Zoonoses III* (eds. GV Hillyer, CE Hopla). Boca Raton, Florida: CRC Press Inc. pp. 89–97.

Rim H-J. 1986. The current pathobiology and chemotherapy of clonorchiasis. *Korean Journal of Parasitology* **24**, Monograph series 3. pp. 1–141.

Rim H-J. 1997. Epidemiology and control of clonorchiasis in Korea. *Southeast Asian Journal of Tropical Medicine and Public Health* **28**, 47–50.

Rim H-J. 2005. Clonorchiasis: an update. *Journal of Helminthology* **79**, 269–281.

Ripert C, Couprie B, Moyou R et al. 1992. Therapeutic effect of triclabendazole in patients with paragonimiasis in Cameroon: a pilot study. *Transactions of the Royal Society of Tropical Medicine and Hygiene* **86**, 417.

Saowakontha S, Pipitgool V, Pariyanonda S et al. 1993. Field trials in the control of *Opisthorchis viverrini* with an integrated programme in endemic areas of northeast Thailand. *Parasitology* **106**, 283–288.

Scholz T, Ditrich O, and Giboda M. 1991. Differential diagnosis of opisthorchiid and hetero-phyid metacercariae (Trematoda) infecting flesh of cyprinid fish from Nam Ngum Dam Lake in Laos. *Southeast Asian Journal of Tropical Medicine and Public Health* **22** (Supplement, December 1991), 171–173.

Sirisinha S, Chawengkirttikul R, Haswell-Elkins MR et al. 1995. Evaluation of a monoclonal antibody-based enzyme linked immunosorbent assay for the diagnosis of *Opisthorchis viverrini* infection in an endemic area. *American Journal of Tropical Medicine and Hygiene* **52**, 521–524.

Sithithaworn P, Tesena S, Pipitgool V et al. 1991. Quantitative post-mortem study of *Opisthorchis viverrini* in man in north-east Thailand. *Transactions of the Royal Society of Tropical Medicine and Hygiene* **85**, 765–768.

Sormani S. 1987. Control of opisthorchiasis through community participation. *Parasitology Today* **3**, 31–32.

Sukontason KL, Sukontason K, Boonsriwong N et al. 2001. Intensity of trematode metacer-cariae in cyprinoid fish in Chiang Mai province, Northern Thailand. *Southeast Asian Journal of Tropical Medicine and Public Health* **32** (Supplement 2), 214–217.

Thamavit W, Moore MA, Ruchirawat S and Ito N. 1992. Repeated exposure to *Opisthorchis viverrini* and treatment with anthelminthic praziquantel lacks carcinogenic potential. *Carcinogenesis* **13**, 309–311.

Thamavit W, Moore MA, Sirisinha S et al. 1993. Time-dependent modulation of liver lesion development in *Opisthorchis*-infected Syrian hamster by an anthelminthic drug. *Japanese Journal of Cancer Research* **84**, 135–138.

Thienpont D, Rochette F and Vanparijs OFJ. 1979. *Diagnosing Helminthiasis Through Copro-logical Examination*. Beerse, Belgium: Janssen Research Foundation.

Udonsi JK. 1987. Endemic *Paragonimus* infection in Upper Igwun Basin, Nigeria: a prelim-inary report on a renewed outbreak. *Annals of Tropical Medicine and Parasitology* **81**, 57–62.

Ukoli FMA, 1984. *Introduction to Parasitology in Tropical Africa*. Chichester and New York: John Wiley and Sons Limited.

Upatham ES, Vianant V, Kurathong S et al. 1984. Relationship between prevalence and intensity of *Opisthorchis viverrini* infection, and clinical symptoms and signs in a rural community in Northeast Thailand. *Bulletin of the World Health Organization* **62**, 451–461.

Velasquez CC. 1982. Heterophydiasis. In: *Handbook series in Zoonoses. Section C: Parasitic Zoonoses Volume III* (eds. GV Hillyer and CE Hopla). Boca Raton, Florida: CRC Press Inc. pp. 99–107.

Watanapa P and Watanapa WB. 2002. Liver-fluke associated cholangiocarcinoma. *British Journal of Surgery* **89**, 962–970.

WHO. 1980. *Manual of Basic Techniques for a Health Laboratory.* Geneva: World Health Organization.

WHO. 1994. *Bench Aids for the Diagnosis of Intestinal Parasites.* Geneva: World Health Organization.

WHO. 1995. *Control of foodborne trematodes infections.* Report of a WHO Study Group. Technical Report Series 849. Geneva: World Health Organization.

WHO. 2000. *Foodborne Disease: a Focus for Health Education* (Prepared by Y. Motarjemi). Geneva: World Health Organization.

WHO. 2002. *Report of the WHO Informal Consultation on the use of praziquantel during pregnancy/lactation and albendazole/mebendazole in children under 24 months.* *Geneva*: World Health Organization. (WHO/CDS/CPE/PVC/2002.4).

WHO. 2004a. *Food-borne trematodes infections in Asia.* Report of a joint WHO/FAO Workshop. Manila: WHO Regional Office of the Western Pacific.

WHO. 2004b. *WHO Model Formulary 2004.* Geneva: World Health Organization.

Wykoff DE, Harinasuta C, Juttijudata P and Winn MM. 1965. *Opisthorchis viverrini* in Thailand–the life cycle and comparison with *O. felineus. Journal of Parasitology* **51**, 207–214.

Yamaguti S. 1975. *A Synoptical Review of the Life Histories of Digenetic Trematodes of Vertebrates with special reference to their larval forms.* Toyko: Keigaku Publishing Co. (Part 1 is descriptive text and Part 2 contains illustrations.)

Yokogawa, M. 1982. Paragonimiasis. In: *Handbook Series in Zoonoses. Section C: Parasitic Zoonoses Volume III* (eds. GV Hillyer and CE Hopla). Boca Raton, Florida: CRC Press Inc. pp. 123–164.

Yokogawa S, Cort WW and Yokogawa M. 1960. *Paragonimus* and paragonimiasis. *Experimental Parasitology* **10**, 81-137, 139–205.

8 Lymphatic Filariasis

Lymphatic filariasis (LF), sometimes still called elephantiasis, is a disease characterized by invasion of the human blood and lymphatic systems by filarial nematodes. Each of three species, *Wuchereria bancrofti, Brugia malayi,* and *B. timori,* can cause a form of the disease (Table 1.1). *Wuchereria bancrofti* is much the commonest and most widely distributed species. Transmission and survival of the worms depends on the feeding behavior of susceptible female mosquitoes. The disease afflicts poor people in the south. Although few people die from LF, millions suffer from chronic ill health, disfigurement, and disability and many have to endure social rejection and psychological stress. In 1997, the World Health Assembly adopted resolution WHA 50.29, thereby making a commitment to work to eliminate LF as a public health problem. The development of new diagnostic techniques and experience with a safe, single-dose, oral drug regimen persuaded the member states that elimination might be achievable (Ottesen et al., 1997; WHO, 2001). The strategy for the elimination of LF aims (1) to disrupt helminth transmission in endemic countries and (2) control and relieve morbidity caused by LF (WHO, 2005). This chapter seeks to describe and examine the science that underpins those aims that now form the basis for the public health challenge that is the elimination of lymphatic filariasis (Zagaria and Savioli, 2002). Much of the science has been summarized in a collection of reviews edited by Nutman (2000).

8.1 LIFE HISTORY AND HOST RANGE

The life history of each of the three species is indirect, with humans serving as definitive hosts and mosquitoes as intermediate hosts (Figure 8.1; Table 8.1). Filarial worms are long and thin, the adult males measure up to 500×1.5 mm and the females up to 1000×2.5 mm. They live in the lymphatics where they mate. The female worms are ovoviviparous and they release microfilariae (MF) that enter the bloodstream. MF are released about 3 months after the start of an infection with *B. malayi* and after about 9 months in the case of *W. bancrofti*. MF measure from 175 to 300 μm, depending on the species, and each is enclosed in a thin, transparent sheath. The sheath represents the egg shell and each MF is a first-stage larva (L1). When a female mosquito of a susceptible species (Table 8.1) bites a human with MF in circulation, some will be ingested and some will develop via two moults to the L3 stage in the mosquito's tissues during 7 to 14 days depending on the worm species and the ambient temperature (Michael, 2000).

Human infection takes place when a female mosquito carrying L3 stages takes a blood meal. The infective L3 are not injected into the host by the insect's mouthparts and must enter under their own power, probably by crawling through the puncture wound made by the proboscis (Spielman and D'Antonio, 2001). Many L3

TABLE 8.1
Mosquito Hosts of *Wuchereria bancrofti*, *Brugia malayi* and *B. timori*. Based on information in Grove (1982), Muller (2002), and WHO (1992, 2002a, 2005).

Species (Disease)	Endemic Region	Common Vectors and Intermediate Hosts	Nonhuman Definitive Hosts
Brugia malayi (p)[1] (brugian filariasis)	South-east Asia Indian sub-continent	*Anopheles barbirostris, A. campestris, A. donaldi, Mansonia annulata, M. annulifera, M. uniformis*	*Macaca fasciularis* (crab-eating macaque), *Presbytis cristata, P. melalophos, P. obscura* (leaf-eating monkeys), *Felis bengalensis, F. planiceps* (wild cats), *Paradoxus hermaphroditus* and *Arctogalidia trivirgata* (civet
B. malayi (sp)		*Mansonia annulata, M. bonneae, M. dives*	cats), *Manis javanica* (pangolin), domestic cats
B. timori (p)	Lesser Sunda Islands, Timor-Este	*Anopheles barbirostris*	
Wuchereria bancrofti (p) (bancroftian filariasis)	Tropical Americas Caribbean Tropical Africa	*Culex quinquefasciatus Anopheles funestus, A. gambiae, Culex quinquefasciatus*	Experimental infections to complete development in *Macaca cyclopis* (Taiwan macaque) *Presbytis cristata* (silvered leaf-eating monkey)
	Middle East South Asia Far East	*Culex quinquefasciatus Culex quinquefasciatus Culex quinquefasciatus*	
W. bancrofti (sp)	Nicobar Islands Thailand Polynesia	*Ochlerotatus riveus, O. harinasuti Aedes polynesiensis*	

[1] p, periodic; sp, semiperiodic.

probably perish during this phase of the life history. On entering the body, the L3 develop slowly, again via two moults, to become adults in the lymphatics. The adult worms are assumed to live from 5 to 10 years, but this must be a difficult variable to measure under uncontrolled conditions and the likelihood of constant reinfection. Greater longevity has been recorded for *W. bancrofti* by Trent (1963) who invested U.S. veterans who had served in the South Pacific during World War II.

The presence of MF in human blood usually has a characteristic periodicity during the course of the 24-h cycle (Figure 8.2). *Wuchereria bancrofti* is found in most places and its periodicity is usually observed to be nocturnal. This means that most MF will be in circulation between 2200 and 0200 h, a time that happens to correspond with the main feeding activities of the vector mosquitoes. The MF accumulate in the lungs for the rest of the cycle. Apparently, nocturnal periodicity can be reversed if the infected person sleeps during daylight and is active at night (Ukoli, 1984). In fewer places, subperiodic MF are observed. For example, *W. bancrofti*

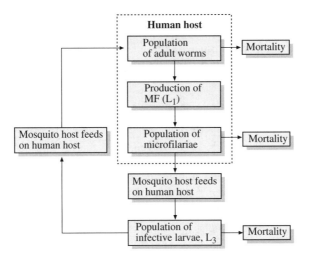

FIGURE 8.1 The life history and population biology of helminths responsible for lymphatic filariasis in humans. From Figure15.3 (Anderson and May, 1992) with permission from Oxford University Press.

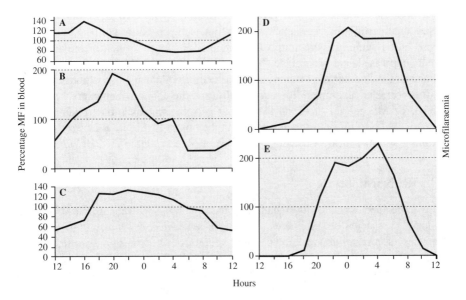

FIGURE 8.2 Plots of microfilarial periodicity for infections of *Wuchereria bancrofti* and *Brugia malayi* in human blood. From Figure 6 (WHO, 1987) with permission from the World Health Organization (www.who.int).

is known from the Philippines where the MF are described as diurnally subperiodic because, although they are always in circulation in the blood, more tend to be present during the day. Similarly *W. bancrofti* from parts of South East Asia is said to be nocturnally subperiodic because more MF are found in the blood at night than at other times in the 24 h-cycle. Periodic and subperiodic populations exist in the case of *B. malayi* (Figure 8.2). The MF of *B. timori* display nocturnal periodicity only (Muller and Wakelin, 1998). It is tempting to conclude that during the evolution of the human-filarial relationship, different strains of the worms have been selected so that infection of the intermediate host and subsequent transmission to the definitive host are enhanced.

Each of the three species of nematode under consideration has low host specificity regarding its need for intermediate hosts (Table 8.1). Information about their definitive hosts is also shown in Table 8.1 (see WHO, 1992). *Wuchereria bancrofti* is highly host specific for humans and *B. timori* is probably the same. *Brugia malayi* may be a zoonotic infection for humans and its range of nonhuman hosts (Nelson, 1966) might pose problems for control activities in places where host distributions overlap and where susceptible mosquitoes are active.

8.2 MICROFILAREMIA

Microfilaremia is a key concept in understanding the relationship between definitive and intermediate hosts of filarial worms, the course of lymphatic filariasis, and the progress of measures to control and eliminate the disease. The term has a qualitative and a quantitative use. Qualitatively, an individual is said to be microfilaremic once MF can be detected in the blood, urine, or hydrocele fluid. In an epidemiological survey, the microfilarial prevalence is measured as the proportion of people surveyed whose blood samples were found to contain MF. The microfilarial prevalence is usually expressed as a percentage (MF% or mf%). Quantitatively, microfilaremia is defined as the number of MF per unit volume of blood. Usually, a blood sample of 20 µl is collected, examined microscopically, and the count is then expressed as the number of MF per ml of blood. Nocturnal periodicity makes blood sampling an antisocial activity for both the observed and the observer (Figure 8.2). Protocols for sampling and measuring variables related to MF are available in WHO (2005).

8.3 TRANSMISSION

Many more species of mosquito than those listed in Table 8.1 transmit filarial worms. At least 31 species of *Anopheles*, 15 species of *Aedes*, 5 species of *Culex,* and 2 species of *Mansonia* are known to be vectors of *W. bancrofti* (Muller, 2002). Transmission of filarial worms between definitive and intermediate hosts is determined by factors that facilitate intimate contact between humans and vector mosquitoes in search of blood meals. Crucial to this contact is the breeding behavior of the mosquitoes and the manner in which human activity contributes to sustaining mosquito populations. The history of public health should teach us that humans generally come in second in conflicts with mosquitoes (Brockington, 1958; Spielman and D'Atonio, 2001). Sometimes we win. Gorgas's remarkable achievements during the construction of the Panama Canal demonstrated that mosquito-borne disease can be controlled by public health measures (Bennett, 1915). By 1906, yellow fever had been eliminated from the construction site and as building continued, the proportion of the workforce hospitalized at any one time from malaria fell from the earlier French rate of 30% to the American rate of 2% (Spielman and D'Antonio, 2001).

Southgate (1984) calculated that thousands of bites from vector mosquitoes carrying L3 stages would be needed to establish a filarial infection. Hairston and De Meillon (1968) reckoned that 15,000 bites annually by infected *C. quinquefasciatus* would be required for a successful infection with *W. bancrofti* in a human host. These estimates may not have appreciated that filariasis begins to be contracted during childhood so fewer bites might be involved than was previously thought (Witt and Ottesen, 2001).

It is remarkable that so many people are infected with filarial worms (Table 1.1) given the conditions under which infection takes place. The L3 stages must find their own way through the skin where they are in danger of desiccation while exposed to the air. Higher filarial prevalences occur in communities living in humid places. The seeming inefficiency of transmission to humans may also have to do with the fact that the prevalence of filarial infections in mosquitoes is usually low, perhaps because many insects are killed while acquiring MF. Any mosquito feeding on a person with high microfilaremia will swallow many MF. Recently ingested MF must penetrate the stomach wall before the peritrophic membrane is formed and the scale of the subsequent damage may prove fatal to the insect (Dye, 1992).

The variety of water sources in areas where LF is endemic influences the distribution of mosquitoes and so determines which species acts as vectors of filarial worms for a particular community. Many breeding sites for mosquitoes are provided by human activity (Smart, 1943; Spielman and D'Antonio, 2001). For example, *C. quinquefasciatus* (Table 8.1) breeds in the polluted, stagnant water that accumulates around the damaged septic tanks, pit latrines, and open drains that are seen all too frequently in urban slums and shanty towns. Other species breed in the water that collects in discarded cans, broken pots, and abandoned rubber tires. Human behavior perpetuates mosquitoes. *Mansonia* spp. (Table 8.1) breed in swamps and so affect the health of rural communities. These mosquitoes obtain oxygen by piercing the tissues of aquatic plants rather than by breathing at the water surface. Knowledge of mosquito breeding habits is important in the design of control programs.

There is some evidence to indicate that *W. bancrofti* is able to pass into the placenta. Eberhard et al. (1993) collected samples of umbilical cord blood and placental tissue from 22 Haitian women with demonstrable microfilaremia. MF were detected in two placental samples and one cord sample involving two of the women. Eberhard et al. concluded that this route of transmission might be uncommon, but they stressed that *in utero* exposure to MF or to filarial antigens might induce some degree of immune tolerance and explain how infants born to infected mothers are almost three times more at risk of infection than infants born to uninfected mothers.

8.4 DIAGNOSIS

Several diagnostic tools are available for identifying infections with filarial worms in both humans and mosquitoes. Until recently diagnosis in humans was mainly based on detecting MF in drops of blood and diagnosis in mosquitoes depended on dissection of the insects to find and identify larval stages. These methods remain valuable, but now polymerase chain reaction (PCR) technology and antigen and antibody tests are available to be applied to filarial detection provided that resources are available (WHO, 1992, 2005).

8.4.1 PARASITOLOGICAL DIAGNOSIS

Parasitological diagnosis of LF usually depends on the identification of MF in a drop of fingerprick blood collected at the time of peak periodicity (Figure 8.2) and processed as described in WHO (1987). Exquisite photomicrographs of the MF of medically important filarial worms have been published by Ash and Orihel (1990). These include the MF of *Onchocerca volvulus* and *Loa loa,* which also occur in the blood of infected people in Africa. The preparations were made from thick and thin blood films stained with Giemsa's stain or hematoxylin. Key features for identification include the presence or absence of a sheath, how the sheath is stained by Giemsa's stain, the shape of the tail, and the distribution of nuclei in the body of the stained larva. All authorities refer to the lengths of the MF. Generally MF measure 250 μm, but this is a difficult character to use because live MF wriggle while fixed and stained MF contract, bend, and wrinkle. Identification based on a single MF would be unsatisfactory. Fortunately most infections involve *W. bancrofti* so the important concurrent infections that need to be recognized are *O. volvulus* and *L. loa.*

In areas where LF is endemic people could well be exposed to the bites of insects that transmit filarial worms to nonhuman hosts (Nelson, 1966). There is, therefore, the need to be able to distinguish the filarial worm responsible for LF from several other species of worm that may be present in addition to *O. volvulus* and *L. loa.* For example, an individual presenting with MF of *W. bancrofti* may also have been bitten by mosquitoes that are vectors for *Brugia* spp. that occasionally infect humans and habitually infect nonhuman hosts or by insects that are vectors of *Mansonella* spp. (see Coombs and Crompton, 1991). Essentially the same problem has to be faced when mosquitoes have been trapped and are examined to find out which species of filarial worm are in transit. A key to sort out the plethora of larval filarial worms by means of morphological features has been devised by Bain and Chabaud (1986), but its use will present a formidable challenge to medical technicians in the countries where LF is endemic.

8.4.2 PCR TECHNOLOGY

The PCR is an extremely sensitive method for amplifying DNA sequences (Sullivan, 1994). In practice, a PCR assay can analyze a minute amount of DNA and detect with confidence a sequence specific to an organism from which the DNA had been extracted (WHO, 2002a). An assay has been developed that enables *W. bancrofti* to be identified from filarial DNA present in a mosquito (Chanteau et al., 1994; Farid et al., 2001). The advantage of this procedure is that mosquitoes can be pooled and any filarial DNA in the pool can be found and identified, thereby making easier the checking of mosquito populations for filarial transmission. For example, Chanteau et al. were able to detect the DNA contained in a single L3 (stage infective to humans) contained in 50 mosquito heads. This procedure can give extremely accurate information about transmission rates. Farid et al. used the PCR to detect DNA in MF present in pools of mosquitoes collected in communities with differing micro-filaremia rates. If the PCR procedure does or does not find filarial DNA in a batch of mosquitoes the health worker will have a significant result because the method is entirely reliable for *W. bancrofti*. The sensitivity of the method makes it potentially invaluable for the mass screening of mosquito populations, for following changes in transmission rates and for identifying areas of LF endemicity (WHO, 2002a). In theory, there should be no need to collect blood samples from people in the middle of the night if *W. bancrofti* is present in the local mosquitoes. PCR assays can also be used to detect *W. bancrofti* in human hosts. The technique needs to be extended for dealing with infections of *B. malayi* and *B. timori*. The sensitivity of the PCR, however, means that it is vulnerable to contamination and excellent laboratory facilities and discipline are required when it is applied (Sullivan, 1994).

8.4.3 IMMUNODIAGNOSIS

A highly sensitive and specific immunochromatographic test (ICT) for the rapid detection of *W. bancrofti* has recently become commercially available (Weil et al., 1997). The ICT can be applied when there is no evidence of MF in the blood either because they are not yet present, as may be the case with children, or because of their periodicity (Figure 8.2). A sample of fingerprick blood is placed on a pad impregnated with antibody raised against antigen produced by *W. bancrofti*. A precise color develops within a few minutes if any *W. bancrofti* antigen in the blood sample reacts with the antibody.

8.4.4 CLINICAL DIAGNOSIS AND PATHOGENESIS

Health workers in areas where LF is endemic will have experience of the signs and symptoms of the disease. These diagnostic features are clues to the complex events that are occurring in the unfortunate host (Figure 8.3, Figure 8.4, Figure 8.5, and Figure 8.6). In adults infected with filarial worms the signs and symptoms may not be obvious until MF appear in the blood. Then the signs become more obvious as the disease takes its course (Ukoli, 1984; Dreyer et al., 2000, 2002). Initially, acute inflammation of the lymphatic system occurs accompanied by headaches, high fever, nausea, and the eosinophilia that is also a feature of so many helminth infections. Later, chronic lymphangitis (inflammation of lymphatics), lymphadenitis (inflam-mation of lymph nodes), and lymphedema (swelling caused by the accumulation of fluid) appear. Chyluria (lymph in urine), hydrocele (fluid in the scrotal sac), and impaired lymph flow may be observed. Eventually lymphedema leads to the hard swollen limbs characteristic of elephantiasis, and the genitalia, breasts, and arms may become irreversibly swollen and hardened. If facilities are available, ultrasonog-raphy can be used to reveal live worms in the lymphatic system (WHO, 2001).

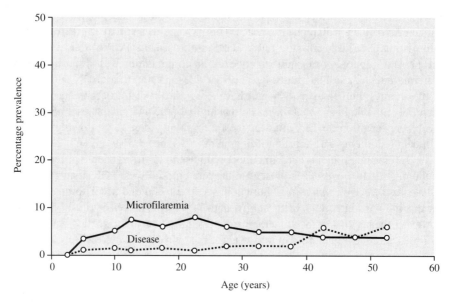

FIGURE 8.3 *Wuchereria bancrofti.* Plot of the prevalence of microfilaremia and morbidity in relation to female host age. From Figure 1a (Pani et al., 1991) with permission from The Royal Society of Tropical Medicine and Hygiene.

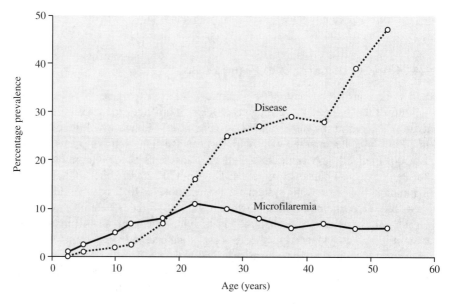

FIGURE 8.4 *Wuchereria bancrofti.* Plot of the prevalence of microfilaremia and morbidity in relation to male host age. From Figure 1b (Pani et al., 1991) with permission from The Royal Society of Tropical Medicine and Hygiene.

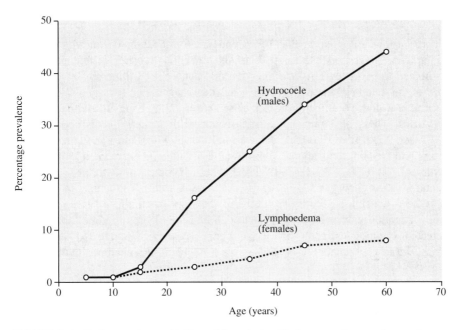

FIGURE 8.5 *Wuchereria bancrofti.* Plot of the relationship between the prevalence of morbidity and host age and gender. From Figure 2a (Pani et al., 1991) with permission from The Royal Society of Tropical Medicine and Hygiene.

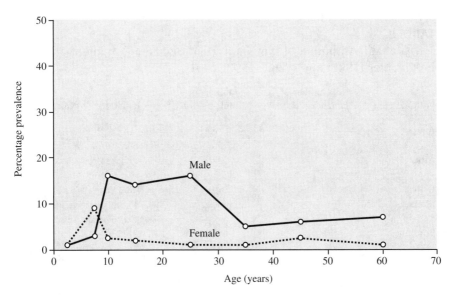

FIGURE 8.6 *Wuchereria bancrofti.* Plot of the relationship between the prevalence of acute episodic lymphangitis and host age and gender. From Figure 2b (Pani et al., 1991) with permission from The Royal Society of Tropical Medicine and Hygiene.

8.5   DISTRIBUTION

There is now a consensus that LF is endemic in at least 80 countries and that *W. bancrofti* is responsible for 90% of the infections. Over a billion people are estimated to be at some risk of infection and about 120 million are infected. The size of the Asian population means that most cases of LF are to be found where the overall prevalence of infection with *W. bancrofti* is 2.25% compared with nearly 9% in Africa. The global prevalence of infection with *Brugia* spp. is about 0.47% (Michael, 2000; WHO, 2002b). Some 16 million, of whom most are women, suffer from lymphedema and about 27 million, most of whom are men, suffer from hydrocele (Table 8.2; WHO, 2001, 2002b). Many people with detectable microfilaremia may not show overt signs of disease (WHO, 1992). The figures in Table 8.2 may need revision especially as results of the work to eliminate LF are reported. More attention may need to be paid to defining the extent of subtle morbidity in infected, symptomless adults and in children who can become infected early in life (WHO, 2001) but do not present at the time with either microfilaremia or the grosser signs of LF.

Knowledge of the distribution of LF has improved recently as part of the program for the elimination of LF. The program depends on the results of accurate epidemiological surveys that can then be used to direct the application of control measures. Gyapong et al. (2002) applied the ICT for the detection of *W. bancrofti* antigen in daytime blood to 50 residents aged 15 years or over in villages in Benin, Burkina Faso, Ghana, and Togo. Geographical information system (GIS) technology was then applied to construct a map predicting the distribution of LF in the four countries.

TABLE 8.2
The Global Distribution of Lymphatic Filariasis. Based on information in Sturchler (1988) and WHO (1992, 2001).

WHO Region	Number of Endemic Countries	Millions of People at Risk	Examples of Prevalence[a]
Americas	7	6.63	Recife, Brazil, 2–15%, Dominican Republic 7–26%[b]
Africa	38	327.5	West Africa (16 countries), 5–40% (> 6 years old)[b]
E Mediterranean	3	30.4	Nile Delta, Egypt, 5–39%[b]
SE Asia[c, d]	8	700.75	India, 45 million cases (12% of people in endemic areas)[b]
W Pacific	24	49.35	Fiji, 0.9–12% [b]

[a] In these cases, prevalence is based on the detection of MF in blood films. Not all cases will present with disease.

[b] *Wuchereria bancrofti*.

[c] *W. bancrofti* and *Brugia malayi* infections may overlap. For example, of 510,000 infected people in Viet Nam, 70,000 have *W. bancrofti* and 440,000 have *B. malayi*.

[d] *Brugia timori* restricted to this region (Table 8.1).

In another survey in Uganda, Onapa et al. (2005) investigated *W. bancrofti* antigen in daytime blood samples from 17,533 school children at 76 sites and then converted the findings into a distribution map. Positive results were obtained from 31 of the sites with prevalence values ranging from 0.4 to 30.7%. This information from Uganda emphasizes the importance of including children in studies of filarial epidemiology.

8.6 FILARIAL POPULATION BIOLOGY AND THE ELIMINATION OF LF

The population biology of *W. bancrofti* is shown as a flow diagram in Figure 8.1 and the population dynamics of filarial nematodes has been analyzed and modeled by Michael (2000). The value of such a model in elucidating human-filarial inter-actions is that it enables a miscellany of field observations to be integrated and certain problems to be overcome. For example, we have little detailed knowledge of how the adult worms behave in the host. We have hardly any direct measurements of the intensity of infection and yet we assume that the bigger the adult worm burden the more severe will be the morbidity. We are usually restricted to working with prevalence estimates and assessments of microfilaremia based on the detection of MF in minute samples of blood. Are we confident that what is found in 50 µl of blood represents what the MF are doing in a volume of 5 million µl?

Transmission between hosts depends on the populations of MF (human to mosquito) and on the populations of L3 (mosquito to human). If the first aim of the LF elimination program is to disrupt transmission with human health as a priority, the best possible information needs to be obtained and understood about MF populations. Experts predict that universal drug application among communities (MDA) with anthelminthic drugs that kill MF will lower the MF population to below a threshold value needed to maintain infections in mosquitoes (Ottesen et al., 1997; Ottesen, 2000). If microfilaremia values below the threshold are achieved, the significance of L3 populations will no longer matter. But several questions need to be investigated:

- What is the microfilaremia threshold value that would need to be achieved to disrupt transmission?
- Given the scale and diversity of the geographical distribution of LF, are present sampling and monitoring methods able to measure the threshold value?
- Is the threshold value the same for all communities in all endemic areas?
- How comprehensive need drug coverage be to lower microfilaremia rates to the level needed to disrupt transmission?
- For how long should drug administration continue?
- Since MF production depends on the reproductive life span of adult worms, is current knowledge of this life span accurate?
- Strains of *W. bancrofti* exist regarding periodicity so is it possible that strains exist regarding reproductive life span?

- Is it possible for MF production to relapse or recrudesce in a host in which microfilaremia had become undetectable?
- What action should be taken if MF populations are found to have become drug resistant?
- Do the different species of vector mosquito behave differently regarding the transmission of filarial worms?
- Do filarial worms change the activities of infected vector mosquitoes?

There is a potential problem in assuming that lowering the MF density to below a certain threshold value will disrupt transmission. Dye (1992) pointed out that if a filarial nematode population is to survive it must not destroy its vector host population. After feeding on human blood containing live MF, at least one MF must penetrate the mosquito's stomach wall and develop to the L3 stage. In some cases facilitation occurs, meaning that within certain limits penetration by MF increases the chances of L3 stages being formed, while in other cases limitation occurs, thereby decreasing the chances of L3 stages being formed (Pichon, 2002). Increased facilitation will put the mosquito at risk of being killed as excessive numbers of MF penetrate its stomach wall. Limitation, however, may serve to reduce mosquito death rates and still support the development of some L3 stages. Pichon (2002) investigated this paradox and concluded that vector control in support of universal anthelminthic treatment (MDA) is desirable depending on the species of vector mosquito. Some species of mosquito transmit malaria parasites in addition to filarial nematodes as occurs in West Africa. Efforts to lower the microfilaremia count in the community without regard to the effect on the lifespan of infected vectors should not be ignored.

8.7 MORBIDITY

The second aim of the global plan to eliminate LF is the control and management of morbidity. This requires some knowledge of how the disease arises and progresses. There is no cure, certainly not at the community level, and so it is important that health workers teach individuals how to help infected relatives, friends, and neighbors manage their condition.

In direct contrast to onchocerciasis where the MF are the main pathogenic agents, LF is initiated by the presence of living, adult filarial worms in the lymphatic system. Acute inflammatory attacks (ADL, adenolymphangitis) occur followed by dilatation of the lymph vessels and lymphatic insufficiency (Dreyer et al., 2000, 2002). Lymphatic insufficiency means that the drainage and circulation of lymph in the body is impaired. One of the consequences of this malfunction is weaker resistance to invading microorganisms from the skin. Under normal circumstances bacteria that have entered the body through tiny abrasions or cuts in the skin's surface will normally be destroyed by leucocytes circulating in lymph. When lymph flow is disturbed, bacterial or fungal entry through skin lesions, known as entry lesions in the filariasis literature, stimulates a form of inflammation that leads to lymphedema. In areas where LF is endemic, the common entry lesions are usually found between the toes because poor people cannot afford to buy shoes.

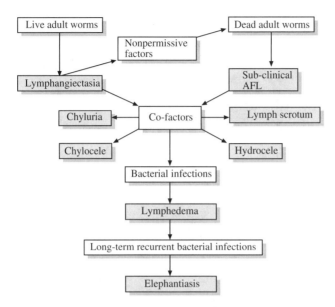

FIGURE 8.7 A concept of the origin and progression of the pathology associated with lymphatic filariasis. From Box 1 (Dreyer et al., 2000) with permission from Elsevier.

For a time it was assumed that lymphedema occurred because the worms had blocked the lymph vessels thereby causing lymphatic insufficiency. That explanation is no longer accepted. Live adult worms release molecules that cause lymphangiectasia (dilatation of the lymph vessels) so that the flow of lymph is disrupted. A conceptual framework for understanding the complex pathology of LF is shown in Figure 8.7 (Dreyer et al., 2000). An important feature of this scheme is the view that attacks of acute filarial lymphangitis (AFL) are caused by host responses to dead adult worms; host responses are no longer thought to be responsible for killing adult worms (Dreyer et al., 2000). During the evolution of the human-filarial relationship the worms appear to have acquired adaptations that facilitate survival despite host immune responses (Maizels et al., 2000). The permanent damage to the lymphatic system leads to the gross pathology in the limbs, breasts, and genitalia. In the case of the lower limbs, the subsequent bacterial and fungal invasions through entry lesions and the hardening of the tissues explains why chronic LF persists after the death of adult worms as indicated by the absence of microfilaremia. Episodes of bacterial invasion account for the bouts of fever and acute inflammation experienced by people with LF.

An important and practical contribution to the management of LF in the community has been made by Dreyer et al. (2002) who have devised a scheme for identifying seven stages in the progression of lymphedema. Their scheme is set out in Figure 8.8 in the form of a chart that describes the course of lymphedema in legs. The progression of lymphedema can also be classified into three grades; grade I, pitting lymphedema that is spontaneously reversible on elevation; grade II, nonpitting lymphedema that is not spontaneously reversible on elevation; grade III, gross

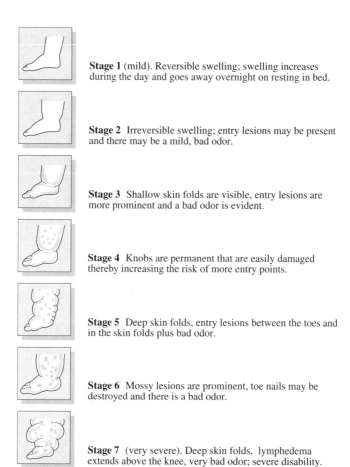

Stage 1 (mild). Reversible swelling; swelling increases during the day and goes away overnight on resting in bed.

Stage 2 Irreversible swelling; entry lesions may be present and there may be a mild, bad odor.

Stage 3 Shallow skin folds are visible, entry lesions are more prominent and a bad odor is evident.

Stage 4 Knobs are permanent that are easily damaged thereby increasing the risk of more entry points.

Stage 5 Deep skin folds, entry lesions between the toes and in the skin folds plus bad odor.

Stage 6 Mossy lesions are prominent, toe nails may be destroyed and there is a bad odor.

Stage 7 (very severe). Deep skin folds, lymphedema extends above the knee, very bad odor; severe disability.

FIGURE 8.8 Stages in the progress of lymphedema in the leg. From Dreyer et al. (2002) with permission from Hollis Publishing Company.

increase in volume with dermatosclerosis and papillomatous lesions (WHO, 2003). Similar pathological events apply to lymphedema in the arms, although complications caused by the entry of microorganisms would be expected to be less common. Dreyer et al. also provided detailed practical advice about the need to exercise personal hygiene to reduce opportunities for bacterial invasion. For sufferers incapacitated by the later stages of lymphedema (Figure 8.8), there is a need for others in the community to assist with washing affected limbs and providing personal support. This form of lymphedema management has short-term psychological benefits. Individuals gain motivation to overcome their condition, they feel better, and they find better acceptance by family and friends. Personal hygiene reduces the bad smell that accompanies lymphedema (Figure 8.8). Lymphedema management has long-term benefits. Daily cleansing reduces the number of entry lesions so acute inflammatory attacks become less frequent and the extent of mossy foot declines. Patients display a sense of well-being and self-esteem improves.

Age and gender analysis of LF data shows that in the many areas where *W. bancrofti* is endemic hydrocele is the major manifestation in men and lymphedema is the major manifestation in women. In Pondicherry, South India, significantly more disease was evident in men than in women where 12% of the men had hydrocele and 2% of the women had lymphedema (Pani et al., 1991). Some aspects of the disease are clearly age dependent (Figure 8.4), some are gender dependent (Figure 8.5), and some may not show clear-cut relationships. Occurrence of hydrocele usually increases as men age (WHO, 2002b). Episodes of ADL, during which the patient experiences fever and malaise, occur more frequently with lymphedema than hydrocele. In an LF endemic area hydrocele is usually assumed to be due to LF, and since even early cases of hydrocele are readily detected with the minimum of equipment, the prevalence of hydrocele in an area may be used as an indicator of the prevalence of LF in the area. Some research evidence, however, indicates that about 30% of hydrocele cases may not result from filarial infection (see WHO, 2000b).

Hydrocele does not respond to any anthelminthic treatment; surgery is needed to correct the condition. Plans have been made and tested to demonstrate that surgery can be provided as a public health measure for the management of hydrocele. The protocol was devised during an informal consultation of experts under the auspices of the World Health Organization and details are included in the report of that meeting (WHO, 2002b). There are three levels of activity in the procedure. First (Level I), a community health worker in consultation with the patient confirms the presence of a scrotal swelling. Second (Level II), the patient is referred to a health center where minor surgical procedures are performed under local anesthesia. This requires the services of a general physician with additional training for the job. Such a center, according to the experts, requires the presence of facilities for basic resuscitation and the observation of patients for up to 48 h. Third (Level III), patients with complicated hydroceles should be referred to a district hospital for more advanced surgery. How many governments in countries of the south, where hydrocele is a serious health problem, already have the resources to offer this service to their citizens?

Relatively few deaths occur as a direct result of LF. Stage 7 lymphedema (Figure 8.8) sometimes forces a patient to be confined to bed for a long time. Dreyer et al. (2002) warn that such patients should not attempt to move or walk without supervision because of the risk of breaking a bone in the leg with lymphedema, which may lead to a blood clot in the leg followed by a fatal pulmonary embolism.

8.8 TROPICAL PULMONARY EOSINOPHILIA

Tropical pulmonary eosinophilia, sometimes called occult filariasis, is an unpleasant condition resulting from invasion of the lungs by eosinophils that are assumed to have been produced in response to the presence of MF in the lungs (Muller, 2002). Patients present with symptoms similar to those of a severe bronchial asthma attack. They may not have detectable MF in their blood but will be serologically positive for *W. bancrofti* or *B. malayi*. The condition is more common in men than in women and is extremely rare in children (Addiss and Dreyer, 2000).

8.9   ANTHELMINTHIC TREATMENT

The global plan adopted for the control and eventual elimination of LF now depends on the universal application of either diethylcarbamazine citrate (DEC) in combination with albendazole (WHO, 2000a) or ivermectin in combination with albendazole (WHO, 2000b; GPELF, 2005; Table 14.1). Such doses of albendazole are effective against infections with soil-transmitted helminths (Chapter 11). There is now evidence that albendazole is even more effective against *Ascaris* and especially against *Trichuris* when used in combination with ivermectin (Belizario et al., 2003). The treatment procedure in LF control programs is referred to as mass drug administration (MDA) and application occurs through implementation units (IU). An implementation unit is defined as a designated administrative area where, if LF is endemic or filarial transmission is known to occur, the entire population should be offered treatment with recommended anthelminthic drugs. An advantage of the IU scheme is it should enable dosage, coverage, side effects, and outcomes to be monitored.

Household salt (sodium chloride) fortified with DEC is proving to be an effective means of providing treatment for LF (GPELF, 2005). One year's use of fortified salt might be sufficient for the disruption of transmission as long as all in need of it actually use it. Trials in India and PR China have found that DEC added to salt at a concentration of 0.1% will eliminate *W. bancrofti* after at least 6 months and when added at a concentration of 0.3% will eliminate *Brugia* spp. after at least 3 months (WHO, 1992; 2004). The use of fortified salt is easier said than done. Community compliance is essential; production must be assured; logistics must be secure to ensure distribution. Government legislation will be required to prohibit the availability of unfortified salt during the duration of the program.

8.9.1   DRUG COMBINATION FOR LF IN COUNTRIES WHERE ONCHOCERCIASIS AND LOIASIS ARE NOT CO-ENDEMIC

The combination of DEC (single, yearly, oral dose of 6 mg/kg body weight) with albendazole (single, yearly, oral dose of 400 mg) is recommended for universal application in communities where *O. volvulus* and *L. loa* are judged to be absent. Current knowledge of the population biology of *W. bancrofti* indicates that MDA for 4 to 6 years should be sufficient to disrupt transmission. Some transient side effects are inevitable when a drug is given to kill MF, which will then disintegrate in the body and challenge the immune system. This annual discomfort is fully justified when compared with the chronic consequences of lymphedema or hydrocele.

Care must be taken to establish the absence of *Onchocerca* and *Loa* if the DEC is a component of the drug combination to be used (WHO, 2000a). A safe rule would be to avoid use of DEC for the treatment of LF in Africa. If *Onchocerca volvulus* is present in a patient for whom DEC has been prescribed, serious adverse effects, known as Mazzotti reactions, may occur, characterized by itching, rash, edema, pain, fever, swelling of the lymph nodes, and even severe eye lesions (see Piessens and Mackenzie, 1982; WHO, 2004). These drug-induced reactions are probably allergic in nature and occur as the immune system is temporarily overwhelmed by the mass of antigen released from dead or dying worms. Although DEC is effective against

Loa loa, its use can be dangerous when the microfilaremia is high (> 50,000 *Loa* MF/ml blood). A form of meningoencephalitis occurs as the DEC kills the MF, which then accumulate in the cerebral capillaries. This condition arises in about 1% of such cases and proves fatal in about half of them (WHO, 2004).

8.9.2 DRUG COMBINATION FOR LF IN COUNTRIES WHERE ONCHOCERCIASIS AND LOIASIS ARE CO-ENDEMIC

The combination of ivermectin (single, yearly, oral dose of 200 μg/kg body weight) and albendazole (single, yearly, oral dose of 400 mg) is recommended for universal application in communities where LF occurs concurrently with infections of *Onchocerca volvulus* or *Loa loa.* This is the drug combination to be used in Africa (WHO, 2000b). The recommended dose of ivermectin will serve to kill MF of *O. volvulus* if that helminth is also present in an individual being treated for LF. The usual dose for onchocerciasis is 150 μg/kg body weight (WHO, 2004) so side effects such as a milder version of the Mazzotti reaction described above may be expected in some people.

Infection with *Loa loa* remains a problem wherever DEC or ivermectin are intended to be used for the treatment of either LF or onchocerciasis. The possibility of severe adverse events such as Mazzotti reactions following DEC use are well known (see above), but occasionally a form of encephalopathy may occur in people treated with ivermectin while harboring a high-intensity infection of *Loa loa.* Program managers should always endeavor to determine the distribution of *Loa loa* in areas endemic for LF and for onchocerciasis (Addiss et al., 2003) and be prepared to deal with a serious adverse event if necessary.

8.10 PREVENTION AND CONTROL IN SUPPORT OF MASS DRUG ADMINISTRATION

Filarial transmission can be reduced if the risk of being bitten by vector mosquitoes can be reduced. Measures to kill mosquitoes, prevent their breeding opportunities, and impede their feeding activities are summarized in Table 8.3 (WHO, 2002a). The greater the sustained compliance with the implementation of these measures in the community, the greater the degree of protection for all living there. Successful control depends on accurate identification of the mosquito species involved in transmission, knowledge of its biology, choice of the most effective method, and funds to cover the costs.

Culex quinquefasciatus is the chief vector for *W. bancrofti.* The control of this mosquito, accompanied by a decrease in the transmission of *W. bancrofti,* has been achieved in parts of East Africa and India by the use of expanded polystyrene beads (EPBs). The mosquito breeds in stagnant water so a layer of EPBs disrupts egg laying by the adults and prevents breathing by the larvae. The addition of EPBs to pit latrines has proved to be a particularly effective means of controlling *C. quinquefasciatus* and has been shown to prevent the resurgence of LF after rounds of MDA had ended. Shredded waste polystyrene (SWAP) has been used with equal success in pit latrines in the West Indies. References to how to apply the methods displayed in Table 8.3 are to be found in WHO (2002a).

TABLE 8.3
Control Methods for Use Against Mosquito Vectors of Lymphatic Filariasis. From Table 1 (WHO, 2002a) with permission from the World Health Organization (ww.who.int).

| | | Type of Filariasis | | Control Method | | | | | | |
| | | | | Environmental/Physical | | | | Chemical/Insecticidal | | |
Mosquito Genus	Peak Time of Biting	Brugia	Wuchereria	Source Reduction[a]	Larval Control	House Screening	Bednets Curtains	Indoor Residual Spray	Larvacidal Chemicals, Biopesticides and MLOs
Anopheles	Night	Nocturnally periodic	Nocturnally periodic	++/-	Predators, e.g., fish	+++[b]	+++	+++	++/-
Culex	Night	Nocturnally periodic	Nocturnally periodic	+++	Expanded polystyrene	+++	+++	++/-	+/-
Mansonia	Night	Nocturnally periodic	NA	+/-	Remove aquatic vegetation	+++	+++	++/-	–
Aedes	Day	Sub-periodic	Sub-periodic	+++	Eliminate containers	++	+/-	+/-	+/-
Ochlerotatus	Day and night	NA	Nonperiodic	+/-	Water management	++	+/-	+/-	++

[a] Source reduction means preventing potential breeding-sites by drainage, flushing, filling of pools, elimination of water-filled receptacles (discarded cans, tires) covering of water-storage jars and basins, installation and maintenance of sanitation systems that do not produce mosquitoes.

[b] +++ most effective; ++ fairly effective; ++/- effective for many but not all situations or vectors; +/- possibly effective for some species/sites; — not appropriate; NA not applicable; MLOs = mosquito larvicidal oils.

The possibility of biological control of mosquitoes exists because of discoveries about *Wolbachia,* a genus of bacteria that live as obligate endosymbionts in arthropods and nematodes (Townson, 2002). These bacteria manipulate the reproduction of their hosts by inducing a process known as cytoplasmic incompatibility (CI), which is maternally inherited. Cytoplasmic incompatibility serves as a transmission adaptation for *Wolbachia,* but CI also causes some of the insects to become sterile. There is even the possibility of using *Wolbachia* as a means of introducing genes into mosquito populations to render them refractory to MF. Much research would be needed, however, before this form of intervention could be adopted. Other research has indicated that *Wolbachia* have a role in the pathogenicity of *Onchocerca volvulus* (Chapter 9).

Although the implementation of measures to control vector mosquitoes may seem an obvious course of action to take in support of universal drug treatment (MDA) in a community where LF prevails, the cost has to be considered. Do the benefits to public health justify the necessary expenditure and are the benefits sustainable? Krishnamoorthy et al. (2002) compared the cost-effectiveness of MDA with the costs of MDA plus vector control in three groups of villages in South India. One group of villages served as a control during the period of the investigation and the mosquito species of concern was *C. quinquefasciatus.* From a cost perspective, MDA plus vector control was not worth the expense. The *per capita* costs were USD 1.49 for MDA alone and USD 3.19 for MDA plus vector control. With MDA alone the cost of stopping an infected mosquito from biting a villager was USD 1.80 compared with USD 3.32 when vector control was included. The cost of reducing the MF% in a group of villages by MDA alone was USD 96.62, but rose to USD 201.16 when vector control was included. The authors stress that their results apply to the communities and conditions encountered in the villages they studied.

8.11 SOCIAL AND ECONOMIC COSTS OF LYMPHATIC FILARIASIS

The social and economic burden of LF is inextricably linked and the linkage will not be weakened or broken until the morbidity due to LF is reduced and the disease is eliminated. LF is considered to be responsible for more than 1% of the DALYs due to infectious and parasitic disease (Table 1.2). Those who make decisions about the use of resources in the public health sector should consider the costs of relieving affected individuals from the impact of physical disfigurement, social stigma and ridicule, loss of self-esteem, impairment of sexual activity, and family discord. The hidden social impact of LF on women in India has been investigated in detail by Bandyopadhyay (1996). Women may not seek relief from LF because of embarrassment, shame, social taboos, and cultural attitudes. The economic losses stem from physical disability, attacks of ADL, lowered employment opportunities, and the costs of treatment. Sufferers who are employed in paid work are less productive and those in unpaid work, who care for their families and households, are equally less productive.

Studies in India demonstrate how LF adversely affects family income and manufacturing output. The productivity of 39 male weavers with overt LF was compared with that of 39 uninfected weavers (Ramu et al., 1996). The weavers were observed for 184 days; the payment for work was 12 rupees per meter of cloth produced. Both groups worked the same number of hours, but the uninfected weavers (controls) produced significantly more cloth and so earned more money for family support than their infected colleagues. Ramaiah et al. (2000) reckoned that ADL puts a worker in India, where 60% of LF cases are men, out of action for an average of 3.5 days per attack. This results in a loss of 59 million working days due to ADL and a loss of USD 700 million in wages.

REFERENCES

Enquiries about the availability of publications of the World Health Organization from health professionals concerned with the control of lymphatic filariasis should be made to the Global Programme for the Elimination of Lymphatic Filariasis (GPELF), World Health Organization, 1211 Geneva 27, Switzerland.

Addiss DG and Dreyer G. 2000. Treatment of lymphatic filariasis. In: *Lymphatic Filariasis* (ed. TB Nutman). London: Imperial College Press. pp. 151–199.
Addiss DG, Rheingans R, Twum-Danso NA and Richards FO. 2003. A framework for decision-making for mass distribution of Mectizan (R) in areas endemic for *Loa loa*. *Filaria Journal* **2**, Suppl. 1:S9.
Anderson RM and May RM. 1992. *Infectious Diseases of Humans*. Oxford University Press (paperback edition).
Ash LR and Orihel TC. 1990. *Atlas of Human Parasitology*, 3rd edition. Chicago: American Society of Clinical Pathologists Press.
Bain O and Chabaud AG. 1986. Atlas des larves infestantes de filaires. *Tropical Medicine and Parasitology* **37**, 301–340.
Bandyopadhyay L. 1996. Lymphatic filariasis and the women of India. *Social Sciences and Medicine* **42**, 1401–1410.
Belizario YV, Amarillo ME, de Leon WU et al. 2003. A comparison of the efficacy of single doses of albendazole, ivermectin, and diethylcarbamazine alone or in combinations against *Ascaris* and *Trichuris* spp. *Bulletin of the World Health Organization* **81**, 35–42.
Bennett IE. 1915. *History of the Panama Canal its Construction and Builders*. Washington DC: Historical Publishing Company.
Brockington F. 1958. *World Health*. Harmondsworth, Middlesex: Penguin Books Ltd.
Chanteau S, Luquiaud P, Failloux A and Williams SA. 1994. Detection of *Wuchereria bancrofti* larvae in pools of mosquitoes by the polymerase chain reaction. *Transactions of the Royal Society of Tropical Medicine and Hygiene* **88**, 665–666.
Coombs I and Crompton DWT. 1991. *A Guide to Human Helminths*. London, New York and Philadelphia: Taylor and Francis.
Dreyer G, Noroes J, Figueredo-Silva J and Piessens WF. 2000. Pathogenesis of lymphatic disease in bancroftian filariasis: a clinical perspective. *Parasitology Today* **16**, 544–548.
Dreyer G, Addiss D, Dreyer P and Noroes J. 2002. *Basic Lymphoedema Management*. Hollis, NH: Hollis Publishing Company.
Dye C. 1992. Does facilitation imply a threshold for the eradication of lymphatic filariasis? *Parasitology Today* **8**,109–110.
Eberhard ML, Hitch WL, McNeeley DF and Lammie PJ. 1993. Transplacental transmission of *Wuchereria bancrofti* in Haitian women. *Journal of Parasitology* **79**, 62–66.

Farid HA, Hammad RE, Hassan MM et al. 2001. Detection of *Wuchereria bancrofti* in mosquitoes by the polymerase chain reaction: a potentially useful tool for large-scale control programs. *Transactions of the Royal Society of Tropical Medicine and Hygiene* **95**, 29–32.

GPELF (Global Programme to Eliminate Lymphatic Filariasis). 2005. *Annual Report on Lymphatic Filariasis 2003.* Geneva: World Health Organization.

Grove DI. 1982. Filariasis. In: *Handbook Series in Zoonoses.* Section C: *Parasitic Zoonoses* Volume II (ed. MG Schultz. Boca Raton, Florida: CRC Press. pp. 123–146.

Gyapong JO, Kyelem D, Kleinschmidt I et al. 2002. The use of spatial analysis in mapping the distribution of bancroftian filariasis in four West African countries. *Annals of Tropical Medicine and Parasitology* **96**, 695–705.

Hairston NG and De Meillon B. 1968. On the inefficiency of transmission of *Wuchereria bancrofti* from mosquito to human host. *Bulletin of the World Health Organization* **38**, 935–941.

Krishnamoorthy K, Rajendran R, Sunish IP and Reuben R. 2002. Cost-effectiveness of the use of vector control and mass drug administration, separately or in combination, against lymphatic filariasis. *Annals of Tropical Medicine and Parasitology* **96** (Supplement), S77–S90.

Maizels RM, Allen JE and Yazdanbakhsh M. 2000. Immunology of lymphatic filariasis: current controversies. *Lymphatic Filariasis* (ed. TB Nutman). London: Imperial College Press. pp. 217–243.

Michael E. 2000. The population dynamics and epidemiology of lymphatic filariasis. In: *Lymphatic Filariasis* (ed. TB Nutman). London: Imperial College Press. pp. 41–81.

Muller R. 2002. *Worms and Human Disease,* 2nd edition. Wallingford, Oxfordshire: CABI Publishing.

Muller R and Wakelin D. 1998. Lymphatic filariasis. In: *Topley and Wilson's Microbiology and Microbial Infections*, 9th edition. Volume 5. *Parasitology* (eds. FEG Cox, JP Kreier and D Wakelin). London: Arnold. pp. 609–619.

Nelson GS. 1966. The pathology of filarial infections. *Helminthological Abstracts* **35**, 311–336.

Nutman TB. (ed.). 2000. *Lymphatic Filariasis.* London: Imperial College Press.

Onapa AW, Simonsen PE, Baehr I and Pedersen EM. 2005. Rapid assessment of the geographical distribution of lymphatic filariasis in Uganda, by screening schoolchildren for circulating filarial antigens. *Annals of Tropical Medicine and Parasitology* **99**, 141–153.

Ottesen EA. 2000. Towards eliminating lymphatic filariasis. In: *Lymphatic Filariasis* (ed. TB Nutman). London: Imperial College Press. pp. 201–215.

Ottesen EA, Duke BOL, Karam L and Behbehani K. 1997. Strategies and tools for the control/elimination of lymphatic filariasis. *Bulletin of the World Health Organization* **75**, 491–503.

Pani SJ, Balakrishnan N, Srividya A et al. 1991. Clinical epidemiology of bancroftian filariasis: effect of age and gender. *Transactions of the Royal Society of Tropical Medicine and Hygiene* **85**, 260–264.

Piessens WF and Mackenzie CD. 1982. Lymphatic filariasis and onchocerciasis. In: *Immunology of Parasitic Infections,* 2nd edition (eds. S. Cohen and KS Warren). Oxford and London: Blackwell Scientific Publications. pp. 622–653.

Pichon G. 2002. Limitation and facilitation in the vectors and other aspects of the dynamics of filarial transmission: the need for vector control against *Anopheles*-transmitted filariasis. *Annals of Tropical Medicine and Parasitology* 96 (Supplement), S143–S152.

Ramaiah KD, Das PK, Michael E and Guyatt H. 2000. The economic burden of lymphatic filariasis in India. *Parasitology Today* **16**, 251–253.

Ramu K, Ramaiah KD, Guyatt H and Evans D. 1996. Impact of lymphatic filariasis on the productivity of male weavers in a south Indian village. *Transactions of the Royal Society of Tropical Medicine and Hygiene* **90**, 669–670.

Smart J. 1943. *A Handbook for the Identification of Insects of Medical Importance.* London: British Museum (Natural History).

Spielman A and D'Antonio M. 2001. *Mosquito.* London: Faber and Faber Limited.

Southgate BA. 1984. Recent advances in the epidemiology and control of filarial infections including entomological aspects of transmission. *Transactions of the Royal Society of Tropical Medicine and Hygiene* **78**, 19–28.

Sturchler D. 1988. *Endemic Areas of Tropical Infections,* 2nd edition. Toronto, Lewiston NY, Bern, Stuttgart: Hans Huber Publishers.

Sullivan KM. 1994. Polymerase chain reaction. In: *The Encyclopaedia of Molecular Biology* (ed. J Kendrew). Oxford: Blackwell Science. pp. 864–866.

Townson H. 2002. *Wolbachia* as a potential tool for suppressing filarial transmission. *Annals of Tropical Medicine and Parasitology* **96**, S117–S127.

Trent SC. 1963. Reevaluation of World War II veterans with filariasis acquired in the South Pacific. *American Journal of Tropical Medicine and Hygiene* **12**, 877–887.

Ukoli FMA. 1984. *Introduction to Parasitology in Tropical Africa.* Chichester and New York: John Wiley and Sons Limited.

Weil GJ, Lammie PJ and Weiss N. 1997. The ICT Filariasis Test: a rapid format antigen test for diagnosis of bancroftian filariasis. *Parasitology Today* **13**, 401–404.

Witt C and Ottessen EA. 2001. Lymphatic filariasis: an infection of childhood. *Tropical Medicine and International Health* **6**, 582–606.

WHO. 1987. *Control of Lymphatic Filariasis. A Manual for Health Personnel.* Geneva: World Health Organization.

WHO. 1992. Lymphatic filariasis: the disease and its control. Fifth Report of the WHO Expert Committee on Filariasis. *Technical Report Series* 821. Geneva: World Health Organization.

WHO. 2000a. Preparing and implementing a national plan to eliminate lymphatic filariasis in countries where onchocerciasis is not co-endemic. Geneva: World Health Organization. WHO/CDS/CPE/CEE/2000.15.

WHO. 2000b. Preparing and implementing a national plan to eliminate lymphatic filariasis in countries where onchocerciasis is co-endemic. Geneva: World Health Organization. WHO/CDS/CPE/CEE/2000.16.

WHO. 2001. Lymphatic filariasis. *Weekly Epidemiological Record of the World Health Organization* **76**, 149–154.

WHO. 2002a. Defining the roles of vector control and xenomonitoring in the Global Programme to Eliminate Lymphatic Filariasis. Report of a WHO Informal Consultation. Geneva: World Health Organization. WHO/CDS/CPE/PVC/2002.3.

WHO. 2002b. Surgical approaches to the urogenital manifestations of lymphatic filariasis. Report of an Informal Consultation. Geneva: World Health Organization. WHO/CDS/CPE/CEE/2002.23.

WHO. 2003. Global programme to eliminate lymphatic filariasis. Annual report on lymphatic filariasis 2002. Geneva: World Health Organization. WHO/CDS/CPE/CEE/2003.38.

WHO. 2004. *WHO Model Formulary 2004.* Geneva: World Health Organization.

WHO. 2005. Monitoring and epidemiological assessment of the programme to eliminate lymphatic filariasis at implementation unit level. Geneva: World Health Organization. WHO/CDS/CPE/CEE/2005.50.

Zagaria N and Savioli L. 2002. Elimination of lymphatic filariasis: a public health challenge. *Annals of Tropical Medicine and Parasitology* **96** (Supplement), S3–S13.

9 Onchocerciasis

Onchocerciasis or "river blindness" is the disease that develops during the course of an infection with *Onchocerca volvulus*. The infection and disease occur in large areas of tropical Africa, in parts of the Americas, and in a small focus in the Arabian Peninsula (WHO, 1995). *Onchocerca volvulus* is a filarial, nematode worm transmitted through the feeding behavior of black flies (*Simulium* spp.), which also serve as intermediate hosts. The pathogenic, first-stage larvae or microfilariae (MF) live in human skin. Consequently onchocerciasis is essentially a disease of skin and skin derivatives including the eyes. The disease causes distress, disfigurement, disability, and blindness in individuals and serious economic loss and social deprivation for communities. Investigations into the pathology of onchocerciasis in Africa have shown that blinding onchocerciasis is commoner in the savannah region than in the forest region. Changes brought about by deforestation have had effects on the distribution and flow of water in which black flies breed (Walsh et al., 1993) so the identification of regions as savannah and forest is no longer clear-cut.

Black flies breed in freshwater. When hunter-gathering declined, people generally became agriculturalists and land near to rivers became the favored sites for settlements. Clearance of forests in river valleys to satisfy the demand for more arable land dispersed many of the animals on which the black flies fed. Inadvertently, humans had increased their exposure to the blood-feeding habits of black flies and onchocerciasis became a major affliction for millions of people, especially in tropical Africa. The nearer people lived to rivers, the more likely they were to become blind. Entire communities were decimated, villages were abandoned, agricultural productivity ceased, and hunger was dominant. By the 1970s, the annual economic loss due to onchocerciasis was estimated to be about USD 30 million (WHO, 2003).

Much progress has been achieved in controlling onchocerciasis and reducing its economic impact. The Onchocerciasis Control Programme (OCP), which began in a selection of countries in West Africa in 1974, has successfully controlled the disease there to the extent that onchocerciasis is no longer seen as a public health problem in much of the region (Samba, 1994; Molyneux, 1995; Figure 9.1). The OCP began with a strategy to control the vector by means of insecticide distribution and later to control and reduce worm infections by mass administration of ivermectin. Aerial spraying with insecticide stopped in 2000 in all but two river basins in the OCP area, but drug administration continues. Control efforts have now been extended to the other parts of Africa where the disease is endemic, again based on ivermectin distribution (Figure 9.1; Remme, 1995). The epidemiological knowledge and practical experience gained from planning, implementing, and managing the control of onchocerciasis on such a large scale have much to teach health workers assigned to dealing with large-scale programs for the control of other forms of helminthiasis.

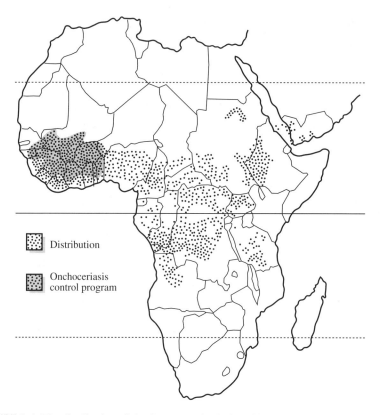

FIGURE 9.1 The distribution of *Onchocerca volvulus* in Africa. From map 10 (Muller, 2002) with permission from CAB International.

9.1 LIFE HISTORY AND HOST RANGE

9.1.1 LIFE HISTORY

Adult *O. volvulus* are thin white worms measuring about 1 mm in width with males being 30 to 50 mm long and females 300 to 800 mm long. Sexually mature male and female *O. volvulus* live coiled around each other in nodules of human tissue (onchocercomata). Nodules formed where bones are located near the surface of the body are easily seen or felt. Detection of nodules provides useful if incomplete information about the prevalence and distribution of *Onchocerca* infection. After insemination, each female releases L1 (MF) that migrate from the nodules to the skin and other sites. The females are ovoviviparous so that embryonation takes place *in utero*. The MF in the skin are observed to be unsheathed, having escaped from their thin eggshells either in or on leaving their parent worm. Absence of a sheath is an important feature because it helps to distinguish the MF of *Onchocerca* from those of *Wuchereria* (see Chapter 8). MF have a lifespan of from 6 to 24 months. Adult female *Onchocerca* may live for up to 16 years although the average length of a female worm's reproductive life is from 9 to 11 years. During this time each fertilized female is reckoned to release from 700 to 1500 MF daily, amounting to an average fecundity of about 3.5 million MF per female (WHO, 1995; Bradley and Unnasch, 1996).

Once in the skin, MF are vulnerable to the feeding activity of female black flies. Having been ingested, the MF leave the insect's gut and reach the thorax where they moult to become L2 stages. After another moult, the L3 stages, which are infective to humans, move to the insect's head and mouth parts. From 6 to 12 days are required for the development of the L1 (MF) to the infective L3 at the ambient temperatures where black flies live. When the black fly feeds, the L3 pass on to the skin and enter the body through the wound made by the intermediate host. Having entered the body two more moults take place before the young adult worms congregate to form new nodules or join the older worms at existing nodules. Little is known about the activities of the immature worms in the human host, but it appears that about 10 to 15 months must elapse before MF are released by females in sufficient numbers for detection by routine methods. Other than mating, little is known about the activities of adult worms in the nodules, which may be occupied for years since live worms and dead, calcified worms may be found in the same nodule at the same time (WHO, 1995; Bradley and Unnasch, 1996).

There appear to be no significant definitive reservoir hosts for *O. volvulus*. A natural infection has been reported from a gorilla and a possible infection has been reported from a spider monkey (see Engelkirk et al., 1982). After an extensive review of *Onchocerca* spp. and their interactions with nonhuman and human hosts, Engelkirk et al. concluded that human onchocerciasis is not a zoonosis. Experimental infections have been established in chimpanzees (*Pan troglodytes*) by injection of L3 stages recovered from black flies (Duke, 1980) and in mangabey monkeys (*Cercocebus atys*) by the same method (Eberhard et al., 1991), but there is no evidence to suggest that either species is a reservoir of infection for human hosts.

9.1.2 INTERMEDIATE HOSTS AND VECTORS

The survival of *Onchocerca* spp. and the distribution of onchocerciasis depend on the vitality and distribution of aggressive, anthropophilic black flies. These small, humped-back flies (up to 5 mm in length) are relatives of mosquitoes and also belong to the suborder Nematocera of the order Diptera (Smart, 1943). Several groups or complexes of sibling species of the genus *Simulium,* in addition to individual species, are the main intermediate hosts and vectors for the development and transmission of *Onchocerca.* The members of the complexes are recognized by such characteristics as chromosome morphology and may be referred to as cytospecies. Any member of one of the complexes is assumed to be capable of supporting *O. volvulus* if it is identified in an area where onchocerciasis is endemic. The identification of black flies to cytospecies or sibling species level requires specialist knowledge and will remain important if the progress achieved by control programs is to be sustained.

Tabulated information about the distribution of the *Simulium* sibling species complexes and other species of *Simulium* involved in onchocerciasis is given by WHO (1995). *Simulium damnosum* complex and the closely related *S. sanctipauli* and *S. squamosum* subcomplexes carry out transmission across West Africa (Figure 9.1).

Simulium rasyani, a sibling of the *S. damnosum* complex, is the host and vector in the Arabian Peninsula. *Simulium naevi* complex is the main vector in East Africa and *S. albivirgulatum* is the vector in the central part of the Democratic Republic of the Congo (Figure 9.1). *Simulium ochraceum* complex is responsible for transmission in the Americas.

Nine species of *Simulium* are recognized as intermediate hosts and vectors of *Onchocerca* in the Americas (WHO, 1995). Most information is known about *S. ochraceum,* which is reckoned to be responsible for most transmission in Guatemala and Mexico. Apparently the annual biting rate of this species is higher than that of members of the *S. damnosum* complex in Africa. The species of black fly that serve as intermediate hosts and vectors for *O. volvulus* have the same role for species of *Onchocerca* that infect wild and domesticated animals (Engelkirk et al., 1982). This situation complicates efforts to monitor the relationship between *O. volvulus* and its vectors. For example, the determination of transmission rates will need the accurate identification of infective L3 of *O. volvulus;* misidentification will give erroneous information.

Black flies of the *S. damnosum* complex in West Africa breed in fast-flowing water that is well oxygenated (Samba, 1994). The larvae become attached to stones and vegetation on the river bed. Black flies of the *S. naevei* complex become attached to freshwater crabs of the genus *Potamonautes* that live in forest streams in East Africa (Williams, 1991). Larvae of the *S. ochraceum* complex in the Americas breed attached to leaves in tiny trickles of water in rugged terrain that is found there (Minter and White, 1982).

9.2 TRANSMISSION

9.2.1 Vector-Borne Transmission

Numerous factors involving black flies affect the transmission of *O. volvulus* from human to black fly and therefore from black fly to human. The fate of *O. volvulus* in the black fly host population is as important for its survival as is its fate in the human host population (Basanez et al., 1995, 1996). The factors include the breeding success of the black fly population, the *Simulium* group to which the black fly belongs, the climate and environment occupied by humans and black flies, the feeding habits and structure of the mouthparts of the black flies and the circumstances that bring humans and black flies into contact (see Ukoli, 1984). The nearer people live to black fly breeding sites, the greater the risk should be of their contracting onchocerciasis. But adult black flies, which may live for up to 4 weeks, are now known to be capable of journeys of several hundred kilometers (Samba, 1994) so distance from breeding sites may not protect people from infection.

Results from a recent investigation by Opara et al. (2005) in the Lower Cross River Basin, Nigeria, after ivermectin had been in use for 5 years, illustrates several of the key determinants of transmission. By means of human bait stationed at two known breeding sites (A and B) of *S. damnosum,* 9,287 flies were caught and dissected to reveal that 313 (3.46%) were infected with *O. volvulus.* Annual biting rates (ABR) were calculated to be 42,419 per person and 28,346 per person at sites

A and B respectively and the annual transmission potential (APT) was calculated to be 419 infective L3 per person and 427 L3 per person at sites A and B, respectively. (The ATP is the number of L3 calculated as being received by a person stationed at an insect capture point during a year.) On a monthly basis, the ABR and the ATP varied significantly between the sites. Transmission was found to be highly seasonal, being greater between April and September than in the rest of the year. In the study area, the vectors had a diurnal feeding pattern with most activity taking place in the early morning and late afternoon, times that corresponded with human activities of farming, fishing, hunting, and timber cutting. No data about ABRs or ATPs were available before the ivermectin intervention began so the results do not show what the impact of mass drug application had been up to this point in the control program.

What was happening in the black fly populations that would have affected the calculations about the transmission rates of *O. volvulus* that were made by Opara et al. (2005)? The survival of black flies during the penetration of their gut walls by MF to produce infective L3 depends on the number of MF that are ingested from the host's skin, the form of the flies' mouthparts that must be negotiated by the MF, and the formation of the flies' peritrophic membranes that must be avoided by the MF. From an investigation of these variables involving field studies, experiments, and mathematical models, Basanez et al. (1995, 1996) proposed that flies with simple, unarmed mouthparts would be more likely to survive and support the formation of L3 stages when feeding on hosts with low numbers of MF in the skin. If such flies ingested many MF, overwhelming gut penetration would kill them. Conversely, flies with complex, armed mouthparts would be better able to survive feeding on skin containing high numbers of MF, presumably because fewer of the MF would pass unscathed through the mouthparts and so fewer would be available to cause the fly's death on penetrating its gut wall. This elegant work reveals something of the variety of adaptations that have evolved to ensure that enough intermediate and vector hosts of *O. volvulus* survive so that the worm itself survives.

9.2.2 Evidence for Transplacental Transmission

Brinkman et al. (1976) obtained evidence to indicate that MF of *O. volvulus* might be transmitted from mother to fetus *in utero*. Since then, convincing circumstantial evidence has accumulated to suggest that transplacental transmission may have considerable significance. Prost and Gorim de Ponsay (1979) found MF of *O. volvulus* within a week of birth in 4 out of 210 babies born in Upper Volta (Burkina Faso). In two other cases, MF were found in the tissues of the umbilical cord. From a cohort of 5,757 children less than 2 years old, 1% under 12 months and 2% between 12 and 23 months were observed to harbor dermal MF (Prost and Gorim de Ponsay, 1979). Since MF appear not to be released until female worms have been established for at least 10 months (Section 9.1.1), the possibility of transplacental transmission seems strong. Ufomadu et al. (1990) detected MF in a 7-month-old baby in Nigeria and also concluded that the MF must have crossed the placenta. More recently, Nwaorgu and Okeibunor (1999) collected skin snips from 642 children aged from 0 to 4 years old in a forest region of southeast Nigeria. In this group, the MF prevalence was found to be 15.7%, but no positive skin snips were found in children under 1 year old. The significance of transplacental transmission remains to be clarified.

9.3 MICROFILAREMIA

Microfilaremia is a key concept in understanding the relationship between humans, black flies, and *Onchocerca volvulus*, the course of onchocerciasis, and the progress of measures to control and eliminate the disease. The term has a qualitative and a quantitative use. Qualitatively, an individual is said to be microfilaremic once MF can be detected in the skin. In an epidemiological survey, the microfilarial prevalence is measured as the proportion of people surveyed whose skin samples were found to contain MF. The microfilarial prevalence is usually expressed as a percentage (MF% or mf%). Quantitatively, microfilaremia is defined as the number of MF per unit mass of skin. The community microfilarial load (CMFL) is the mean number of MF per skin snip among sampled people aged 20 years and over, including those in whom no MF were observed. The CMFL is a particularly useful variable in epidemiological surveys, in monitoring the progress of control measures, and in surveillance once control stops. Usually a small skin biopsy is taken, weighed, left in saline, examined microscopically, and the number of MF to emerge from the skin sample is expressed per milligram of skin (Section 9.4.1) and the mean value for those samples is calculated.

9.4 DIAGNOSIS

Diagnosis of an *O. volvulus* infection or onchocerciasis is made through a combination of parasitological evidence and clinical signs and symptoms. Communities located where onchocerciasis is entrenched will be well aware of the presence of the disease and its impact on their health and prosperity. Diagnostic skills need to be learned and practiced to facilitate the care of individual patients and the management of control programs.

Progress toward achieving the targets chosen for control activities will depend on following changes in the rates of microfilaremia and common clinical signs of onchocerciasis. In the course of a diagnostic survey, some individuals may complain of muscular pain, present with epilepsy, or be underweight for age. The cause of these conditions should not be assumed to be directly due to onchocerciasis.

9.4.1 PARASITOLOGICAL DIAGNOSIS

Diagnosis can be made by examining adult *O. volvulus* present in nodules removed from a subject following local anesthesia or examined *in situ* by means of ultrasonography (WHO, 1995). These reliable methods are time-consuming procedures, are restricted to small numbers of subjects, and require expensive equipment and highly trained staff. In practice, in a public health setting the detection of microfilariae in skin snips is a long-established means of reliable diagnosis provided that certain conditions are noted (Figure 9.2).

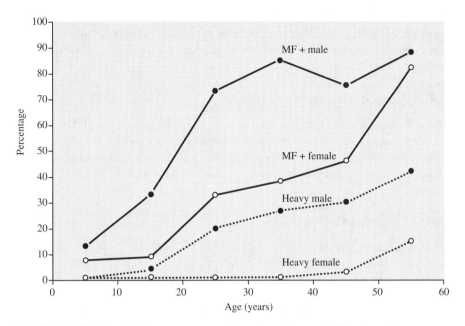

FIGURE 9.2 Relationships between age and gender of hosts in Nigeria and percentage prevalences of *Onchocerca* MF positive skin snips and heavy infections. From Figure 2 (Wyatt, 1971), reproduced from *Annals of Tropical Medicine and Parasitology* 65, page 517, with the permission of the Liverpool School of Tropical Medicine.

In an epidemiological survey or control program in which regular monitoring is required, skin snips should be obtained from the same sites of the body, at the same time of day, and by the same method. This discipline helps to ensure that data comparisons are valid. Snips are best taken in the afternoon, from skin over the iliac crest in African subjects, from skin of the upper back in American subjects, and from skin of the lower calf in Arabian subjects (Muller, 2002). When seeking skin snips care must be taken to respect cultural traditions and personal privacy and allay anxiety. The calf is probably the least intrusive site for fearful individuals. Some technicians use a needle and razor blade, but a corneoscleral punch is to be preferred because a snip of standard size is obtained. All instruments must be clean before use and contamination must be avoided. Some technicians take one snip while others take two, one from either side of the iliac crest or wherever (Bradley and Unnasch, 1996).

The skin snip must be seen to be free from blood otherwise MF that live in blood might be present and might interfere with the reliability of the diagnosis. The skin snip should be incubated for a set time under defined conditions and then the MF should be identified and counted. In West Africa there is the possibility that the human host is harboring a concurrent infection of *Mansonella streptocerca*. The MF of *M. streptocerca* are also unsheathed and live in skin (Meyers et al., 1977).

Searching for MF in skin snips has limitations. The MF population in the skin may be below the sensitivity of the method. The female worms may not yet be producing MF. Finding and correctly identifying MF of *O. volvulus* means that the subject is infected. Failure to find MF means failure to find MF and not that the subject is free from infection. More extensive examinations may reveal the presence of MF in body fluids and organs, but such examinations are beyond the scope of public health activity.

9.4.2 CLINICAL DIAGNOSIS

Clinical diagnosis relies on the detection of pathological changes in the skin and eyes, on finding palpable nodules, and the presence of lesions such as hanging groin. Any or all of these signs and symptoms are evidence of disease and morbidity. Individuals may be encountered with detectable microfilaremia, but no detectable disease.

9.4.2.1 Skin Appearance

Several schemes have been published to aid in the diagnosis of onchocerciasis by reference to changes in the condition of the skin (see Wyatt, 1971; WHO, 1995). In outline the chronic effects of increasing numbers of MF in the skin are as follows:

- rash and intense itching often accompanied by localized bacterial invasion
- lichenification, the skin being thickened and darkened with pustules and ulcers
- leopard skin appearance, the skin over the shins showing prominent patches of uneven pigmentation
- atrophy, the skin of a young person appearing like that of an old person as elasticity is lost

There are other well known manifestations of skin pathology that relate to particular endemic areas. For example, in the focus of onchocerciasis in the Yemen a condition known locally as sowda is seen in which the skin becomes deeply pigmented. These observed skin changes may not always be caused by the MF of *O. volvulus*. The skin is always vulnerable to the effects of other pathogens and nutritional deficiencies.

The presence of leopard skin (LS) has been widely used in West Africa as a cardinal sign of onchocerciasis. The advantages of determining rates of leopard skin are (1) the lesion is easily recognized and can be detected by untrained staff; (2) no equipment is needed and no tissue samples have to be taken; (3) from the subject's perspective there is hardly any intrusion; and (4) diagnosis is quick for both subject and observer. Measurements of the extent of leopard skin have contributed to understanding the epidemiology of the disease in rural Nigeria (Edungbola and Asaolu, 1984; Edungbola et al., 1987). These authors have shown that leopard skin rates correlate satisfactorily with the intensity of onchocerciasis in the community. In a survey of 8,426 villagers, the presence of LS was recorded from the 3,765 found to have MF positive skin snips. The LS results were interpreted against the arbitrary convention that MF prevalences of < 10%, 10 to 39%, 40 to 69%, and > 70% indicate that onchocerciasis is sporadic, hypoendemic, mesoendemic, and hyperendemic, respectively (Table 4.2). Edungbola et al. (1987) concluded that LS prevalences of < 1%, 1 to 6%, and > 6% indicate that onchocerciasis is sporadic, hypoendemic, and mesoendemic or more, respectively. In contrast, Fischer et al. (1993), who examined 6,271 people for onchocerciasis in Uganda where transmission depends on *S. naevei*, decided that the prevalence of leopard skin was not a useful indicator of the extent of onchocerciasis in that region.

9.4.2.2 Nodules (Onchocercomata)

Subcutaneous nodules containing adult worms are easily found on palpation by an experienced health worker, but the privacy of subjects must be respected during examinations. Nodules vary in size from about 10 to 60 mm and consist of a matrix of fibrous tissue in which the worms are embedded. In African onchocerciasis, most palpable nodules are found around the pelvic region and lower limbs while in American onchocerciasis most are found on the head and upper part of the body (Eddington and Gilles, 1976). Nodules located between bone and thin soft tissue such as is found on the head are directly visible. Generally, nodules are found to be more abundant in men than in women (Figure 9.3).

Nodules may be located in deeper tissues so failure to find them during super-ficial examination should not be assumed to mean that the subject is free from infection with *O. volvulus*. Nodules may arise from other causes. For example, during the course of an infection with *Loa loa,* transient "Calabar" swellings may arise and be seen and felt at the surface of the body (Muller, 2002). Lymph nodes may become enlarged to give the sensation of onchocercomata.

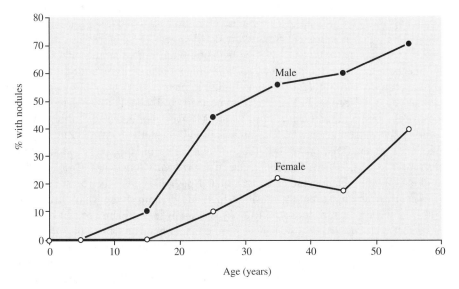

FIGURE 9.3 Relationships between the age and gender of hosts in Nigeria and the percentage prevalence of *Onchocerca* nodules. From Figure 3 (Wyatt, 1971), reproduced from *Annals of Tropical Medicine and Parasitology* 65, page 517, with the permission of the Liverpool School of Tropical Medicine.

9.4.2.3 Ocular Signs and Symptoms

If slit lamp equipment and trained staff are available, live MF and the damage initiated by dead MF may be detected in the eye. In the front of the eye, live MF may be visible in the cornea and anterior chamber. Dead MF and evidence of their death (keratitis, iritis, and secondary glaucoma) may be present in the same sites. Cataracts may have developed in the lens. In the rear of the eye, lesions may be seen in the retina, choroid, chorioretina, and optic nerve. Appropriate tests will reveal various levels of visual impairment including night blindness, field loss, and permanent blindness (WHO, 1995). Accurate recording of these symptoms is important if the impact and sustainability of control measures are to be demonstrated.

9.4.2.4 Lymphatic Signs and Symptoms

Enlarged lymph nodes in the groins, some degree of lymphedema, and a condition known as "hanging groin" are detectable features of onchocerciasis. Superficially, hanging groin has similarities to the scrotal swelling that is diagnostic of lymphatic filariasis. Hanging groin, however, is not normally associated with any hydrocoele.

9.4.3 IMMUNOLOGICAL AND MOLECULAR DIAGNOSIS

A cocktail of three recombinant antigens has been developed to allow the early detection of infections of *O. volvulus* during the prepatent period when the numbers of MF in the skin are too low to be found with the standard skin snip procedure (WHO, 1995). The method should also be useful for screening communities for evidence of recrudescence, for the detection of suspected drug resistance, and for examining the infection status of migrants. Research has also been done with the polymerase chain reaction (PCR) and the development of DNA probes to characterize the strains of *Onchocerca* carried by black flies, to give more sensitive information about the epidemiology of *O. volvulus* infections, and to monitor the severity of onchocerciasis in different bioclimes (Zimmerman et al., 1992, 1994). In a skin snip survey of 94 people in Ecuador, Zimmerman et al. (1994) found 60 to be MF positive for *O. volvulus*. Of the 34 in whom MF were not detected, 13 more were found to be infected when a PCR assay was used. In a study in Africa, Zimmerman et al. (1992) used DNA probes specific to *O. volvulus* from the forest and savannah bioclimes, to demonstrate that the strain of *O. volvulus* found in the savannah is significantly more pathogenic as regards ocular pathology than the forest strain.

The PCR is an extremely sensitive method for amplifying DNA sequences (Sullivan, 1994). In practice, a PCR assay can analyze a minute amount of DNA and detect with confidence a sequence specific to an organism from which the DNA had been extracted. The PCR is complex (Fung, 2004) and depends on access to modern facilities run by highly trained technicians. The procedure is expensive and requires conditions that minimize risks of contamination. Bradley and Unnasch (1996), after a review of progress in the development of molecular diagnostic tests, concluded that although many of the earlier problems of specificity have been resolved, there is still considerable work remaining to calibrate the tests in different infection situations. Perhaps the international community should set up specialized reference laboratories to which samples could be sent for accurate immunological and molecular investigation.

9.4.4 DEC Patch Test

Diethylcarbamazine (DEC) is no longer recommended for the treatment of patients with *O. volvulus* infections because severe Mazzotti reactions may occur (Greene et al., 1985; WHO, 2004). DEC kills the MF, thereby stimulating itching, rash, pain, edema, fever, and eye lesions as the larvae disintegrate. Stingl et al. (1984) devised a diagnostic test that exploits the Mazzotti reaction. A minute amount of DEC is incorporated into a patch that is then applied to the skin and, on removing the patch, any subsequent minor change in the appearance of the skin would be assumed to denote the presence of MF of *O. volvulus*. When tested in the Sudan, Stingl et al. obtained 69 positive results from 75 trials of the test. Histological examination of the skin from a positive test showed dead MF in the upper dermis surrounded by eosinophils. Boatin et al. (2002) tested the patch test on a larger scale and compared its sensitivity with results from skin snips and from the PCR method. The subjects were 313 villagers from a region of Guinea that had been exposed to OCP activities for 10 years. Both PCR and the DEC patch test were found to be more sensitive than skin snips for the detection of onchocerciasis in this situation. The advantages of the DEC patch are its noninvasiveness, its simplicity, and its low cost.

9.5 DISTRIBUTION AND EPIDEMIOLOGY

When the WHO Expert Committee met in late 1993 to review global onchocerciasis the disease was known to be endemic in 34 countries, 26 in sub-Saharan Africa, 6 in the Americas, and 2 in the Arabian Peninsula (WHO, 1995). About 17.7 million people were considered to be infected with *O. volvulus* and about 268,000 were considered to be blind as a result of the disease. Globally, 123 million people were judged to be at risk of infection. Over 99% of those infected or blind were living in the tropical region of sub-Saharan Africa. There has been a significant reduction in the extent and public health significance of onchocerciasis in the countries included in the OCP; most of the problem now is to be found in the 15 African countries that were not included in the OCP (Figure 9.1).

9.5.1 *Onchocerca volvulus* in the Savannah and Forest Regions (Bioclimes) of Africa

Onchocerciasis, like other forms of helminthiasis, should not be expected to be distributed evenly in an endemic region or to have the same effects and characteristics wherever it is found. Environmental and climatic conditions, the properties of the vector black flies and the nature of the host community influence the course of onchocerciasis. Evidence for that view is provided by the marked differences seen in the form of onchocerciasis across the territory once considered as savannah and forest in West Africa (WHO, 1995), but now changing in response to deforestation and accompanying climatic change as dryness extends throughout the region (Walsh et al., 1993).

9.5.2 ONCHOCERCA VOLVULUS IN THE COMMUNITY

A survey carried out in Nigeria by Wyatt (1971) provides a picture of a community living in an area where onchocerciasis was endemic before any control measures had been introduced. On the basis of single skin snips taken from the right calf, the MF% was found to be higher in men than in women and the intensity of infection or microfilaremia (MF/mg skin) was found to be higher in men than in women (Figure 9.2). That more nodules were detected in men than in women (Figure 9.3) was to be expected given this finding, which probably accounts for more blindness eventually occurring in men than in women (WHO, 1995). The higher prevalence rates in men than in women are considered to be due to increased exposure of men to the bites of black flies rather than to some gender-related difference in resistance to infection. Health workers should not expect all epidemiological surveys of *O. volvulus* infections to give results like those obtained by Wyatt (1971). Helminth control would be straightforward if little variation occurred. As yet there are no known gender differences in the establishment or acquisition of infection (WHO, 1995; Nwaorgu and Okeibunor, 1999).

A longitudinal investigation lasting 18 years has discovered that a dominant risk factor for children to acquire infection with *O. volvulus* is having a mother who is infected with *O. volvulus*. This important conclusion was drawn by Kirch et al. (2003) who worked in West Africa with a population of 15,290 people of whom 9,640 were children. Children with infected mothers became infected themselves earlier than others and developed higher infection levels. Maternal infection was shown to be a greater risk for children than the intensity of transmission measured as the ATP. Infection in the children with infected mothers continued at a higher level than in other children despite the vector control activities of the OCP.

9.6 MORBIDITY

There is now consensus that the common skin and ocular manifestations of onchocerciasis (Section 9.4.2) result from the host reaction to disintegrating MF (Little et al., 2004a,b). The higher the number of MF in the skin the more severe the disease. Furthermore, more MF are generally found in men than women so men suffer more from the disease than women and the disease worsens with age as still more MF accumulate and die. The clinical signs and symptoms of onchocerciasis involving the skin reflect the course of the disease in that organ progressing from itching to prematurely aged atrophied tissue. The pathological ocular changes referred to in the literature are complex and include, in alphabetical rather than chronological order, cataract, choroidoretinitis, glaucoma, iridocyclitis, iritis, keratitis, optic atrophy, punctate keratitis, uveitis, and white intraretinal deposits. These may all progress through time to blindness in one or both eyes if the MF numbers continue to increase. In West Africa, statistical analysis of data collected during the OCP showed the incidence of blindness due to onchocerciasis to be significantly and positively correlated with MF burden and the risk to women there of becoming blind to be 40% less than the risk to men (Little et al., 2004a).

The eye lesions were thought to be due to either direct immune responses to MF antigens or to cross reactions between MF antigens and human antigens (see Viney, 2002). A further explanation of eye pathology has emerged that involves microorganisms that are endosymbionts of arthropods and nematodes. Endosymbionts, identified as belonging to the *Wolbachia* group, influence the reproductive activity of their partners. If *O. volvulus* are exposed to doxycycline (broad spectrum antibiotic), sterility follows for the worms (Hoerauf et al., 2000). If extracts of *Onchocerca* are injected into the eyes of mice, a predominant inflammatory response and corneal pathology are observed, but not if the extract is first treated with antibiotic (Saint Andre et al., 2002). The conclusion is that the *Wolbachia* living in the MF induce the eye pathology and speculation that antibiotic therapy for onchocerciasis should be investigated seems justified. Maybe research on the *Wolbachia* that inhabit black flies merits support if it resulted in control through impairment of their reproduction. This proposition has been discussed in some detail by Townson (2002) as a potential control method for the mosquitoes that transmit *Wuchereria bancrofti*.

There have been debates about links between onchocerciasis and epilepsy and mortality. A case-control study conducted by Boussinesq et al. (2002) compared microfilaremias between 72 epilepsy cases and 72 nonepilepsy cases in an area of onchocerciasis endemicity in Cameroon. The skin snip MF burdens in the epileptic group were statistically significantly higher than in the nonepileptic group, a relationship not readily demonstrated in areas subjected to repeated rounds of ivermectin treatment, but a result that further supported the case for community treatment. The question of whether death rates increase as a consequence of onchocerciasis has been investigated by reference to the fate of 297,756 people covered by the OCP during 1971 to 2001 (Little et al., 2004b). The deaths of 24,517 people were recorded in this time frame and 1,283 were reckoned to be due to onchocerciasis. Death was related not directly to blindness but to increasing MF burden. Women were found to have a 7.5% less risk of death attributable to onchocerciasis than men.

9.7 PREVENTION AND CONTROL

9.7.1 THE ONCHOCERCIASIS CONTROL PROGRAM (OCP) IN WEST AFRICA

The OCP began in 1974 when the decision was taken to relieve the magnitude of the public health problem caused by onchocerciasis. Rhetoric was replaced by action concerning the dreadful impact of the disease on the health, social well-being, and productivity of individuals and their communities. The history of the remarkable partnership between governments, UN agencies, and donors that supported the OCP and a detailed account of the management structure of the program has been recorded by Samba (1994). A brief summary of the achievements of the program has been published by Molyneux (1995).

The OCP initially involved Benin, Burkino Faso, Cote d'Ivoire, Ghana, Mali, Niger, and Togo. In 1986, Guinea, Guinea-Bisseau, Senegal, and Sierra Leone were added so that the OCP was covering 11 countries where onchocerciasis was endemic. By means of helicopters and low-flying aircraft the insecticide temephos was sprayed over vast tracts of terrain to kill black fly larvae in their breeding sites. Ecological fears were eased when monitoring showed that this organophosphorous compound was not destroying fish or aquatic invertebrates. Almost inevitably, temephos-resistant populations of black flies were selected by this process and reinvasion of areas with renewed *Onchocerca* transmission occurred. Temephos was replaced with a biocide, Bt H14, developed from *Bacillus thuringensis,* with specific toxicity for dipteran larvae (IPCS, 1999). When susceptibility to temephos returned, the OCP reverted to using it in place of Bt H14.

By 1994, some achievements of the program included:

- As early as 1978 the ATP had dropped from 800 to 100 in two thirds of the OCP territory. Experts had concluded that holding the ATP at 100 and reducing it still further would allow people to reoccupy abandoned river valleys without the risk of developing serious ocular lesions.
- By 1987, over 90% of stations for ATP assessment were reporting ATPs of zero.
- The CMFL had declined to close to zero in most of the OCP territory apart from in some of the reinvaded areas.
- 30 million people have been protected from contracting onchocerciasis.
- 9 million children born since the OCP began have been protected from developing blinding onchocerciasis and the prediction at the time of Samba's evaluation of the achievements of the program was that by 2000 some 15 million children would have had the same protection.
- 1,250,000 people had been spared the misery on ocular onchocerciasis.
- Over 100,000 cases of blindness had been averted.
- Ivermectin distribution has reduced many other aspects of the disease.
- An impressive return of people to abandoned farming land has taken place so that about 25 million hectares have been reclaimed.
- Many health professionals have been recruited and trained during the OCP, thereby acquiring transferable skills for other spheres of service in public health.

The OCP has cost millions of U.S. dollars, but when the millions of beneficiaries are noted, the cost per person over the duration of the OCP is a few dollars each, not to mention the increased productivity that is estimated as being equivalent to a 20% return on investment (Molyneux, 1995).

9.7.2 Mass Distribution and Use of Ivermectin in the Community

In 1987 efforts to control onchocerciasis were boosted by the decision of Merck and Co. Inc., a major producer of quality drugs in the research-based pharmaceutical industry, to donate Mectizan (ivermectin, see Chapter 14) for as long as necessary and for as many people as necessary for the treatment of onchocerciasis. Since ivermectin is not effective against adult worms, the view has been that mass distribution in areas where onchocerciasis is endemic would be needed for at least 15 years (Section 9.1.1). Over 250 million doses of the drug have been used in the effort to control onchocerciasis (Tielsch and Beeche, 2004). Ivermectin at the dose used in communities against *O. volvulus* (single oral dose annually of 150 µg/kg body weight) is a suppressive drug that kills the MF and inhibits their release from parent female worms (WHO, 2004). The instructions for ivermectin advise against use in children less than 5 years of age or in children weighing 15 kg or less. The instructions also advise against use during pregnancy and recommend that ivermectin treatment should be delayed until after delivery (WHO, 2004). A physician may decide otherwise having considered an individual's state of health.

Five points should be noted regarding ivermectin, the treatment of young children, and pregnancy. First, the manufacturer's instructions probably reflect the likelihood that clinical trials were not carried out on these groups of subjects. Second, onchocerciasis occurs in children less than 5 years of age (Nwaorugu and Okeibunor, 1999). Third, Gyapong et al. (2003) investigated circumstances in which 50 out of 343 pregnant women had inadvertently been treated with both albendazole and ivermectin in an area of Ghana where lymphatic filariasis and onchocerciasis were coendemic. After analysis of the birth outcomes, authors concluded that there was no evidence of a higher risk of congenital malformations or abortions in those who were inadvertently exposed to the anthelminthic drugs. Doumbo et al. (1992) had earlier expressed concern about the fact that pregnant women were likely to take ivermectin during the distribution of the drug for onchocerciasis control in Mali, but they also noted no adverse outcomes. Fourth, there is strong evidence to show that onchocerciasis is an even more debilitating disease during the lives of children who are born to mothers already enduring onchocerciasis (Kirch et al., 2003). Fifth, there is some evidence that transplacental transmission may occur (Section 9.2.2). What would an impartial quantitative risk/benefit analysis recommend about the deliberate inclusion of young children and pregnant women in mass drug administration for the treatment of onchocerciasis? If experts were to conclude that ivermectin is not to be used to relieve pregnant women from onchocerciasis, then some system should be found for ensuring that treatment is offered as soon as possible after delivery (Maduka et al., 2004).

Concern has been expressed about the use of ivermectin in communities where infections of *Loa loa* occur concurrently with *O. volvulus*. If the *Loa* microfilaremia is of high intensity there is a risk of a potentially serious adverse event known as *Loa* encephalopathy. If an individual experiencing a high intensity infection with

Loa MF were to be given the recommended dose of ivermectin for onchocerciasis, there is the remote possibility (1 per 10,000 according to Muller [2002]) of fatal neurological complications developing in the brain (Addiss et al., 2003). Health professionals in Africa dealing with onchocerciasis and lymphatic filariasis (Chapter 8) should not overlook problems and adverse events linked to the presence of *Loa loa*.

9.7.3 EXIT STRATEGY

The control of onchocerciasis, based on the distribution of ivermectin, is continuing across sub-Saharan Africa and the supply of the drug is assured for as long as it needed. Molyneux (2005) has pointed out that several major matters will have to be resolved if control is to continue, if the disease is not to regain ground, and if elimination is to be attained. Governments of the countries where onchocerciasis is endemic may not be able to rely indefinitely on the donor support that covers the expense of drug delivery and coverage. An exit strategy will be needed to enable the benefits of control to be sustained. Onchocerciasis control now depends on high coverage to subdue populations of MF, thereby protecting people and reducing transmission. If coverage declines, the prevalence of the infection will increase. Coverage could be strengthened by the integration of ivermectin distribution with existing health services such as vitamin A distribution (Molyneux, 2005) or the recruitment of local women who are aware of the threat of onchocerciasis to their community and who are trusted by their neighbors (Katabarwa et al., 2002). Once the CMFL has been lowered to the level that is viewed as acceptable (Samba, 1994) surveillance will be even more important to detect any signs of recrudescence and any evidence of the emergence of ivermectin-resistant *O. volvulus*. Molyneux (2005) also makes the case for the provision of a safety-tested anthelminthic drug that will kill adult *O. volvulus* in the human host. If such a drug became available at an affordable price, the global elimination of onchocerciasis would become a realistic aim.

REFERENCES

Addiss DG, Rheingans R, Twum-Danso NA and Richards FO. 2003. A framework for decision-making for mass distribution of Mectizan (R) in areas endemic for *Loa loa*. *Filaria Journal* **2**, Suppl.1, S9.

Basanez MG, Remme JH, Alley ES et al. 1995. Density-dependent processes in the transmission of human onchocerciasis: relationship between the numbers of microfilariae ingested and successful larval development in the simuliid vector. *Parasitology* **110**, 409–427.

Basanez MG, Townson H, Williams JR et al. 1996. Density-dependent processes in the transmission of human onchocerciasis: relationship between microfilarial intake and mortality of the simuliid vector. *Parasitology* **113**, 331–355.

Boatin BA, Toe L, Alley ES et al. 2002. Detection of *Onchocerca volvulus* infection in low prevalence areas: a comparison of three diagnostic methods. *Parasitology* **125**, 545–552.

Boussinesq M, Pion SD, Demanga-Ngangue and Kamgno J. 2002. Relationship between onchocerciasis and epilepsy: a matched case-control study in the Mbam valley, Republic of Cameroon. *Transactions of the Royal Society of Tropical Medicine and Hygiene* **96**, 537–541.

Bradley JE and Unnasch TR. 1996. Molecular approaches to the diagnosis of onchocerciasis. *Advances in Parasitology* **37**, 57–106.

Brinkman UK, Kramer P, Presthus GT and Sawadogo B. 1976. Transmission *in utero* of microfilariae of *Onchocerca volvulus*. *Bulletin of the World Health Organization* **54**, 708–709.

Doumbo O, Soula G, Kodio B and Perrenoud M. 1992. Ivermectin and pregnancy in mass treatment in Mali. *Bulletin Societe Pathologie Exotique* **85**, 247–251. [In French, abstract available in English via PubMed.]

Duke BOL. 1980. Observations on *Onchocerca volvulus* in experimentally infected chimpanzees. *Tropenmedicine und Parasitologie* **31**, 41–54.

Eberhard ML, Dickerson JW, Boyer AE et al. 1991. Experimental *Onchocerca volvulus* infections in mangabey monkeys (*Cercocebus atys*) compared to infections in humans and chimpanzees (*Pan troglodytes*). *American Journal of Tropical Medicine and Hygiene* **44**, 151–160.

Eddington GM and Gilles HM. 1976. *Pathology in the Tropics*, 2nd edition. London: Edward Arnold (Publishers) Ltd.

Edungbola LD and Asaolu SO. 1984. Parasitologic survey of onchocerciasis (river blindness) in Babama District, Kwara State, Nigeria. *American Journal of Tropical Medicine and Hygiene* **33**, 1147–1154.

Edungbola LD, Alabi TO, Oni GA et al. 1987. "Leopard skin" as a rapid diagnostic index for estimating the endemicity of African onchocerciasis. *International Journal of Epidemiology* **16**, 590–594.

Engelkirk PG, Williams JF, Schmidt, GM and Leid RW. 1982. Zoonotic onchocerciasis. In: *Handbook Series in Zoonoses*. Section C: *Parasitic Zoonoses* Vol. II (ed. MG Schultz). Boca Raton, Florida: CRC Press. pp. 225–250.

Fischer P, Kipp W, Bamuhiga J et al. 1993. Parasitological and clinical characterization of *Simulium naevei*-transmitted onchocerciasis in western Uganda. *Tropical and Medical Parasitology* **44**, 311–321.

Fung DYC. 2004. Rapid methods for the detection and enumeration of microorganisms in water. In: *Waterborne Zoonoses* (eds. JA Cotruvo, A Dufour, G Rees et al.). Geneva: World Health Organization. pp. 367–375.

Greene BM, Taylor HR, Cupp EW et al. 1985. Comparison of ivermectin and diethylcarbamazine in the treatment of onchocerciasis. *New England Journal of Medicine* **313**, 133–138.

Gyapong JO, Chimbuah MA and Gyapong M. 2003. Inadvertent exposure of pregnant women to ivermectin and albendazole during mass drug administration for lymphatic filariasis. *Tropical Medicine and International Health* **8**, 1093–1101.

Hoerauf A, Volkmann L, Hamelmann C et al. 2000. Endosymbiotic bacteria in worms as targets for a novel chemotherapy in filariasis. *Lancet* **355**, 1242–1243.

ICPS (International Programme on Chemical Safety). 1999. Environmental Health Criteria 217. *Bacillus thuringiensis*. Geneva: World Health Organization.

Katabarwa MN, Habomugisha P and Agunyo S. 2002. Involvement and performance of women in community-directed treatment with ivermectin of onchocerciasis in Rukungiri District, Uganda. *Health and Social Care in the Community* **10**, 382–393.

Kirch AK, Duerr HP, Boatin B et al. 2003. Impact of parental onchocerciasis and intensity of transmission on development and persistence of *Onchocerca volvulus* infection in offspring: an 18 year follow-up study. *Parasitology* **127**, 327–335.

Little MP, Basanez MG, Breitling LP et al. 2004a. Incidence of blindness during the onchocerciasis control programme in western Africa. *Journal of Infectious Diseases* **189**, 1932–1941.

Little MP, Breitling LP, Basanez MG et al. 2004b. Association between microfilarial load and excess mortality in onchocerciasis; an epidemiological study. *Lancet* **363**, 1514–1521.

Maduka CU, Nweke LN, Miri ES et al. 2004. Missed treatment opportunities, for pregnant and breast-feeding women, in onchocerciasis mass-treatment programmes in south-eastern Nigeria. *Annals of Tropical Medicine and Parasitology* **98**, 697–702.

Manson-Bahr PEC and Apted FIC. 1982. *Manson's Tropical Diseases,* 18th edition. London: Balliere Tindall.

Meyers WM, Neafie RC, Moris R and Bourland J. 1977. Streptocerciasis: observation of adult male *Dipetalonema streptocerca* in man. *American Journal of Tropical Medicine and Hygiene* **26**, 1153–1155.

Minter DM and White GB. 1982. Medical entomology. Appendix III. In: Manson-Bahr and Apted (1982) pp. 734–823.

Molyneux DH. 1995. Onchocerciasis control in West Africa: current status and future of the Onchocerciasis Control Programme. *Parasitology Today* **11**, 399–402.

Molyneux DH. 2005. Onchocerciasis control and elimination: coming of age in resource constrained health systems. *Trends in Parasitology* **21**, 525–529.

Muller R. 2002. *Worms and Human Disease,* 2nd edition. Wallingford, Oxfordshire: CABI Publishing.

Nwaorgu OC and Okeibunor JC. 1999. Onchocerciasis in pre-primary school children in Nigeria: lessons for onchocerciasis country control programme. *Acta Tropica* **73**, 211–215.

Opara KN, Fagbemi OB, Ekwe A and Okenu DM. 2005. Status of forest onchocerciasis in the Lower Cross River Basin, Nigeria: entomologic profile after five years of ivermectin intervention. *American Journal of Tropical Medicine and Hygiene* **73**, 371–376.

Prost A and Gorim de Ponsay E. 1979. The epidemiological significance of neo-natal parasitism with microfilariae of *Onchocerca volvulus. Tropenmedizin und Parasitologie* **30**, 477–481. [In French; abstract available in English via PubMed]

Remme JHF. 1995. The African programme for onchocerciasis control: preparing to launch. *Parasitology Today* **11**, 403–406.

Saint Andre A, Blackwell NM, Hall LR et al. 2002. The role of endosymbiotic *Wolbachia* bacteria in the pathogenesis of river blindness. *Science* **295**, 1892–1895.

Samba EM. 1994. *The Onchocerciasis Control Programme in West Africa.* Geneva: World Health Organization.

Smart J. 1943. *Insects of Medical Importance.* London: British Museum (Natural History).

Stingl P, Ross M, Gibson DW et al. 1984. A diagnostic "patch test" for onchocerciasis using topical diethylcarbamazine. *Transactions of the Royal Society for Tropical Medicine and Hygiene* **78**, 254–258.

Sullivan KM. 1994. Polymerase chain reaction. In: *The Encyclopaedia of Molecular Biology* (ed. J Kendrew). Oxford: Blackwell Science. pp. 864–866.

Tielsch JM and Beeche A. 2004. Impact of ivermectin on illness and disability associated with onchocerciasis. *Tropical Medicine and International Health* **9**, 45–56.

Townson H. 2002. *Wolbachia* as a potential tool for suppressing filarial transmission. *Annals of Tropical Medicine and Parasitology* **96**, S117–S127.

Ufomadu GO, Sato Y and Takahashi H. 1990. Possible transplacental transmission of *Onchocerca volvulus*. *Tropical and Geographical Medicine* **42**, 69–71.

Ukoli FMA. 1984. *Introduction to Parasitology in Tropical Africa*. Chichester and New York: John Wiley and Sons Ltd.

Viney M. 2002. *Wolbachia* and river blindness. *Trends in Parasitology* **18**, 244.

Walsh JF, Molyneaux DH and Birley MH. 1993. Deforestation: effects on vector-borne disease. *Parasitology* **106**, S55–S75.

Williams TR. 1991. Freshwater crabs and *Simulium naevei* in east Africa. III. Morphological variation in *Potamonautes loveni* (Decapoda: Potamidae). *Annals of Tropical Medicine and Parasitology* **85**, 181–188.

WHO. 1995. Onchocerciasis and its control. Report of a WHO Expert Committee on onchocerciasis control. *Technical Report Series* 852. Geneva: World Health Organization.

WHO. 2003. *Global Defence against the Infectious Disease Threat* (ed. MK Kinderhauser). Geneva: World Health Organization.

WHO. 2004. *WHO Model Formulary 2004*. Geneva: World Health Organization.

Wyatt GB. 1971. Onchocerciasis in Ibarapa, Western State, Nigeria. *Annals of Tropical Medicine and Parasitology* **65**, 513–523.

Zimmerman PA, Dadzie KY, De Sole G et al. 1992. *Onchocerca volvulus* DNA probe classification correlates with epidemiologic patterns of blindness. *Journal of Infectious Diseases* **165**, 964–968.

Zimmerman PA, Guderian RH, Aruajo E et al. 1994. Polymerase chain reaction-based diagnosis of *Onchocerca volvulus* infection: improved detection of patients with onchocerciasis. *Journal of Infectious Diseases* **169**, 686–689.

10 Schistosomiasis

Species belonging to the family schistosomatidae are digenean flukes with some characteristics that are not typical of their many relatives. Schistosomes are elongate worms, uncharacteristically having separate sexes and living in the blood vessels of their definitive hosts. Their development omits the redial and metacercarial stages typical of many digenean species (Dawes, 1968). Like nearly all digenean flukes, however, schistosomes develop in snails. Twelve species of schistosome have been found to infect humans (List, Chapter 1), but in some cases infection does not progress beyond penetration of the skin followed by cercarial dermatitis (see Coombs and Crompton, 1991). Five species have public health significance. *Schistosoma haematobium, S. japonicum,* and *S. mansoni* are the most widely distributed and are responsible for most morbidity, while *S. intercalatum* and *S. mekongi* have localized distributions but still cause disease (Table 1.1 and Table 10.1; WHO, 1993; Chitsulo et al., 2000; Attwood, 2001). Comprehensive accounts of schistosomiasis covering established knowledge and research developments have been compiled and edited by Jordan et al. (1993) and Mahmoud (2001).

10.1 LIFE HISTORY AND HOST RANGE

The schistosome life history is summarized in Figure 10.1. High-quality photomicrographs of the species of schistosome described in Table 10.1 have been published by Ash and Orihel (1990). Eggs produced by sexual reproduction in the human host are released in either stools or urine (Table 10.1). The embryos inside eggs that reach fresh water develop into miracidia and these penetrate susceptible snails (Figure 10.2). Once inside the snail, miracidia transform into sporocysts and daughter sporocysts. After some time (Table 10.2), cercariae formed by asexual reproduction in the daughter sporocysts are released from the snails into the water and these stages swim actively until they encounter human skin, which is then penetrated to begin the establishment of infection. Miracidia and cercariae are short lived (Table 10.2) and the survival and persistence of schistosomes depends on their finding susceptible snails and susceptible humans. The aquatic environment determines the discontinuous distribution of snail hosts, and human contact with this environment determines the extent and severity of schistosomiasis.

Humans are not the only definitive hosts of schistosomes (Table 10.1). *Schistosoma japonicum* is known to infect many species of domesticated and wild mammals and is probably a genuine zoonotic infection when present in humans. The other four species of schistosome listed in Table 10.1 have fewer mammalian hosts, but the significance of these as reservoirs of infection should not be overlooked by health workers concerned with the control of schistosomiasis. The diversity and susceptibility of snail hosts is more complicated than would appear from Table 10.1 and

TABLE 10.1
Features of the Life History and Biology of Schistosome Species Infecting Humans. Based on information from Rollinson and Simpson (1987), Chitsulo et al. (2000), Ohmae et al. (2004), SW Attwood (2001), and VR Southgate (personal communications).

	S. haematobium	S. intercalatum[a]	S. japonicum	S. mansoni	S. mekongi
General distribution	Confined to Africa and adjacent islands	Cameroon, Dem. Rep. of Congo, Gabon	China, Philippines, Indonesia	Africa, especially south of the Sahara, Brazil, Surinam, Venezuela, certain Caribbean Islands	Cambodia, Laos (Thailand)
Numbers of infections (millions)	114	1.73	1.6	83	0.19
Site of adult worms in human host	Venous system of the bladder	Blood vessels of the hepatic portal system	Blood vessels of the hepatic portal system	Blood vessels of the hepatic portal system	Blood vessels of the hepatic portal system
Worm size (mm)					
Male	10–15		12–20	10–15	
Female	13.5–22.5	13–28	15–30	7.2–17	14.5–20.1
Life span (years)	3.3–6	3–4	up to 47	1.5–10	up to 33
Prepatent period in human host (days)	56	41	34	34	43

Fecundity estimate (eggs/worm pair/day)	150–250	150–250	2,000	150–250	95
Common species of intermediate hosts: freshwater snails (Figure 10.2)[b]	*Bulinus africanus* *B. truncatus* *B. forskalii*	*Bulinus africanus* *B. forskalii*	*Oncomelania hupensis*	*Biomphalaria glabrata* *B. alexandrina* *B. pfeifferi* *B. straminea* *B. tenagophila*	*Neotricula aperta*
Nonhuman, mammalian hosts (reservoirs of infection)	Sheep, pigs, vervet monkey, chimpanzee, baboons	Rats	Cattle, dogs, wild pig, wild deer, macaque, mice, rats, vole, hare, squirrel, mongoose, badger, fox, civet, weasel, muntjac, water deer, hedgehog	Cattle, sheep, dogs, rats, baboons, chimpanzee, monkeys, shrews, waterbuck, guinea pig, crab-eating raccoon, anteater	Dogs, pigs

[a] *Schistosoma guineensis* (Pages et al., 2003) is included with *S. intercalatum.*
[b] Biodiversity and systematics reviewed by Rollinson and Southgate (1987).

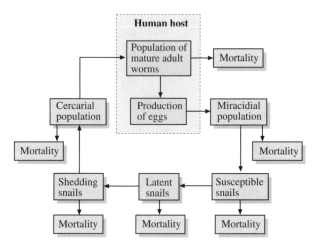

FIGURE 10.1 Flow chart of the main features in the life history of *Schistosoma* spp. infecting humans. From Figure 15.2 (Anderson and May, 1992) with permission from Oxford University Press.

Table 10.2. The snails mentioned in the tables are representatives of species groups and the reader is referred to publications by Brown (1994) and Southgate and Rollinson (1987) for detailed information.

10.2 TRANSMISSION AND ESTABLISHMENT OF INFECTION IN HUMANS

The transmission of human schistosomes is influenced by (1) the population dynamics of susceptible freshwater snails, (2) contamination of the snails' environment with human stools and urine containing schistosome eggs, (3) the ecology of the snails, (4) numbers and properties of infective cercariae in water, (5) human contact with water and exposure to infective cercariae, and (6) the development of some degree of protective immunity in the human host (Wilkins, 1987; Anderson and May, 1992). Human schistosomiasis would become a disease of historical interest if people had access to safe water and if those responsible for development projects involving water in endemic areas, such as irrigation schemes and hydroelectric power generation, implemented measures to keep people and snails away from water-based enterprise.

10.2.1 WATER CONTACT

Most of the people at risk of contracting schistosomiasis do not have access to safe water and appropriate sanitation. Natural water bodies provide them with water for drinking, cooking, bathing, washing vegetables and utensils, religious duties, personal hygiene, laundry, fishing, playing, and tending animals and crops. Water contact is unavoidable and, if cercariae are present in the water, infection for some people may be inevitable. Jordan et al. (1980) investigated the relationship between

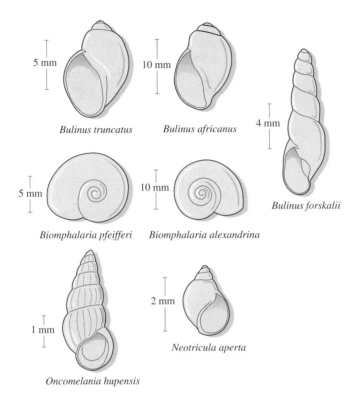

FIGURE 10.2 Sketches of the shells of snails that are important in the development of schistosomes, redrawn from illustrations in Brown (1994), images traced through Google and material supplied by SW Attwood. *Bulinus africanus* and *B. truncatas* support *Schistosoma haematobium*. *Bulinus forskalii* supports *S. intercalatum*. *Biomphalaria pfeifferi* and *Bio. alexandrina* support *S. mansoni*. *Oncomelania hupensis* supports *S. japonicum*. *Neotricula aperta* supports *S. mekongi*. Snail-schistosome relationships vary from place to place (Rollinson and Southgate, 1987).

host age, gender, water contact, and infection with *S. mansoni* in St Lucia (Figure 10.3). Most cases of infection were observed to occur in children (5 to 14 years), presumably because this group had most contact with water (Figure 10.3a, b), although they may not have developed as strong an immune response as adults. Thereafter, women were observed to have more water contact than men (Figure 10.3b), presumably because of their household responsibilities, and so to harbor more infections than men. This correlation of water contact and infection may have general application for *S. mansoni* elsewhere and for other species of schistosome (Table 10.1), but various factors will also be involved in the degree of water contact including the form of the water body and the amount of water it contains. Water levels will vary according to seasons and that will affect the biology of the snails so that transmission may be a seasonal event. There are exceptions to the trends shown in Figure 10.3a, b. For example, freshwater fishermen are an obvious high-risk group for contracting schistosomiasis (Kabatereine et al., 2004). Tourists may be at risk

TABLE 10.2
Features of the Life History and Biology of Human Schistosomes during Development in Snail Hosts. Based on information from Rollinson and Simpson (1987), Anderson and May (1992), and VR Southgate (personal communication).[a,b]

Variable	Biomphalaria glabrata	Biomphalaria pfeifferi	Bulinus globosus	Oncomelania hupensis	Neotricula aperta
Schistosome species	mansoni	mansoni	haematobium	japonicum	mekongi
Development time in snail host (days)	20–35	20–35	20–35	> 40	> 40
Prepatent/latent period (days)	17–45	15–43	25–119	39–70	43
Cercarial output/snail/day[c]	Values range from 42–16,000, depending on many factors[c]				
Life span of infected snail (weeks)	Values range from 2–11 weeks, depending on extent of infection				
Prevalence of infection (%)	0.6–25	0.8–27	0.4–7.1	0.3–17.4	2.7–60[d]

[a] The data have been selected to indicate typical values. Development in the snails is dependent on environmental factors such as temperature and water quality and on the number of miracidia that enter and survive in a snail.

[b] Free-living miracidia and cercariae have life expectancies of 4–16 hours and 8–20 hours respectively, subject to environmental conditions.

[c] Depends on the size of the snail, number of invading miracidia, and the water temperature.

[d] Depends on the strain of N. aperta.

as members of a group discovered after being diagnosed with *S. mansoni* infection caught during a rafting trip on the Omo River in Ethiopia (Schwartz et al., 2005).

10.2.2 SKIN PENETRATION AND TISSUE MIGRATION

Information about the process of infection has been obtained from elegant experiments involving *S. mansoni* and mice (Wilson, 1987). Freshly released cercariae (Figure 10.4) do not feed but depend on stored energy reserves to facilitate survival in water, finding a host and penetrating its skin. These processes are dependent on the ambient temperature and must usually be accomplished within 20 hours of cercarial shedding (Table 10.2). Human hosts are most vulnerable to cercarial skin penetration when their water contact activity corresponds with the time of maximum shedding. Having made contact with human skin, cercariae attach, search for a site for penetration, and then enter through a combination of enzymic secretions and muscular activity. Cercariae of *S. mansoni* reach the base of the epidermis of mouse skin within 30 min of attachment. Once penetration begins, the tail is lost and the body of the cercaria transforms to the schistosomulum.

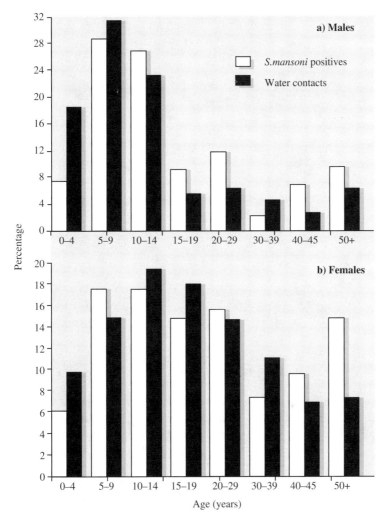

FIGURE 10.3 Relationship between host age, water contact, and infection with *Schistosoma mansoni* in St Lucia for (a) males and (b) females. From Figure 20.13 (Anderson and May, 1992) with permission from Oxford University Press.

During experiments, schistosomula have been found to remain in rodent skin for from 48 to 96 hours. The presence of the schistosomula in skin stimulates an inflammatory response that occurs more quickly and powerfully following repeated penetrations. Some schistosomula survive their time in rodent skin and they reach the lungs via the blood and lymphatic systems. On leaving the lungs after a few days, schistosomula move into blood vessels and travel against the flow to their final site. In the case of *S. mansoni* in mice, the accumulation of schistosomula in the liver appears to be completed within three weeks and some eventually reach the portal vessels. Extrapolation of results from mice to the human situation should be

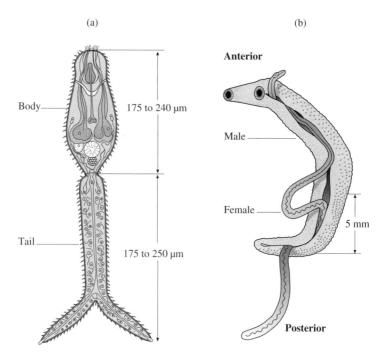

FIGURE 10.4 (a) A schistosome cercaria and (b) a pair of adult schistosomes. From Figure 64 (Chandler and Read, 1961) with permission from the Chandler family.

accepted with care since the skin, metabolic rate, and size of a mouse are so markedly different. He et al. (2005) have compared human skin penetration between the three common species of schistosome and found major differences in the timing and behavior once the schistosomula have entered the skin. The schistosomula of *S. japonicum* complete skin migration and reach the lungs one day after penetration, whereas those of *S. haematobium* and *S. mansoni* take three days. Also far fewer *S. japonicum* die in the skin as compared with the other two species.

10.3 DIAGNOSIS

10.3.1 PARASITOLOGICAL DIAGNOSIS

Unequivocal diagnosis of a schistosome infection still depends on finding eggs in stool, urine, or tissue samples. The Kato-Katz procedure is now widely promoted as the method of choice for the diagnosis of each of the four schistosome infections during which eggs are released in stools (WHO, 1994; Appendix 3). The method also provides an indirect estimate of the intensity of infection so it is useful for identifying individuals most at risk of severe disease, tracking the effectiveness of anthelminthic treatment, and following the progress of control interventions. The procedure cannot detect prepatent infections or those in which egg release may be irregular and so not found when a single sample is collected. Also, the sensitivity of the method has been questioned in the case of light infections. Eggs can be detected in tissue samples taken by intrusive biopsy procedures that are not suitable for epidemiological surveys.

Tests of the reliability of the Kato-Katz procedure have been carried out by Engels et al. (1996, 1997) in an area of Burundi where *S. mansoni* is endemic and by Yu et al. (1998) in China where *S. japonicum* is endemic. Engels et al. found that light infections were most likely to be missed, particularly if a single slide was examined, and that as the intensity of infection increased the schistosome eggs were not distributed evenly throughout the stool. Yu et al. examined Kato-Katz preparations on seven consecutive days from 570 people. Some (356) were from a village of high endemicity and some (214) from a village of low endemicity. Examination of one slide gave a prevalence value of 42.4% in the high village that increased to 68.3% when seven slides were checked. Equivalent values for the low village were 17.0 to 36%. In practice, a single examination is likely to be all that a public health program can afford to perform, but managers need to be aware of the limitations of the Kato-Katz procedure.

Facilities for the diagnosis of schistosomes whose eggs are discharged in stools could be facilitated by use of the Meade Readiview handheld microscope, manufactured by the Meade Instruments Corporation (Irvine, CA, U.S.). This microscope is a portable, robust, light-weight instrument that costs about a tenth of the price of a conventional compound microscope. Stothard et al. (2005) have undertaken a thorough evaluation of the use of the instrument in an epidemiological survey of the distribution of *S. mansoni* infections in school children in Uganda. The results were satisfactory, especially when the costs of the equipment were taken into account.

The usual method of parasitological diagnosis for *S. haematobium* infections has been the detection of eggs in samples of urine and to express the findings as the number of eggs per 10 ml of urine, preferably collected between 1100 and 1700 hours (WHO, 1980). Savioli et al. (1990) noted that variability occurred in the discharge of eggs in urine and were able to show that visible hematuria meant that the egg count would be at least 50 per 10 ml of urine. This finding is of considerable importance for the control of schistosomiasis haematobium.

10.3.2 DIAGNOSIS OF *SCHISTOSOMIASIS HAEMATOBIUM* BY QUESTIONNAIRE

The fact that hematuria is a cardinal feature of disease caused by infection with *S. haematobium* has enabled a rapid and inexpensive method to be developed for surveying and assessing the extent of morbidity at national, regional, and district levels (Chitsulo et al., 1995). The procedure begins with the use of a questionnaire that is answered by primary-school children under the supervision of their teachers. Operational research conducted by the Red Urine Study Group (1995) in eight African countries has shown that children can be relied upon to remember whether they have seen blood in their urine or not. Chitsulo et al. explain how to design and prepare to use the questionnaire, how to distribute it to obtain the most accurate information, how to carry out the necessary interviews with children, and how to collate and analyze the findings. The scheme has to be adapted to prevailing culture and attitudes in the places where it is to be used.

10.3.3 IMMUNODIAGNOSIS

Concerns over problems encountered in the methods for the detection of schistosome eggs in stool and urine samples, and the intrusive nature of taking tissue samples, have stimulated a vast effort to develop simple, reliable, and cheap immunodiagnostic tests (Ikeda and Akao, 2001). The principle behind the effort has been to detect either schistosome-specific antibodies or circulating antigens released by schistosomes. A comparison of 25 immunodiagnostic systems carried out by Mott and Dixon (1982) indicated that at that time the best results were to be gained from enzyme-linked immunosorbent assays (ELISAs) and indirect hemagglutination assays (IHAs) using antigens from schistosome eggs.

Sorgho et al. (2005) investigated the performance of a series of immunodiag-nostic tests by comparing their sensitivity for the detection of *S. mansoni* infections using 450 serum samples from Burkino Faso. The research involved an IHA diag-nostic kit (SMIHA) from Dade Behring, Germany, designed to detect *S. mansoni*, and an IHA diagnostic kit (SjIHA) from China, based on soluble *S. japonicum* egg antigen. Also tested were three ELISA tests involving soluble adult *S. mansoni* anti-gens. Sorgho et al. concluded that further work should be done to refine IHA kits based on soluble egg antigens. Good progress is being made with the application of procedures to detect *S. japonicum* antibody with a dot immunogold filtration assay (DIGFA) test and *S. mekongi* antibody with a dipstick dye immunoassay (DDIA) test (Zhu et al., 2005; Wen et al., 2005).

A simple and easy to read diagnostic dipstick has been developed for the detection of infections of *S. haematobium* and *S. mansoni*. The dipstick (schistoso-miasis one step test) came on to the market in 2003 and is available from EVL (European Veterinary Laboratory), Woerden, The Netherlands (www. evlonline.nl). Stothard et al. (2006) have carried out a thorough appraisal of the science on which immunodiagnostic tests are based and have tested the sensitivity and specificity of the EVL dipstick in Burkina Faso, Niger, Uganda, and Zanzibar (Tanzania) in places where *S. haematobium* and *S. mansoni* were known to be endemic. The test, which does not discriminate between the different species of schistosome, works by the dipstick's capturing schistosome circulating cathodic antigen (CCA, immunogenic glycoprotein) in urine. Both CCA and circulating anodic antigen (CAA) are produced by schistosomula and adult worms *in vitro* and *in vivo* with more CCA being produced by female worms than males (van Dam et al., 1996).

For *S. mansoni*, the test was found to have 83% sensitivity and 81% specificity and these results agreed well with cases of infection determined by Kato-Katz examination of stool samples. For *S. haematobium*, however, the results of the dipstick test were unsatisfactory and did not serve to confirm the presence of infection in children who had been shown to be passing eggs in their urine. Zanzibar was chosen for the investigation of *S. haematobium* to avoid cross-reactivity with CCA from *S. mansoni*. Subsequently, similar results for *S. haematobium* were obtained in Burkino Faso and Niger. Stothard et al. concluded that the CCA dipstick is a useful tool to support the diagnosis of *S. mansoni* infections. The current cost of the dipsticks may put them beyond the financial means of health services in much of Africa.

10.3.4 CLINICAL DIAGNOSIS

Signs and symptoms of each form of schistosomiasis are recognizable to health workers in areas where infections are endemic provided that the infection is not in a phase of being asymptomatic. Infected people are usually aware of the disease even if they have little knowledge of blood flukes (see Chapter 15). When resources for diagnosis are minimal, key pointers to infection include, for schistosomiasis haematobium, hematuria (visible blood in urine), microhematuria (blood detectable in urine with reagent sticks such as the Bayer *Haemastix®*), and dysuria (painful or difficult urination); for schistosomiasis japonicum and mansoni, bloody diarrhea, abdominal pain, and palpable hepatosplenomagaly. Ultrasonography has transformed clinical diagnosis, has increased our understanding of morbidity, and has facilitated monitoring the effects of anthelminthic treatment (Hatz, 2001). Portable ultrasound units are now available and can be used in communities that do not have clinics or health care premises.

10.4 EPIDEMIOLOGY

The geographical distribution of species of schistosome has been mapped by Doumenge et al. (1987) and analyzed by Chitsulo et al. (2000) who reckon that schistosome infections are endemic in 74 countries. About 193 million people are currently infected and some 650 million more are at risk of infection. Eighty-five percent of the infected people live in Africa. *Schistosoma mansoni* is found in 54 countries, *S. intercalatum* in 10 African countries, *S. japonicum* in China, Indonesia, and the Philippines, *S. mekongi* in Cambodia and Laos (there may be cases in Thailand), and *S. haematobium* in 53 countries, most of which are in Africa. Estimates of the numbers of cases in each of the endemic countries and the numbers of people at risk are given by Chitsulo et al. (2000). Generalizations and predictions about schistosomiasis are difficult to make because the problem depends on the patchy distribution of susceptible species of snail, which in turn depends on the distribution of their microhabitats, and the amount of water contact to which people are exposed.

When resources permit, in-depth epidemiological surveys aid in making decisions and enable public health managers to plan, deliver, and monitor control interventions. Surveys involving over 89,000 people across Egypt have provided data based on questionnaires, stool and urine samples, and physical and ultrasound examinations (El-Khoby et al., 2000). Information was obtained about the distribution of *S. haematobium* and *S. mansoni,* risk factors for infection and morbidity, and the extent and severity of morbidity, all on a demographic basis. Readers are recommended to review this work as an example of what comprehensive surveys can achieve.

10.4.1 PREVALENCE

The results of epidemiological surveys to detect schistosome infections in communities often reveal that more children are infected than adults (Figure 10.5 and Figure 10.6). This is not always the case (Figure 10.7). In terms of water contact, children are more

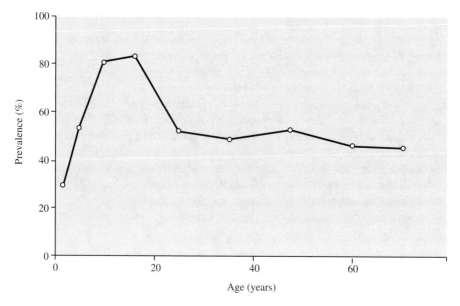

FIGURE 10.5 Relationship between host age and the prevalence of *Schistosoma haematobium* infection in Tanzania. From Figure 20.12b (Anderson and May, 1992) with permission from Oxford University Press and The Royal Society of Tropical Medicine and Hygiene.

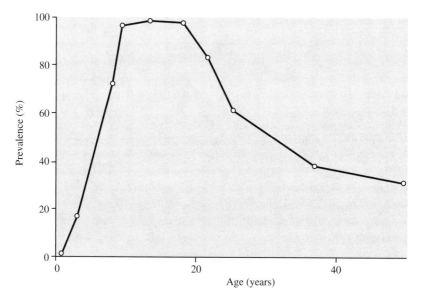

FIGURE 10.6 Relationship between host age and the prevalence of *Schistosoma japonicum* infection in the Philippines. From Figure 20.12a (Anderson and May, 1992) with permission from Oxford University Press and the World Health Organization (www.who.int).

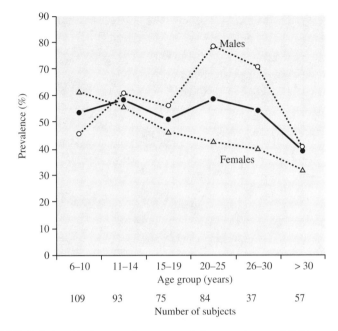

FIGURE 10.7 Relationship between host age and the prevalence of *Schistosoma mansoni* infection in Burkina Faso in 2002. From Figure 1 (Sorgho et al., 2005) with permission from Elsevier.

likely to swim and so expose more skin to cercariae than adults. Adults may expose less skin, but for longer times if they are occupied as fishermen or rice growers. The finding that prevalence is higher in children means that anthelminthic treatment for the control of morbidity can readily be part of the school-based deworming strategy advocated at the 54th World Health Assembly in May 2001 that also involves soil-transmitted helminthiasis (Engels and Savioli, 2005; Chapter 11 and Chapter 14).

10.4.2 INTENSITY

Surveys designed to measure the intensity of infection have almost always been dependent on egg counts; adult schistosomes cannot be counted after anthelminthic treatment as is the case with soil-transmitted helminth infections passed in host stools. Measuring intensity, even if indirectly through egg counts, remains an important task if interventions are to be used efficiently for the reduction and control of morbidity. Results of three intensity surveys are shown in Figure 10.8, Figure 10.9, and Figure 10.10. Again, intensity, like prevalence, is often found to be higher in children than adults. Care is needed in accepting this conclusion since many eggs remain trapped in a host's tissues (Section 10.5.1) so that the number of eggs per unit of sample may not always correlate with worm burden. The problem of squaring observed egg counts from an individual's stool sample with the worm burden responsible for their production has been explored by Gryseels and de Vlas (1996). Despite the difficulties they identify, all the evidence shows that the severity of morbidity

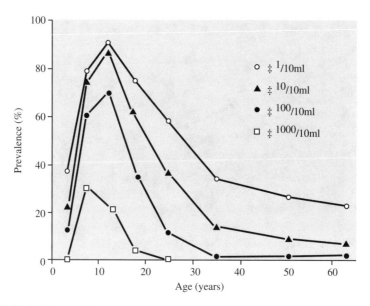

FIGURE 10.8 The prevalence and intensity of *Schistosoma haematobium* in relation to age in a Gambian community. From Figure 1 (Wilkins, 1987) with permission from Elsevier and The Royal Society of Tropical Medicine and Hygiene.

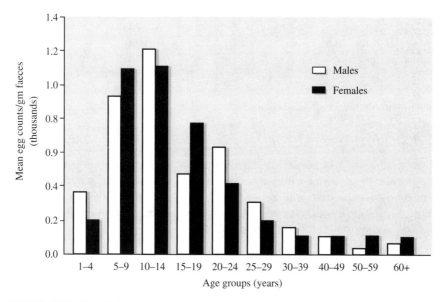

FIGURE 10.9 Plot of the relationship between host age and gender and the mean intensity (epg) of *Schistosoma mansoni* infection in Machakos District, Kenya. From Figure 20.17 (Anderson and May, 1992) with permission from Oxford University Press and from a histogram (Mahmoud and Warren, 1980) with permission from BC Decker.

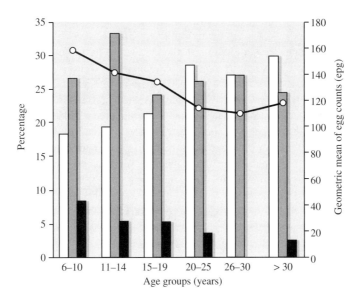

FIGURE 10.10 Relationship between host age and the intensity (epg) of *Schistosoma man-soni* infection in Kou valley, Burkina Faso. From Figure 2 (Sorgho et al., 2005), with permission from Elsevier.

during schistosomiasis is related to the intensity of infection (Table 4.1). Intervention to reduce intensity will reduce morbidity.

In developing a mathematical model of the dynamics influencing schistosome population biology Anderson and May (1992) point out that the frequency distribution of numbers of flukes per human host is aggregated or overdispersed, meaning that in any community most worms will be found in a few hosts. Consequently those few hosts will be at risk of most morbidity and death. Should they be identified as a top priority for treatment in a control program? The problem with that approach is the fact that schistosomiasis is a chronic disease and others in the community, with apparently low infection intensities, will be accumulating large numbers of eggs in their tissues. Anderson and May also point out that there is some evidence to suggest that predisposition to heavy or light intensities of infection occurs. This comes from studies of patterns of reinfection and is stronger in children than adults, perhaps as a result of some degree of adaptive immunity.

10.5 PATHOGENESIS AND MORBIDITY

The pathological consequences of schistosome infections in humans have been described in detail by von Lichtenberg (1987) and investigated *in situ* through the application of ultrasonography (see Hatz, 2001; Richter, 2003). The extent and severity of the observed clinical morbidity prompted an Expert Committee of the World Health Organization (WHO, 2002a) to suggest that the disability value then being assigned to schistosomiasis was an underestimate in need of review. King et al. (2005) have undertaken a meta-analysis of disability-related outcomes of schistosomiasis based on 482 published and unpublished reports. The importance of this approach is that the subtle morbidity accompanying schistosomiasis was recognized and quantified. In addition to the damage to tissues and organs initiated by infection with schistosomes, the infection contributes to iron-deficiency anemia, chronic abdominal and genitourinary pain, diarrhea and undernutrition. Establishing causality in the case of anemia is difficult because of the effects of concurrent infections, dietary intake, and other factors (Friedman et al., 2005). Nevertheless the chronic blood loss that occurs during schistosomiasis cannot be ignored.

The literature that can be accessed through these four articles should convince decision makers in the public health arena that schistosomiasis is a much more serious disease than was previously thought. King et al. (2005) note that at one time schistosomiasis had been given age-specific DALY weights ranging from 0.005 to 0.006, almost as if the disease were no more serious than facial discoloration! Their meta-analysis indicated that the total cost to an individual with detectable schistosomiasis could range from 2 to 15% chronic disability, values of more concern than the 0.5% accepted elsewhere.

10.5.1 SCHISTOSOME EGGS AND GRANULOMAS

Much of the morbidity of schistosomiasis can be attributed to the fate of the eggs after they leave the female worms. Having been released into blood vessels occupied by their parent worms, schistosome eggs, which are relatively large measuring from 60 to 170 µm long, must pass to either the gut or the bladder if they are to reach the microhabitats of susceptible snails. Little is known about how eggs move through tissues, but at least six days are required for those that succeed (Jourdane and Theron, 1987). While the spines on the egg shells (Appendix 3) may aid in passage through tissues, they cause the bleeding that occurs during schistosomiasis. Chronic, egg-related hemorrhage will be largely responsible for the anemia discussed by King et al. (2005).

Many eggs remain trapped in the tissues, stimulating the formation of granulomas around them and the migrations of cells contributing to hepatomegaly and splenomegaly. Trapped eggs continue to accumulate in the tissues as long as patent infections persist. An infection becomes established, reaches maturity, produces eggs, and dies, to be replaced by more worms and so on. The build-up of trapped eggs and the pathology they cause accounts for schistosomiasis being such a chronic disease. A granuloma is a tumor-like nodule composed of fibroblasts and several types of white blood cells. This agglomeration of egg and host cells initially measures about 350 µm in diameter and is a form of inflammatory response elicited by soluble egg antigens diffusing out of the miracidium through tiny pores in the egg shell. In time, after death of the miracidia, the granulomas may disperse and the disintegrated egg shell may be removed or calcified. The damage caused by scattered eggs is less of a problem than that occurring at sites where many have become trapped. Fibrous tissue is formed, veins may be obstructed, and ureters blocked. Calcification may also set in. Cheever et al. (1977) established from human autopsy studies that there is a positive correlation between organ egg load and organ pathology.

10.5.2 PROGRESSIVE MORBIDITY DURING SCHISTOSOMIASIS

Experts recognize that each species of schistosome generates species-specific morbidity. For public health intervention, however, distinguishing between lesions is less important than having a broad understanding of the general manifestations and progression of morbidity (Table 10.3). A scheme showing how disease and disability progress during the course of schistosomiasis is set out in Figure 10.11.

TABLE 10.3
Chronic Morbidity Characteristics of Schistosomiasis in Endemic Areas. Abstracted from von Lichtenberg (1987) and Richter (2003).

	S. haematobium	*S. japonicum*	*S. mansoni*
Pathogenic agent	Egg	Egg	Egg
Principal organs affected	Urinary tract	Gut, liver, spleen	Gut, liver, spleen
Important ectopic lesions	CNS	CNS	CNS
Typical pathology	Microhematuria	Bloody diarrhea	Bloody diarrhea
	Hematuria (anemia)	(anemia)	(anemia)
	Ureteric abnormalities	Hepatomegaly	Portal hypertension
	Female genital	Splenomegaly	Hepatomegaly
	schistosomiasis	Periportal fibrosis	Splenomegaly
	Bladder cancer	hematesis	Periportal fibrosis
	Hydronephrosis		hematesis

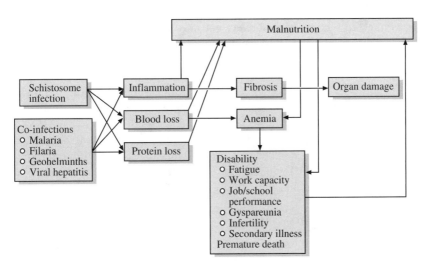

FIGURE 10.11 Proposed pathways in the generation of disease and disability during schistosomiasis. From Figure 5 (King et al., 2005) with permission from Elsevier.

10.5.3 Female Genital Schistosomiasis (FGS)

Female genital schistosomiasis (FGS) is defined as the presence of schistosome eggs and characteristic pathology in any part of the female reproductive tract. *Schistosoma haematobium* mature in blood vessels of the pelvic system, but are known to explore other blood vessels in the region, thereby causing their eggs, with subsequent granuloma formation, to be trapped in many parts of the female reproductive tract (Poggensee and Feldmeier, 2001). Fibrosis and calcification commonly occur. Cases of FGS are found wherever *S. haematobium* infection is endemic. Epidemiological surveys aimed at measuring the population prevalence of FGS are difficult to undertake since accurate diagnosis depends on biopsy samples during gynecological examination in addition to detecting eggs in urine. When urine samples from three consecutive days were collected and examined for eggs, over 20% of women with FGS were not identified (Poggensee and Feldmeier, 2001). Kjetland et al. (1996) decided that the prevalence of FGS in women infected with *S. haematobium* in Malawi was at least 55%. As expected the intensity of infection increases the risk of developing FGS.

FGS is understandably causing concern because it may enhance conditions for the transfer to women of HPV-2 (Herpes simplex virus), which in turn is involved in cervical neoplasia. Damage to the epithelium in the lower reproductive tract during FGS is likely to assist in the entry of HIV in the blood stream. Women in Zimbabwe, participating in a cross-sectional study in the northwestern part of the country, have been found to have a higher risk of contracting HIV infection if suffering from FGS than if not (Kjetland et al., 2006). Evidence in support of these consequences of FGS is discussed in detail by Poggensee and Feldmeier (2001).

10.5.4 Bladder Cancer

Patients infected with *S. haematobium* not infrequently are found to be suffering from bladder cancer (Table 10.3). Epidemiological evidence from Tanzania (Kitinya et al., 1986) and Zimbabwe (Thomas et al., 1990) indicates that squamous cell carcinoma of the bladder occurs more frequently in people infected with *S. haematobium* than in people who are uninfected. In Zimbabwe, patients from provinces with a high prevalence of *S. haematobium* had more cases of squamous-cell bladder carcinoma than patients from elsewhere in the country. Bedwani et al. (1998) analyzed data on Egyptian patients with urinary disorders and, after taking account of confounding variables, concluded that schistosomiasis was responsible for 16% of bladder cancer cases in the Egyptian population. *Schistosoma haematobium* has been recognized as having carcinogenic properties (IARC, 1994).

10.5.5 Ectopic Schistosomiasis

Schistosome eggs may be carried in the blood stream to every part of the body. According to von Lichtenberg (1987) there is virtually no human tissue in which the occurrence of schistosomes has not been reported at least once. Neuroschistosomiasis occurs when eggs have reached the brain and spinal cord. Granulomas develop and neurological symptoms may be observed (Pittella, 1997).

TABLE 10.4
Estimated Number of Individuals With Morbidity or Pathology Due to *Schistosoma haematobium* Infection by Age Group in Sub-Saharan Africa in Millions (90% Confidence Interval). From Table 2 (van der Werf et al., 2003) with permission from Elsevier.

Category	Preschool Children	Schoolchildren	Adults	Total
At risk of infection	71	168	196	436
Infected	14	56	43	112
Hematuria in last 2 weeks	9.5 (6.6–12)	33 (25–39)	28 (20–36)	70 (51–87)
Dysuria in last 2 weeks	3.6 (1.7–7.0)	17 (10–28)	11 (5.5–21)	32 (17–54)
Minor bladder wall pathology	12 (8.5–15)	41 (33–46)	35 (25–42)	88 (67–102)
Major bladder wall pathology	2.3 (0.5–3.5)	9.0 (3.0–13)	6.9 (1.6–10)	18 (5.1–27)
Moderate hydronephrosis	2.4	4.1	2.8	9.3
Major hydronephrosis	0	5.2	4.3	9.6

10.5.6 EXTENT OF MORBIDITY AND MORTALITY DURING SCHISTOSOMIASIS

Estimates of the numbers of people infected with *Schistosoma* spp. (Table 10.1) are unlikely to have as much influence with decision makers as estimates of the numbers suffering from disease. Since most infections occur in sub-Saharan Africa, van der Werf et al. (2003) calculated the numbers of people there who would be likely to be suffering from different aspects of schistosomiasis haematobium (Table 10.4) and schistosomiasis mansoni (Table 10.5) and so would present with clinical morbidity. Estimates by van der Werf et al. indicate that about 150,000 and 130,000 people die annually in sub-Saharan Africa from schistosomiasis haematobium and schistosomiasis mansoni respectively. Death from infection with *S. haematobium* infection is likely to follow from kidney failure or bladder cancer. Death from *S. mansoni* infection probably results from liver failure. These mortality figures should not be ignored when DALYs due to schistosomiasis are revisited (WHO, 2002a).

10.6 IMMUNITY

In immunocompetent hosts, the innate and adaptive elements of the immune system (Chapter 4) respond to the presence of schistosome infections in a complex manner, the details of which are still being unravelled. The responses result in the control of hypersensitive granuloma formation (Section 10.5.1), the slow development of a degree of protective immunity, and a blend of interactions (concomitant immunity) that results in avoidance of destruction of established worms and the regulation of numbers of new arrivals (McLaren and Smithers, 1987; Hagan and Abath, 1992; Pearce and MacDonald, 2002). In humans, immunity requires the action of T helper cell type 2 (Th2) responses characterized by a high production of IgE (immunoglobulin

TABLE 10.5
Estimated Number of Individuals with Morbidity or Pathology Due to
Schistosoma mansoni **Infection by Age Group in Sub-Saharan Africa in**
Millions (90% Confidence Interval). From Table 3 (van der Werf et al., 2003)
with permission from Elsevier.

Category	Preschool Children	Schoolchildren	Adults	Total
At risk of infection	65	152	177	393
Infected	4.7	25	23	54
Diarrhea in last 2 weeks	0.034 (0.00–0.72)	0.42 (0.0–3.6)	0.32 (0.0-3.5)	0.78 (0.0–7.8)
Blood in stool in last 2 weeks	0.24 (0.16–0.69)	2.3 (1.6–4.1)	1.9 (1.3–3.7)	4.4 (3.0–8.3)
Hepatomegaly (MSL)	0.076	4.0	3.8	8.5
Splenomegaly	[0.61]	[2.9]	[2.8]	[6.3]
Ascites	[0]	[0]	[0.29]	[0.29]
Hematemesis	[0]	[0]	[0.93]	[0.93]

Note: Data in square brackets [ ] should be used with caution until more information becomes available.

MSL, mid-sternal level.

E) and eosinophilia. Soluble egg antigens are powerful inducers of Th2 responses, which may initiate antibody production and killing by cytotoxic cells.

The fact that protective immunity occurs has stimulated research to identify molecules that might serve as potential vaccines (Capron et al., 2005). *Schistosoma mansoni* contains an enzyme described as 28 kDa glutathione-*S*-transferase (Sm28GST) that is important for some of the worm's functions including fertility. Might this enzyme serve as the antigen for a vaccine? Would antibody or other effectors released in response to an inoculation of a dose of Sm28GST damage the schistosomes and impede egg production? Even when those questions are answered the vaccine will need to withstand clinical trials and be tested under community conditions, preferably after its precise mode of action and the responses it stimulates have been understood (see Section 10.7.6).

10.7 PREVENTION AND CONTROL

Options for the prevention and control of schistosomiasis have been identified and evaluated in terms of cost-effectiveness by Hotez et al. (2006). The options comprise anthelminthic treatment, access to safe water, provision of appropriate sanitation, health education and the application of chemicals to water to kill snails. There is now a movement to integrate the prevention and control of a wide range of infectious diseases into a single strategy that would tackle HIV/AIDS, malaria, TB, filariasis, schistosomiasis, and soil-transmitted helminthiasis together. Although schistosomiasis is discussed in isolation in this chapter, we urge the public health community

TABLE 10.6
Interventions for the Control of Morbidity in School-Age Children in Areas Where *Schistosoma haematobium* and *S. mansoni* Are Endemic. From Table 1 (Magnussen, 2003) with permission from Elsevier.

Results of Community Diagnosis through Schools	School-Based Interventions	Community Provision and Intervention
S. hematobium ≥ 30% hematuria (questionnaire) or *S. mansoni* ≥ 50% infected (Kato-Katz slide)	Treatment (PZQ) of all enrolled and nonenrolled school children once a year	Access to PZQ for passive case management *and* community-directed treatment (PZQ) for risk groups
S. hematobium 1 to 29% haematuria (questionnaire) or *S. mansoni* 20 to 50% infected (Kato-Katz slide)	Treatment (PZQ) of all enrolled and nonenrolled schoolchildren once every two years	Access to PZQ for passive case management
S. hematobium known to be present but no visual hematuria (questionnaire) or *S. mansoni* < 20% infected (Kato-Katz slide)	Selective case treatment (PZQ) *or* universal treatment (PZQ) at start and end of primary school period for enrolled and nonenrolled children	Access to PZQ for passive case management

Note: PZQ, praziquantel (see WHO, 2004).

to work for a holistic approach to the control of communicable disease. Anthelminthic treatment for the control of morbidity due to schistosomiasis (Table 10.3) has been accepted as the most affordable intervention at present. Treatment strategies for schistosomiasis control are set out in Table 10.6. Re-infection is inevitable in endemic areas and improved and extended access to safe water and effective sanitation will be essential if progress is to be sustained and schistosomiasis is to be struck off the public health agenda. Killing snails by adding poisonous chemicals to their habitats may be an effective option but there are bound to be environmental concerns. Research to find natural molluscides from plants is in progress.

10.7.1 ANTHELMINTHIC TREATMENT WITH PRAZIQUANTEL

The objective of morbidity control due to schistosomiasis can be implemented by ensuring access to essential anthelminthic drugs in all health care facilities and by the regular treatment of population groups identified as being at risk of developing morbidity (WHO, 2002a). How deworming is offered in community settings (universal, targeted, or selective; Chapter 4) depends on epidemiological information (see El-Khoby et al., 2000) and the availability of resources. Treatment targeted at school-aged children is a highly desirable strategy (Bundy and Guyatt, 1996) because of their infection status (Figure 10.5, Figure 10.6, Figure 10.8, and Figure 10.9), vulnerability to disease, and the routine assembly of many of them in schools. Such programs have been successful in Egypt (Curtale et al., 2003) during which praziquantel PZQ administration has had a notable effect on the intensity of *S. mansoni*

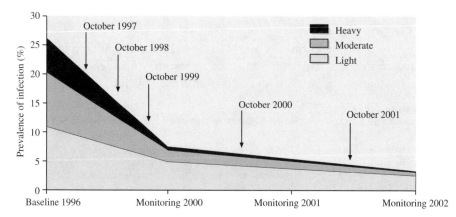

FIGURE 10.12 Plot showing the prevalence of *Schistosoma mansoni*, by different level of intensity (epg), among school children in Behera, Egypt. Vertical arrows indicate campaigns to distribute praziquantel to all school children. From Figure 2 (Curtale et al., 2003) with permission from Elsevier.

infection in children (Figure 10.12). Deworming programs have been started in Burkina Faso, Mali, Niger, Tanzania, Uganda, and Zambia with support from the Bill and Melinda Gates Foundation (Southgate et al., 2005). Every effort has to be made to include children who for whatever reason are not attending school (Useh and Ejezie, 1999).

Praziquantel is the drug of choice for the treatment of all forms schistosomiasis (Fenwick et al., 2003; WHO, 2004). Experience has shown that for school-aged children, a dose of 40 mg/kg body weight is effective and, in schools or communities where scales are lacking, the use of the dose pole is recommended. The results of operational research have shown that a child's height can be safely equated with body weight for purpose of dose calculation (Montresor et al., 2005). Manufacturers of PZQ are encouraged to include in packs of tablets a durable tape that can be pinned on a wall to help teachers and health workers calculate doses.

PZQ is well tolerated and is suitable for use in community treatment programs. Over a period of 27 years of the drug's use in China against *S. japonicum*, more than 50 million people received PZQ and the recorded adverse effects were classified as mild and transient rarely lasting for more than 24 hours (Chen, 2005). During the development of PZQ, no evidence was found for the compound's having mutagenic potential for humans or fetal and maternal toxicity (Dayan, 2003). A single oral dose (40 to 60 mg/kg body weight) results in high cure rates and a reversal of some of the pathology and morbidity (Table 10.6). Although PZQ is available as a suspension, most is supplied in 600 mg tablet form. Half of this amount (the levorotary isomer) has therapeutic activity against the adult stages of schistosomes. The price of PZQ has plummeted since its patent expired. Quality generic versions, costing as little as USD 0.10 when ordered in bulk, are now available and tablets are produced by 20 companies around the world (Fenwick et al., 2003). Nevertheless, managers of control programs are advised to arrange for the quality of the PQZ they intend to purchase to be checked; a fake version of the drug has been uncovered. Despite its good safety record based on 20 years of use, offering PZQ treatment to pregnant and lactating women has been avoided. New advice has emerged recently following a risk/benefit review (Section 10.7.1.2).

10.7.1.1 Impact of Praziquantel Treatment on Schistosomiasis

Richter (2003) has carried out a major review of the consequences of treating infected people with PZQ and some of the findings are shown in Table 10.7. Generally there was a marked reversal in morbidity as detected by biochemical tests, clinical examinations, and ultrasonography, and these improvements correlated with reductions in prevalence and intensity. The following conclusions emerged from this analysis. (1) In most cases, annual treatment is sufficient to achieve much organ pathology. (2) The chances for the reversal of pathology increase if PZQ treatment is given early in life. (3) Children and adolescents should be targeted for PZQ treatment in public health interventions. (4) Reversal of hepatic pathology is slow and may need more than one round of treatment during the year. Olveda et al. (1996) demonstrated that hepatic pathology was slow to decline after treatment for schistosomiasis japonicum (Figure 10.13). (5) PZQ treatment has beneficial effects on childhood development, physical fitness, and working capacity. Richter's survey also covered the impact of oxamniquine and metrifonate on morbidity and he pointed out that information about anthelminthic treatment on genital pathology is inadequate.

Further evidence of the merits of intervention in the community with PZQ (40 mg/kg body weight) is provided by an account of the control of morbidity due to schistosomiasis mekongi in Lao People's Democratic Republic during a program lasting from 1989 to 1999 (Khamkeo and Pholsena, 2003). The account deals with such details as how to target school children and how to ensure that people swallowed the tablets, and emphasizes the importance of health education. At the start, prevalence values in individual villages ranged from 98 to 8% and intensities from 137 to 31 (mean epg). After the rounds of treatment during the program, the pooled prevalence for people living in the villages had fallen to 0.8% and mean village intensity values ranged from 0.4 to 0.03 (epg). Adverse effects in school children following a single dose of PZQ were monitored in a group of 102 school children. Not all the effects can necessarily be attributed to the drug. Headache was recorded in 54 children, diarrhea in 43, and abdominal pain in 20, but most effects had regressed after 6 to 12 hours. Khamkeo and Pholsena concluded their report as follows "Overall, despite limited resources and a lack of experience among the personnel, the program has been largely successful. Coverage of the population by mass treatment was excellent. Health education campaigns had a substantial impact on the population, facilitating progressive behavioral change in terms of contact with the waters of the Mekong and use of latrines."

10.7.1.2 Praziquantel, Pregnancy, and Lactation

Pregnant women suffering from schistosomiasis are likely to be left untreated because clinical tests of PZQ on pregnant women could not be allowed to be carried

TABLE 10.7

Examples of Outcomes of Anthelminthic Treatment (PZQ)[a] on the Prevalence, Intensity, and Morbidity of Infection With *Schistosoma haematobium*, *S. japonicum*, and *S. mansoni*. From Table 1, Table 2, and Table 3 (Richter J. 2003. *Acta Tropica* 86, 161–184), with permission from Elsevier.

Country	Age Group	Treatment	Follow-Up Interval	Assessment[b]	Re-Infection	Outcome and Impact[c]	Reference
Schistosoma haematobium (Urinary Schistosomiasis Haematobium)							
Congo	Children	PZQ 40 mg/kg	11 months	H, MH, P, US	+	Prevalence ↓; intensity ↓; MH ↓; BWA ↓ UTO ↓	Doehring et al. (1986)
Niger	All ages, mainly children	PZQ 40 mg/kg universal then selective treatment	1–4 years	H, MH, P, US	+	Prevalence ↓; intensity ↓; MH ↓; BWA ↓; UTO ↓	Devidas et al. (1989) Laurent et al. (1990)
Mali	All ages	PZQ 40 mg/kg	1 year	H, MH, P, US	+	Prevalence ↓; intensity ↓; H ↓; MH ↓; BWA ↓; UTO ↓ Resolution of ureteric abnormalities	Kardorff et al. (1994)
Ghana	All ages	PZQ 40 mg/kg	12 and 18 months	H, MH, P, US	+	Month 12: BWA ↓; UTO ↓ Month 18: BWA (mild pathology); UTO at 63%	Wagatsuma et al. (1999)
Tanzania	Children	PZQ 40 mg/kg annual, selective treatment of MH cases	1–5 years	CE, H, MH, P	+	Prevalence of H ↓ from 51 to 23% after 5 years	Magnussen et al. (1997)
Schistosoma japonicum (Intestinal Schistosomiasis Japonicum)							
China	All ages	PZQ 60 mg/kg	1 year	CE, P, US and biochem tests	+	Periportal changes ↓; hepatosplenomegaly ↓; biochemical markers ↓	Cai et al. (1997)
China	All ages	PZQ 60 mg/kg risk groups	2 years	CE, P, US and biochem tests	(+)	Prevalence ↓; periportal changes ↓; hepatosplenomegaly ↓	Li et al. (2000)

Country	Age group	Treatment[a]	Follow-up	Method[b]		Outcome[c]	Reference
China	All ages	PZQ 50 mg/kg repeated annually	1, 2 and 3 years	CE, P, US	+	Prevalence ↓; hepatomegaly ↓; splenomegaly ↓	Wiest et al. (1994)
Philippines	All ages	PZQ 40 mg/kg	Yearly, for 8 years	CE, P	+	Prevalence ↓; intensity ↓; hepatomegaly ↓;	Olveda et al. (1996)
Schistosoma mansoni (Intestinal Schistosomiasis Mansoni)							
Madagascar	All ages	PZQ 40 mg/kg	3 years	CE, P, US	–/+	Prevalence ↓; bloody stools ↓; PPF ↓; splenomegaly ↑	Boisier et al. (1998)
Burundi	All ages	annually, selective treatment	3 months to 3 years	CE, P	+	Prevalence ↓; organomegaly ↓	Gryseels et al. (1994)
Sudan	Children	PZQ 20 mg/kg and PZQ 40 mg/kg	7 and 23 months	CE, P, US	+	Month 23: prevalence ↓; intensity ↓; hepatomegaly ↓; splenomegaly ↑	Mohamed et al. (1991) Doehring-Schewrdtfeger et al. (1992)
Uganda	All ages	PZQ 40 mg/kg repeated after one year	2.7 years	CE, P, US	+	Prevalence ↓; PPF = or ↓ in people > 30 years, less reversible in women and girls	Frenzel et al. (1999)
Zambia	Children and adolescents	PZQ 40 mg/kg and PZQ 40 mg/kg twice, 6-monthly	6 months and 1 year	CE, P	+	Prevalence ↓; intensity ↓; bloody diarrhoea ↓; hepatomegaly ↓; splenomegaly ↓	Sukwa (1993)
Brazil	All ages	PZQ 40 mg/kg	3 and 6 months, 5 years	CE, P, US, and Biochem tests	–/(+)	PPF = no progression; liver size =; splenomegaly ↓	Richter (2003)

[a] PZQ, praziquantel (WHO, 2004).

[b] CE, clinical examination; H, hematuria; MH, microhematuria; P, parasitology (urine or stool samples for eggs); US, Ultrasonography:.

[c] ↓ decrease; ↑ increase; = no change; BWA, bladder wall abnormality; PPF, periportal fibrosis assessed by US; UTO, urinary tract obstruction.

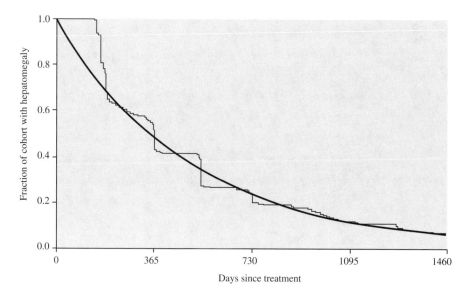

FIGURE 10.13 Evidence of the time taken for the decline in the liver enlargement following treatment with praziquantel for schistosomiasis japonicum. From Figure 3 (Olds, 2003) with permission from Elsevier.

out. Of equal concern is the view that PZQ treatment should not be given during breast feeding and that if nursing mothers are to be treated then breast feeding should be temporarily discontinued for from 24 to 48 hours. Under most circumstances, this advice is impractical causing discomfort and distress on top of the schistosomiasis and disruption to the infant's feeding routine. There are no known pregnancy-related risks from the standard, recommended treatment with PZQ to either mother, growing fetus, or suckling infant (WHO, 2002b; Dayan, 2003) and, given the severity of morbidity caused by schistosomiasis (Siegrist and Siegrist-Obimpeh, 1992; Olds, 2003), members of an Informal Consultation convened by the World Health Organization (WHO, 2002b) decided that treatment with PZQ should not be withheld from women during pregnancy and lactation. The original text of the report of the Informal Consultation should be considered before embarking on control measures during pregnancy. In essence, the recommendations, which apply to the use of PZQ, are:

- women of child-bearing-age should not be excluded from population-based anthelminthic treatment programs (with PZQ) for schistosomiasis;
- pre- and postpubescent females should be included in all schistosomiasis control interventions
- all pregnant and lactating women found to have schistosomiasis should be offered immediate treatment

- pregnant and lactating women living in areas of high schistosomiasis endemicity, where population-based treatment is employed, should be included and offered treatment
- in areas where schistosomiasis is endemic and where universal anthelminthic treatment is not available, women of child-bearing age, including those pregnant and lactating, should be accepted as a high-risk group for morbidity

Since these recommendations, Adam et al. (2004) have asked whether praziquantel therapy is safe during pregnancy. They searched for congenital abnormalities in the children of Sudanese women who had taken PZQ during pregnancy and compared their findings with children born to mothers who had not been treated. In summary they stated that no congenital abnormalities were noted by clinical examination in any of the babies born to either group and they concluded that PZQ treatment is safe during pregnancy.

10.7.1.3 Therapeutic Efficacy of Praziquantel

Many publications confirm high cure rates and impressive egg reduction rates when PZQ is used to treat schistosomiasis in a community setting (Magnussen, 2003). Occasionally reports are received suggesting that the expected effectiveness has not been achieved. For example, Silva et al. (2005) observed poor therapeutic activity in the PZQ they prescribed to treat Brazilian soldiers with urinary schistosomiasis contracted while on duty in Mozambique. Possible explanations for such therapeutic "failures," including the investigation of the emergence of strains of drug-resistant worms, are discussed in Chapter 14.

10.7.1.4 Other Deworming Treatments for Schistosomiasis

Praziquantel is not the sole drug available for the treatment of schistosomiasis. Oxamniquine is recommended for the effective treatment of *S. mansoni* infections, but it must be given as 60 mg/ kg body weight in divided doses over 2 to 3 days (WHO, 2004) so increasing the cost of delivery in community control programs. Currently, oxamniquine and metrifonate, which has good activity against *S. haematobium* infections, are no longer available commercially (Fenwick et al., 2003). There is clearly an urgent need to identify more drugs for the expanding effort to control morbidity due to schistosomiasis. What would happen if PZQ-resistance emerged and spread either through human activity or the evolutionary adaptability of the schistosomes themselves?

Alternative treatments to PZQ, which is also crucial for controlling for cestodiasis (Chapter 5) and foodborne trematodiasis (Chapter 7), may be sought by modifying the PZQ molecule, by developing the use of artemether, and by investigating the antischistosomal compounds such as myrrh (Fenwick et al., 2003) and other substances from plants. Artemether is a derivative of artemesinin (qinghaosu), which is extracted from the leaves of *Artemisia annua* and has been widely used in traditional Chinese medicine for the treatment of fevers. Utzinger et al. (2001) have reviewed the results of trials with artemether to control schistosomiasis. Artemether is prescribed for humans at doses of 6 mg/kg body weight, but several doses are needed over a period of time to achieve most therapeutic activity. The special feature of artemether, in contrast to PZQ, is its activity against schistosomula. Combination therapy with artemether and PZQ, in which both schistosomula and adult schistosomes are killed at the same time, may become a major advance in the control of morbidity due to schistosomiasis. Artemether is currently recommended by the World Health Organization for restricted use in the treatment of falciparum malaria. Artemether must undergo more operational research before it can be considered as a compound for the treatment of schistosomiasis.

Myrrh is a resin extracted from the stem of the plant *Commiphora molmol*. Myrrh's antischistosomal activity is obtained by giving a dose of 10 mg/kg body weight on three consecutive days. Sheir et al. (2001) treated 204 schistosomiasis patients with myrrh by giving them doses on three consecutive days of 10 mg/kg body weight and then recorded a cure rate of nearly 92%. With such a regimen using Mirazid, a preparation now marketed in Egypt, Soliman et al. (2004) observed a 100% cure rate from *Schistosoma mansoni* infection four weeks after treatment, but this trial involved as few as eight children. Other studies have given equivocal results. There is still no general agreement about the merits of myrrh as an antischistosomal agent.

10.7.2 Vaccination Prospects

The identification and development of Sm28GST as a candidate antigen for vaccine production (Section 10.6) is an important breakthrough for the prevention and control of schistosomiasis. Experiments with infected animals showed that the recombinant protein version of Sm28GST had a significant inhibitory effect on the fecundity of female *S. mansoni* and on the fecundity of female *S. haematobium*. Capron et al. (2005) refer to this encouraging effect as being due to a cross-species-specific anti-fecundity vaccine. The first of a series of vaccination trials in humans has been carried out with rSh28GST, the recombinant form of the protein from *S. haematobium*, to begin an evaluation of the safety of the putative vaccine. No local or systemic adverse reactions were detected from this trial in healthy adult volunteers (Capron et al., 2005). Whether vaccination becomes a realistic intervention in areas where the infections are endemic, either alone or in integration with deworming, will depend on resources available to health care systems.

Another promising development concerns the vaccination of bovine hosts of *S. japonicum*, which causes a zoonotic form of schistosomiasis in humans (Table 10.1). Domesticated cattle and water buffalo are the principal reservoir hosts in China and the Philippines and are considered to be responsible for nearly all the contamination of the environment with the eggs of *S. japonicum*. The outline strategy is to identify, prepare, and test antigens that can be used as a vaccine to block transmission via bovine hosts (McManus, 2005). Blocking transmission would have been achieved if the egg output from infected bovines crashed after vaccination. An attraction of this scheme is its benefit for the welfare of valuable animals as well as for their owners. This form of vaccination might also be used in combination with human deworming in the prevention and control of schistosomiasis.

10.7.3 SUSTAINABLE PROGRESS

All those who have studied schistosomiasis agree that people in endemic areas will continue to be at risk of infection and disease until they have safe water and effective sanitation with the support of sewage collection and treatment and waste disposal (Chapter 16). Vaccination may increase protection and support deworming. For the present, however, community deworming supported by health awareness (Chapter 15) is the practical public health intervention that decision makers are encouraged to take and implement.

REFERENCES

Adam I, Elwasila el T and Homeida M. 2004. Is praziquantel therapy safe during pregnancy? *Transactions of the Royal Society of Tropical Medicine and Hygiene* **98**, 540–543.

Anderson RM and May RM. 1992. *Infectious Diseases of Humans.* Oxford University Press (paperback edition).

Ash LR and Orihel TC. 1990. *Atlas of Human Parasitology,* 3rd edition. Chicago: American Society of Clinical Pathologists Press.

Attwood SW. 2001. Schistosomiasis in the Mekong region: epidemiology and phylogeography. *Advances in Parasitology* **50**, 87–152.

Bedwani R, Renganathan E, El Kwhsky F et al. 1998. Schistosomiasis and the risk of bladder cancer in Alexandria, Egypt. *British Journal of Cancer* **77**, 1186–1189.

Boisier R, Ramarokoto CE, Ravaoalimalala VE et al. 1998. Reversibility of *Schistosoma mansoni*-associated morbidity after yearly mass praziquantel therapy: ultrasonographic assessment. *Transactions of the Royal Society of Tropical Medicine and Hygiene* **92**, 451–453.

Brown DS. 1994. *Freshwater Snails of Africa and their Medical Importance,* 2nd edition. London and Philadelphia: Taylor and Francis.

Bundy DAP and Guyatt HL. 1996. Schools for health: focus on health, education and the school-age child. *Parasitology Today* **12**, 1–16.

Cai W, Chen Z, Chen F et al. 1997. Changes of ultrasonography and two serum biochemical indices for hepatic fibrosis in *Schistosoma japonica* patients one year after praziquantel treatment. *Chinese Medical Journal* **110**, 797–800.

Capron A, Riveau G, Capron M and Trottein F. 2005. Schistosomes: the road from host-parasite interactions to vaccines in clinical trials. *Trends in Parasitology* **21**, *143–149.*

Chandler AC and Read CP. 1961. *Introduction to Parasitology,* 10th edition. New York and London: John Wiley & Sons Inc.

Cheever AW, Kamel IA, Elwi AM et al. 1977. *Schistosoma mansoni* and *S. haematobium* infections in Egypt. II. Quantitative parasitologic findings at necropsy. *American Journal of Tropical Medicine and Hygiene* **26**, 702–716.

Chen M-G. 2005. Use of praziquantel for clinical treatment and morbidity control of schistosomiasis japonica in China: a review of 30 years' experience. *Acta Tropica* **96**, 168–176.

Chitsulo L, Lengeler C and Jenkins J. 1995. The schistosomiasis manual. *Methods for Social Research in Tropical Diseases* 3. Geneva: World Health Organization.

Chitsulo L, Engels D, Montresor A and Savioli L. 2000. The global status of schistosomiasis and its control. *Acta Tropica* **77**, 41–51.

Coombs I and Crompton DWT. 1991. *A Guide to Human Helminths.* London and Philadelphia: Taylor & Francis.

Curtale F, Hassanein YAW, Wakeel AE et al. 2003. The school health programme in Behera: an integrated helminth control programme at governorate level in Egypt. *Acta Tropica* **86**, 295–307.

Dawes B. 1968. *The Trematoda.* Cambridge University Press.

Dayan AD. 2003. Albendazole, mebendazole and praziquantel. Review of nonclinical toxicity and pharmacokinetics. *Acta Tropica* **86**, 141–159.

Devidas A, Lamothe F, Develoux M et al. 1989. Ultrasonographic assessment of the regression of bladder and renal lesions due to *Schistosoma haematobium* after treatment with praziquantel. *Annales de la Societe Belge Medicine Tropicale* **69**, 57–65.

Doehring E, Ehrich JHH, Reider F et al. 1986. Reversibility of urinary tract abnormalities due to *Schistosoma haematobium* infection. *Kidney International* **30**, 582–585.

Doehring-Schwerdtfeger E, Abdel-Rahim IM, Kardorff R et al. 1992. Ultrasonographic investigation of periportal fibrosis in children with *Schistosoma mansoni* infection: reversibility of morbidity twenty-three months after treatment with praziquantel. *American Journal of Tropical Medicine and Hygiene* **46**, 409–415.

Doumenge J, Mott KE, Cheung C et al. 1987. *Atlas of the Global Distribution of Schistosomiasis.* Bordeaux: Presses Universitaires de Bordeaux.

El-Khoby T, Galal N, Fenwick A et al. 2000. The epidemiology of schistosomiasis in Egypt: summary of findings in nine governorates. *American Journal of Tropical Medicine and Hygiene* **62**, 88–99.

Engels D and Savioli L. 2005. Public health strategies for schistosomiasis control. In: *World Class Parasites:* Vol. 10. *Schistosomiasis* (eds. WE Secor and DG Colley). New York: Springer Science & Business Media, Inc. pp. 207–222.

Engels D. Sinzinkayo E and Gryseels B. 1996. Day-to-day egg count fluctuation in *Schistosoma mansoni* infection and its operational implications. *American Journal of Tropical Medicine and Hygiene* **54**, 319–324.

Engels D, Sinzinkayo E, De Vlas SJ and Gryseels B. 1997. Intraspecimen fecal egg count variation in *Schistosoma mansoni* infection. *American Journal of Tropical Medicine and Hygiene* **57**, 571–577.

Fenwick A, Savioli L, Engels D et al. 2003. Drugs for the control of parasitic diseases: current status and development in schistosomiasis. *Trends in Parasitology* **19**, 509–515.

Frenzel K, Grigull L, Odongo-Agnya E et al. 1999. Evidence for a long term effect of a single dose of praziquantel on *Schistosoma*-induced hepatosplenic lesions in northern Uganda. *American Journal of Tropical Medicine and Hygiene* **60**, 927–931.

Friedman J, Kanzaria HK and McGarvey ST. 2005. Human schistosomiasis and anaemia: the relationship and potential mechanisms. *Trends in Parasitology* **21**, 386–392.

Gryseels B and de Vlas SJ. 1996. Worm burdens in schistosome infections. *Parasitology Today* **12**, 115–119.

Gryseels B, Nkulikyinka L and Engels D. 1994. Impact of repeated community-based selective chemotherapy on morbidity due to *Schistosoma mansoni. American Journal of Tropical Medicine and Hygiene* **51**, 634–641.

Hagan P and Abath FG. 1992. Recent advances in immunity to human schistosomes. *Memorias do Instituto Oswaldo Cruz* **87**, 95–98.

Hatz CFR. 2001. The use of ultrasound in schistosomiasis. *Advances in Parasitology* **48**, 225–284.

He, Y-X, Salafsky B and Ramaswamy K. 2005. Comparison of skin invasion among three major species of *Schistosoma. Trends in Parasitology* **21**, 201–203.

Hotez PJ, Bundy DAP, Beegle K et al. 2006. Helminth infections; soil-transmitted helminth infections and schistosomiasis. In: *Disease Control Priorities in Developing Countries,* 2nd edition (eds. DT Jamison, G Alleyne, J Breman et al.). Oxford University Press. Chapter 24.

IARC (International Agency for Research on Cancer). 1994. *IARC Monographs on the Evaluation of Carcinogenic Risks to Humans.* Schistosomes, Liver flukes and *Helicobacter pylori.* Geneva: World Health Organization.

Ikeda T and Akao N. 2001. Immunodiagnosis of helminthic diseases. In: *Perspectives in helminthology* (eds. N Chowdhury and I Tada). Enfield (NH), U.S.: Science Publishers Inc. pp. 397–418.

Jordan P, Christie JD and Unrau GO. 1980. Schistosomiasis transmission with particular reference to possible ecological and biological methods of control. *Acta Tropica* **37**, 95–138.

Jordan P, Webbe G and Sturrock RF (eds). 1993. *Human Schistosomiasis.* Wallingford Oxon: CABI Publishing.

Jourdane J and Theron A. 1987. Larval development: eggs to cercariae. In: *The Biology of Schistosomes. From Genes to Latrines* (eds. D Rollinson and AJG Simpson). London and New York: Academic Press. pp. 83–113.

Kabatereine NB, Kemijumbi J, Ouma JH et al. 2004. Epidemiology and morbidity of *Schistosoma mansoni* infection in a fishing community along Lake Albert in Uganda. *Transactions of the Royal Society of Tropical Medicine and Hygiene* **98**, 711–718.

Kardorff R, Traore M, Doehring-Schwerdtfeger E et al. 1994. Ultrasonography of ureteric abnormalities induced by *Schistosoma haematobium* infection before and after praziquantel treatment. *British Journal of Urology* **74**, 703–709.

Khamkeo T and Pholsena K. 2003. Control of schistosomiasis due to *Schistosoma mekongi* in Khong District, 1989 to 1999. In: *Controlling Disease due to Helminth Infections* (eds. DWT Crompton, A Montresor, MC Nesheim and L. Savioli). Geneva: World Health Organization. pp. 171–181.

King CH, Dickman K and Tisch DJ. 2005. Reassessment of the cost of chronic helminth infection: a meta-analysis of disability-related outcomes in endemic schistosomiasis. *Lancet* **365**, 1561–1569.

Kitinya JN, Lauren PA, Eshleman LJ et al. 1986. The incidence of squamous and transitional cell carcinomas of the urinary bladder in northern Tanzania in areas of high and low levels of endemic *Schistosoma haematobium* infection. *Transactions of the Royal Society of Tropical Medicine and Hygiene* **80**, 935–939.

Kjetland EF, Ndhlovu PD, Gomo E et al. 2006. Association between genital schistosomiasis and HIV in rural Zimbabwean women. *AIDS* **20**, 593–600.

Kjetland EF, Poggensee G, Helling-Giese G et al. 1996. Female genital schistosomiasis due to *Schistosoma haematobium.* Clinical and parasitological findings in women in rural Malawi. *Acta Tropica* **62**, 239–255.

Laurent C, Lamothe F, Develoux D et al. 1990. Ultrasonographic assessment of urinary tract lesions due to *Schistosoma haematobium* in Niger after four consecutive years of treatment with praziquantel. *Tropical Medicine and Parasitology* **41**, 139–142.

Li YS, Sleigh AC, Ross AGP et al. 2000. Two-year impact of praziquantel treatment for *Schistosoma japonicum* infection in China: reinfection, subclinical disease and fibrosis marker measurements. *Transactions of the Royal Society of Tropical Medicine and Hygiene* **94**, 191–197.

McLaren DJ and Smithers SR. 1987. The immune response to schistosomes in experimental hosts. In: *The Biology of Schistosomes. From Genes to Latrines* (eds. D Rollinson and AJG Simpson). London and New York: Academic Press. pp. 233–263.

McManus DP. 2005. Prospects for development of a transmission blocking vaccine against *Schistosoma japonicum. Parasite Immunology* **27**, 297–308.

Magnussen P. 2003. Treatment and retreatment strategies for schistosomiasis control in different epidemiological settings: a review of 10 years' experience. *Acta Tropica* **86**, 243–254.

Magnussen P, Muchiri E, Mungai P et al. 1997. A school-based approach to the control of urinary schistosomiasis and intestinal helminth infections in children in Matugua, Kenya: impact of a two-year chemotherapy programme on prevalence and intensity of infections. *Tropical Medicine and International Health* **2**, 825–831.

Mahmoud AAF (ed.). 2001. *Schistosomiasis.* London: Imperial College Press.

Mohamed AR, Ali Q, Doehring-Schwerdfteger E et al. 1991. Ultrasonographical investigation of periportal fibrosis in children with *Schistosoma mansoni* infection: reversibility of morbidity seven months after treatment with praziquantel. *American Journal of Tropical Medicine and Hygiene* **44**, 444–451.

Montresor A, Odermatt P, Muth S et al. 2005. The WHO dose pole for the administration of praziquantel is also accurate in non-African populations. *Transactions of the Royal Society of Tropical Medicine and Hygiene* **99**, 78–81.

Mott KE and Dixon H. 1982. Collaborative study on antigens for immunodiagnosis of schistosomiasis. *Bulletin of the World Health Organization* **60**, 729–753.

Ohmae H, Sinuon M, Kirinoki M et al. 2004. Schistosomiasis mekongi: from discovery to control. *Parasitology International* **53**, 135–142.

Olds GR. 2003. Administration of praziquantel to pregnant and lactating women. *Acta Tropica* **86**, 185–195.

Olveda RM, Daniel BL, Ramirez BD et al. 1996. *Schistosomiasis japonica* in the Philippines: the long-term impact of population based chemotherapy on infection, transmission, and morbidity. *Journal of Infectious Diseases* **174**, 163–172.

Pages JR, Jourdane J, Southgate VR and Tchuem Tcheuente LA. 2003. Reconnaissance de deux especes jumelles au sien du taxon *Schistosoma intercalatum* Fisher, 1934, agent de le schistosomose rectale humaine en Afrique. Description de *Schistosoma guineensis* n. sp. In: *Taxonomy, Ecology and Evolution of Metazoan Parasites* Tome II (eds. C Combes and J Jourdane). Bordeaux: Presses Universitaires de Bordeaux.

Pearce EJ and MacDonald AS. 2002. The immunology of schistosomiasis. *Nature Reviews Immunology* **2**, 499–511.

Pittella JE. 1997. Neuroschistosomiasis. *Brain Pathology* **7**, 649–662.

Poggensee G and Feldmeier H. 2001. Female genital schistosomiasis; facts and hypotheses. *Acta Tropica* **79**, 193–210.

Red Urine Study Group. 1995. Identification of high risk communities for schistosomiasis in Africa: a multi-country study. *Social and Economic Research Projects* 15. Geneva: World Health Organization.

Richter J. 2003. The impact of chemotherapy on morbidity due to schistosomiasis. *Acta Tropica* **86**, 161–183.

Rollinson D and Simpson AJG (eds). 1987. *The Biology of Schistosomes. From Genes to Latrines.* London and New York: Academic Press.

Rollinson D and Southgate VR. 1987. The genus *Schistosoma*: a taxonomic appraisal. In: *The Biology of Schistosomes. From Genes to Latrines* (eds. D Rollinson and AJG Simpson). London and New York: Academic Press. pp. 1–49.

Savioli L. Hatz C and Dixon H et al. 1990. Control of morbidity due to *Schistosoma haematobium* on Pemba Island: egg excretion and hematuria as indicators of infection. *American Journal of Tropical Medicine and Hygiene* **43**, 289–295.

Schwartz E, Kozarsky P, Wilson M and Cetron M. 2005. Schistosome infection among river rafters on Omo River, Ethiopia. *Journal of Travel Medicine* **12**, 3–8.

Sheir Z, Nasr AA, Massoud A et al. 2001. A safe, effective, herbal antischistosomal therapy derived from myrrh. *American Journal of Tropical Medicine and Hygiene* **65**, 700–704.

Siegrist D and Siegrist-Obimpeh P. 1992. *Schistosoma haematobium* infection in pregnancy. *Acta Tropica* **50**, 317–321.

Silva IM, Thiengo R, Conceicao MJ et al. 2005. Therapeutic failure of praziquantel in the treatment of *Schistosoma haematobium* infection in Brazilians returning from Africa. *Memorias do Instituto Oswaldo Cruz* **100**, 445–449.

Soliman OE, El-Arman M, Abdul-Samie ER et al. 2004. Evaluation of myrrh (Mirazid) therapy in fascioliasis and intestinal schistosomiasis in children: immunological and parasitological study. *Journal of the Egyptian Society of Parasitology* **34**, 941–966.

Sorgho H, Bahgat M, Poda J-N et al. 2005. Serodiagnosis of *Schistosoma mansoni* infections in an endemic area of Burkino Faso: performance of several immunological tests with different parasite antigens. *Acta Tropica* **93**, 169–180.

Southgate VR and Rollinson D. 1987. Natural history of transmission and schistosome interactions. In: *The Biology of Schistosomes. From Genes to Latrines* (eds. D Rollinson and AJG Simpson). London and New York: Academic Press. pp. 347–378.

Southgate VR, Rollinson D, Tchuem Techuente LA and Hagan P. 2005. Towards control of schistosomiasis in sub-Saharan Africa. *Journal of Helminthology* **79**, 181–185.

Stothard JR, Kabatereine NB, Tukahebwa FM et al. 2005. Field evaluation of the Meade Readiview handheld microscope for diagnosis of intestinal schistosomiasis in Ugandan school children *American Journal of Tropical Medicine and Hygiene* **73**, 949–955.

Stothard JR, Kabatereine NB, Tukahebwa EM et al. 2006. Use of circulating cathodic antigen (CCA) dipsticks for detection of intestinal and urinary schistosomiasis. *Acta Tropica* **97**, 219–228.

Sukwa TY. 1993. A community based randomised trial of praziquantel to control schistoso-miasis morbidity in school children in Zambia. *Annals of Tropical Medicine and Parasitology* **87**,185–194.

Thomas JE, Bassett MT, Sigola LB and Taylor P. 1990. Relationship between bladder cancer incidence, *Schistosoma haematobium* infection, and geographical region in Zimbabwe. *Transactions of the Royal Society of Tropical Medicine and Hygiene* **84**, 551–553.

Useh MF and Ejezie GC. 1999. School-based schistosomiasis control programmes: a com-parative study on the prevalence and intensity of urinary schistosomiasis among Nigerian school-age children in and out of school. *Transactions of the Royal Society of Tropical Medicine and Hygiene* **93**, 387–391.

Utzinger J, Shuhua X, N'Goran EK et al. 2001. The potential of artemether for the control of schistosomiasis. *International Journal for Parasitology* **31**, 1549–1562.

van Dam GJ, Bogitsh BJ, van Zeyl RJ et al. 1996. *Schistosoma mansoni: in vitro* and *in vivo* excretion of CAA and CCA by developing schistosomula and adult worms. *Journal of Parasitology* **82**, 557–564.

van der Werf MJ, de Vlas SJ, Brooker S et al. 2003. Quantification of clinical morbidity associated with schistosome infection in sub-Saharan Africa. *Acta Tropica* **86**, 125–139.

von Lichtenberg F. 1987. Consequences of infections with schistosomes. In: *The Biology of Schistosomes. From Genes to Latrines* (eds. D Rollinson and AJG Simpson). London and New York. pp. 185–232.

Wagatsuma Y, Aryeetey, Sack DA et al. 1999. Resolution and resurgence of *Schistosoma haematobium*-induced pathology after community-based chemotherapy in Ghana as detected by ultrasound. *Journal of Infectious Diseases* **179**, 1515–1522.

Wen LY, Chen JH, Ding JZ et al. 2005. Evaluation of the applied value of the dot immunogold filtration assay (DIGFA) for rapid detection of anti-*Schistosoma japonicum* antibody. *Acta Tropica* **96,** 142–147.

Wiest PM, Wu G, Zhong S et al. 1994. Impact of annual screening and chemotherapy with praziquantel on schistosomiasis japonica on Jishan Island, People's Republic of China. *American Journal of Tropical Medicine and Hygiene* **51**, 162–169.

Wilkins HA. 1987. The epidemiology of schistosome infections in man. In: *The Biology of Schistosomes. From Genes to Latrines* (eds. D Rollinson and AJG Simpson). London and New York: Academic Press. pp. 379–397.

Wilson RA. 1987. Cercariae to liver worms: development and migration in the mammalian host. In: *The Biology of Schistosomes. From Genes to Latrines* (eds. D Rollinson and AJG Simpson). London and New York: Academic Press. pp. 115–146.

WHO. 1980. *Manual of Basic Techniques for a Health Laboratory.* Geneva: World Health Organization.

WHO. 1993. The control of schistosomiasis. *Technical Report Series* 830. Geneva: World Health Organization.

WHO. 1994. *Bench Aids for the Diagnosis of Intestinal Parasites.* Geneva: World Health Organization.

WHO. 2002a. Prevention and control of schistosomiasis and soil-transmitted helminthiasis. *Technical Report Series* 912. Geneva: World Health Organization.

WHO. 2002b. Report of the WHO Informal Consultation on the use of praziquantel during pregnancy/lactation and albendazole/mebendazole in children under 24 months. Geneva: World Health Organization. WHO/CDS/CPE/PVC/2002.4.

WHO. 2004. *WHO Model Formulary 2004.* Geneva: World Health Organization.

Yu JM, de Vlas SJ, Yuan HC and Gryseels B. 1998. Variations in fecal *Schistosoma japonicum* egg counts. *American Journal of Tropical Medicine and Hygiene* **59**, 370–375.

Zhu YC, Socheat D, Bounlu K et al. 2005. Application of dipstick dye immunoassay (DDIA) kit for the diagnosis of schistosomiasis mekongi. *Acta Tropica* **96**, 137–141.

11 Soil-Transmitted Helminthiasis

Four species of nematode (Chitwood and Chitwood, 1974), *Ascaris lumbricoides* (roundworm), *Ancylostoma duodenale* and *Necator americanus* (hookworms), and *Trichuris trichiura* (whipworm), rank among the commonest and most persistent of human pathogens (Table 1.1). Nematode development involves an egg, four juvenile or larval stages (L1, L2, L3, and L4), and four molts before maturity is reached. These four species of soil-transmitted helminth (STH) generate a burden of neglected disease, which was sidelined for many years in the north but not by the impoverished millions in the south (Figure 1.1). At last, the STHs or geohelminths are beginning to get long overdue attention (WHO, 2004a; Hotez et al. 2006). Comprehensive reviews covering the biology of STHs have been compiled by Crompton (2001) for *A. lumbricoides*, by Brooker et al. (2004) for hookworms, and by Bundy and Cooper (1989) for *T. trichiura*. The results of operational research relating to the control of STH infections is to be found in the series of *Collected Papers on Control of Soil-transmitted Helminthiases*. This valuable resource is published by The Asian Parasite Control Organization and the first volume appeared in 1980.

11.1 LIFE HISTORY AND HOST RANGE

Each species has a direct life history pattern and each shows high host specificity for humans. Each matures and undergoes sexual reproduction in the intestinal tract. Information about their biology is given in Table 11.1 and their life histories in Figure 11.1. There is little evidence of significant reservoir hosts for STHs. There are reports of *A. lumbricoides* infections in chimpanzees, gibbons, monkeys, gorillas and orangutans (Habermann and Williams, 1957; Orihel, 1970). The future for STHs is likely to be more secure if tied to *Homo sapiens* rather than endangered primate hosts.

11.2 TRANSMISSION AND ESTABLISHMENT OF INFECTION

Soil-transmitted helminths are so called because their transmission depends on contamination of the environment with fertile eggs (eggshells protecting zygotes) passed in feces from infected humans (Table 11.1; Figure 11.2). In most parts of the south where STHs are endemic, transmission occurs without seasonal disruption. STH eggs are not infective until the zygotes they contain have undergone embryonation to reach the appropriate larval stage. Embryonation is temperature dependent and requires moisture, oxygen, and shade.

TABLE 11.1

Biological Features of *Ascaris lumbricoides*, *Ancylostoma duodenale*, *Necator americanus*, and *Trichuris trichiura*

	Ascaris[1,2]	*Ancylostoma*[3–5]	*Necator*[3–5]	*Trichuris*[3,6,7]
Life history	Direct	Direct	Direct	Direct
Host specificity	High	High	High	High
Life span (years)	1–2	1[a]	3–5[a]	1–2
Reproduction	Sexual diecious	Sexual diecious	Sexual diecious	Sexual diecious
Prepatency (days)	70	50	50	60
Fecundity (egg production/ female/day)	250,000	18,000	7,500	5,000
Embryonation to infectivity (days)[b]	14 at 30°C 50 at 17°C	6 at 25°C	6 at 30°C	15 at 30°C
Infective stage	L2 (?)	L3	L3	L1
Route of infection	Oral	Skin, oral, and transplacental	Skin	Oral
Tissue phase (days)	21	7	7	7[c]
Location of adults	Small intestine (lumen)	Small intestine (mucosa)	Small intestine (mucosa)	Cecum and colon (mucosa)
Adult length and width (mm)				
Male	200 × 3	10 × 0.5	8 × 0.3	30[d]
Female	300 × 6	12 × 0.6	10 × 0.4	45

[1] Crompton (1994).
[2] Crompton (2001).
[3] Little (1985).
[4] Hoagland and Schad (1978).
[5] Pawlowski et al. (1991).
[6] Bundy and Cooper (1989).
[7] Stephenson et al. (2000a).
[a] Ash and Orihel (1990) state that the life spans of *Ancylostoma* and *Necator* are 5 to 10 years.
[b] Eggs, as seen in human stool samples, are illustrated in Appendix 3.
[c] After escaping from the egg shell, the larval *Trichuris* are intimately associated with the mucosa, but this may not represent a true tissue phase.
[d] Tapered (whiplike) bodies.

11.2.1 *Ascaris lumbricoides*

Establishment of infection depends on infective eggs being swallowed or inhaled (Table 11.1) after embryonation has produced the infective larval stage in its tough and sticky eggshell (Appendix 3). For years we have accepted that formation the second larval stage (L2), involving a single molt, completes embryonation and the development of infectivity. Fagerholm et al. (2000), working with *A. suum*, have obtained evidence to suggest that two molts occur during embryonation so that the infective stage of *Ascaris* that emerges from the eggshells is a third larval stage (L3).

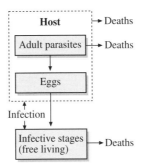

FIGURE 11.1 Life histories of soil-transmitted helminths. From Figure 1 (Crompton, 1994) with permission from Elsevier.

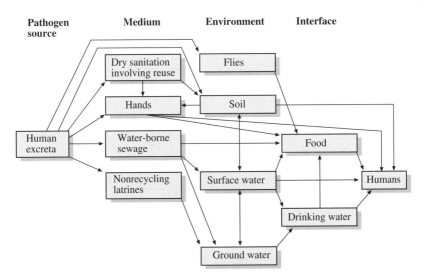

FIGURE 11.2 Transmission pathways of soil-transmitted helminths. Based on Figure 16.1 (Pruss-Ustun et al., 2004) with permission from the World Health Organization (www.who.int).

Infective eggs can be recovered from soil, houses, school rooms, door handles, chopping boards, fingers, insects, money, and vegetables. The practice of using fresh human feces (night soil) as a fertilizer for vegetables promotes the persistence of ascariasis. Crompton (1989a) summarized details of embryonation, contamination of the environment, structure of the eggshells, and the transmission process. Although *Ascaris* eggs may survive for years in the soil, the variable most affecting transmission is the length of the period during which infective eggs and humans may make contact. The fact that a female *Ascaris* produces so many eggs daily (Table 11.1) suggests that many either perish or are dispersed without ever encountering a susceptible human host.

Conditions in the alimentary tract activate larval *Ascaris,* which leave the eggshells, bore through the intestinal wall, enter the hepatic portal system, and reach the liver. After a few days they move to the lungs. Having molted the larvae crawl out of the lungs, up the trachea into the mouth from where they are swallowed to regain access to the intestine in preparation for maturity, mating, and reproduction (Table 11.1). Tissue migration has three important consequences. First, immature *Ascaris* and the human immune system make contact. Secondly, if *A. lumbricoides* in humans undergo the same events as *A. suum* in pigs, many larvae will perish in the tissues (Murrell et al., 1997). Perhaps this high larval death rate serves to reduce competition for limited resources including space in the intestine. Thirdly, acute morbidity is triggered by the migration in some individuals (Section 11.5.4).

11.2.2 ANCYLOSTOMA DUODENALE AND NECATOR AMERICANUS

Embryonation involves release of first-stage larvae (L1) from the eggshells followed by free-living development in the surface of the soil to the infective third-stage larvae (L3). These minute larvae, usually enclosed in the discarded cuticle from the last molt, stop feeding and climb up shoots of vegetation. Hookworm infection is acquired mainly when bare feet or bare hands make contact with soil or vegetation carrying infective larvae that then enter the body by penetrating the skin. If poor people living in endemic areas were able to afford shoes, hookworm infections would decline. After skin penetration, the larvae arrive in the lungs within about 10 days and subsequently reach the intestine by means of the trachea and mouth (Hotez et al., 2004; Table 11.1). *Necator* L3 larvae must penetrate human skin if the helminth's development is to continue (Salafsky et al., 1990). *Ancylostoma* L3 larvae can enter the body and develop when swallowed, particularly with contaminated vegetables (Yu Sen-hai and Shen Wei-xia, 1990). Such larvae are not known to undergo tissue migration (Pawlowski et al., 1991). *Ancylostoma* larvae may pass the placenta to establish an infection in an unborn child leading to neonatal hookworm disease (Yu Sen-hai and Shen Wei-xia, 1990) and may migrate from an infected mother to an infant during breast-feeding. Immature *Ancylostoma* may also undergo arrested development, an adaptation ensuring that egg production is synchronized with good conditions for larval development (Schad et al., 1973). Larval activity in the skin and during migration has consequences for human health (Section 11.5.4).

11.2.3 TRICHURIS TRICHIURA

After egg release, embryonation produces infective first-stage larvae (L1) contained in protective eggshells (Table 11.1). As with infective eggs of *Ascaris*, extensive contamination of the human environment results in infective eggs being swallowed. On exposure to gastrointestinal conditions, the L1 larvae emerge from the eggshells and gradually arrive in the epithelial tissue of the cecum and colon. There is no migratory phase such as that experienced by *Ascaris* larvae (Bundy and Cooper, 1989). Under some circumstances, and for as yet unexplained reasons, children eat soil (pica or geophagia). Cooper and Bundy (1987) recorded that in every case of severe, symptomatic trichuriasis they encountered in a child there was a history of geophagia.

11.3 DIAGNOSIS

Cheesbrough (1998) describes methods for the collection, management, and examination of samples intended for the diagnosis of STH infections. In the public health context with its concern for morbidity control, collecting stool samples and examining them by the Kato Katz technique offers a well tried means of diagnosis (WHO, 1994; Appendix 3). Kato Katz facilitates identification of patent STH infections and enables reasonably accurate estimations to be made of infection intensity expressed as the number of eggs per gram of feces (epg). The assumption is made that the higher the egg count the greater the number of mature female worms in the gut. If the sex ratio of the STHs is roughly 1:1 then a similar number of male worms will be present. The special merit of the Kato Katz technique is the opportunity it offers to be quantitative. Tracking egg counts (intensity) allows heavily infected individuals to be identified, enables the progress of control measures to be monitored, and serves to check on the effectiveness of anthelminthic drugs (Section 11.7).

The Kato Katz technique and all diagnostic methods based on stool examination do not distinguish between the eggs of *Ancylostoma* and *Necator*. Infection with hookworm is the only safe conclusion to be drawn. If preparations of adult worms are not available for examination (Figure 11.3), identification may be attempted by obtaining third-stage larvae with the Harada-Mori *in vitro* culture method (Figure 11.4) described by Pawlowski et al. (1991).

If resources are too limited to allow use of the Kato Katz technique, the direct fecal smear procedure is recommended. This procedure is generally satisfactory for qualitative diagnosis but it contributes little to the management of control programs. Both techniques are described in Appendix 3 together with accurate illustrations of helminth eggs found in human stool samples. Further information about the morphology of STH eggs is to be found in WHO (1980). Examples of how to keep records of the results of stool examinations are set out by Montresor et al. (2002). Thienpont et al. (1979) produced a comprehensive monograph on diagnosing helminthiasis through fecal examination.

11.4 EPIDEMIOLOGY

Ascaris lumbricoides, T. trichiura, Anc. Duodenale, and *N. americanus* share similar geographical distributions across the countries of the South (Sturchler, 1988). Further information at the national level can be obtained by reference to Chan et al. (1994), Crompton (1989b), Janssens (1985), Peng et al. (1998), and Xu Long-Qi et al. (1999). The application of Geographical Information Systems offers a more detailed insight into the distribution of STHs, both actual and predicted (Brooker et al., 2000). Most cases of hookworm infection are due to *N. americanus,* which is much more common than *Anc. duodenale* in tropical and subtropical countries. Mixed hookworm infections occur and the distribution of *Anc. duodenale* extends into temperate regions (Pawlowski et al., 1991). The transmission process for hookworms tends to confine them and hookworm disease to rural communities. *Ascaris* and *Trichuris* thrive equally well in the crowded, unplanned slums that arise as the consequences

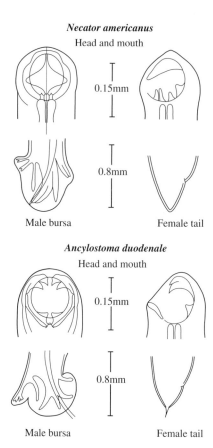

FIGURE 11.3 Morphological features of the adult stages of *Necator americanus* and *Ancylostoma duodenale*. From Figures 177 and 178 (Zaman and Keong, 1989) with permission from Elsevier.

of urbanization overtake the cities of the South (Crompton and Savioli, 1993; Beall, 2000).

11.4.1 PREVALENCE

Estimates of prevalence provide information about the distribution of helminth infections within territories, the numbers and types of people infected, and the course of infection. Consider the case of *A. lumbricoides* infection in Nigeria (Crompton, 1989b). Regardless of misgivings about pooling data obtained by different observers at different times and by different procedures, it was possible to claim that *A. lumbricoides* infected 18.6% of the Nigerian population. However, this prevalence rate does not indicate an even distribution of infection across the nation. Prevalence rates were recorded as varying from 0.9 to 98%. Knowledge of prevalence indicates the patchy distribution of STHs and points to where resources for control measures should be provided. Knowledge of prevalence also contributes to estimating the cost

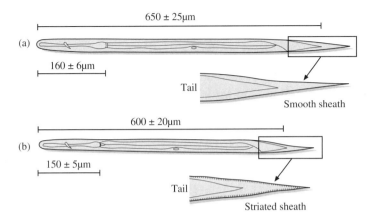

FIGURE 11.4 Diagnostic features of third-stage (infective) larvae of (a) *Ancylostoma duodenale* and (b) *Nectator americanus* obtained by the Harada-Mori coproculture technique. From Figure A1-5 (Pawlowski et al., 1991) with permission from the World Health Organization (www.who.int).

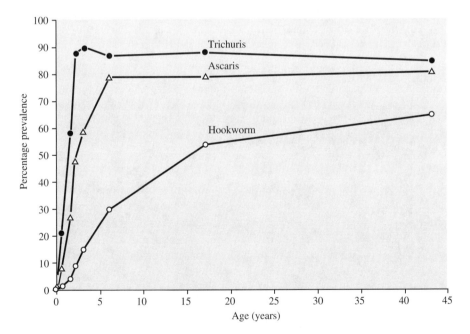

FIGURE 11.5 Relationship between host age and the prevalences of *Ascaris lumbricoides*, hookworm and *Trichuris trichiura* infections. From Figure 10.1 (Bundy, 1990) with permission from Routledge/Taylor & Francis Group, LLC.

resources for the numbers of people in need of attention. Plots of changes in the prevalences of STHs against host age are shown in Figure 11.5 to Figure 11.10. Each infection is acquired during childhood and each persists through the host's life.

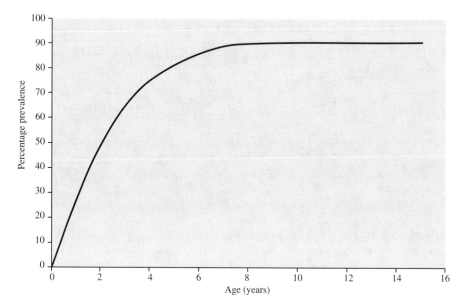

FIGURE 11.6 Relationship between age of children in rural Bangladesh and the prevalence of *Ascaris lumbricoides* infection. From Figure 1 (Crompton, 1987) with permission from Elsevier and The Royal Society of Tropical Medicine and Hygiene.

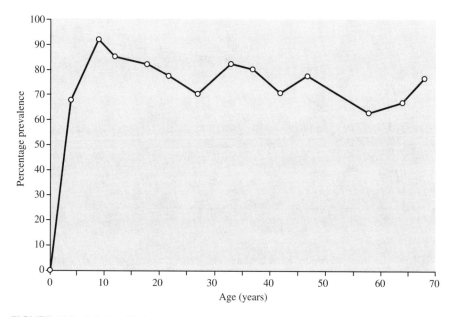

FIGURE 11.7 Relationship between host age and prevalence of *Ascaris lumbricoides* infection in rural Myanmar. From Figure 18 (Crompton, 1987) with permission from Elsevier and Routledge/Taylor & Francis Group, LLC.

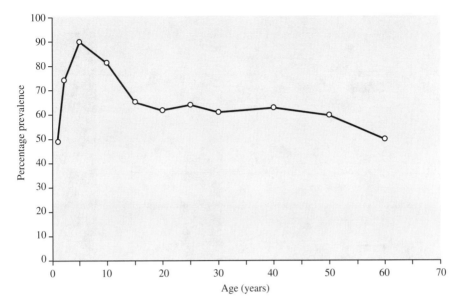

FIGURE 11.8 Relationship between host age and the prevalence of *Ascaris lumbricoides* infection in rural Jiangxi Province, P. R. China. From Figure 2 (Peng Weidong et al., 1996) with permission from Cambridge University Press.

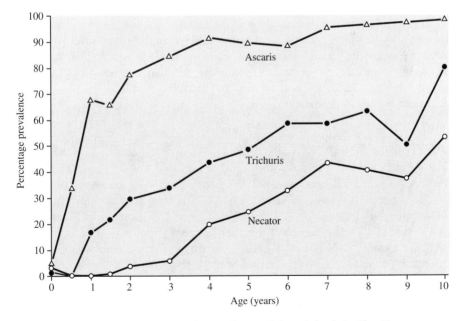

FIGURE 11.9 Changes with host age in prevalences of *Ascaris lumbricoides*, *Necator americanus*, and *Trichuris trichura* in children from rural Madagascar. From data published by Kightlinger et al. (1995).

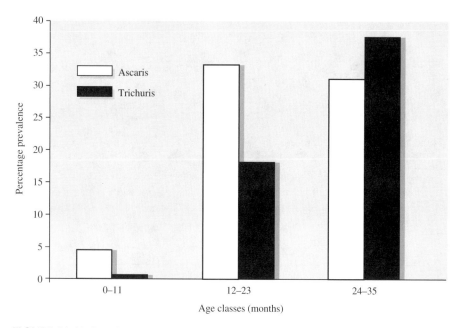

FIGURE 11.10 Prevalences of *Ascaris lumbricoides* and *Trichuris trichura* infections in preschool children from St Lucia during 1977 to 1979. From data in Henry (1988).

11.4.2 INTENSITY

Plots of changes in the intensities of STHs against host age are shown in Figure 11.11 to Figure 11.15. Intensity of infection is responsible for the severity of morbidity experienced by an infected person. Intensity should be used to identify groups in need of morbidity control. It is clear from the plots that, on average, children will suffer more from infection with *Ascaris* and *Trichuris* than adults. It is also clear that intensity of *Ancylostoma* and *Necator* infections does not decline during adulthood (Figure 11.11 and Figure 11.15). When intensity plots (Figure 11.12 and Figure 11.14) are compared with prevalence plots (Figure 11.5 and Figure 11.8) some explanation is needed for the decline in intensity with host age in the case of *Ascaris* and *Trichuris*. Perhaps some degree of immunity has developed in adults or perhaps their behavior has changed or their personal hygiene has become better than that of children. Although the intensity of STHs is invariably measured indirectly by counting eggs, confidence in its reliability comes from comparing Figure 11.12 and Figure 11.14 and noting the correlation between worm burden and egg counts shown in Figure 11.16. The same intensity with age pattern is seen whether eggs or worms are counted, the worms having been retrieved from the stools of children given an anthelminthic drug (Thein Hlaing et al., 1984; Forrester and Scott, 1990).

Concurrent STH infections are common. Interestingly, individuals with heavy *Ascaris* infections often harbor heavy *Trichuris* infections as has been demonstrated in communities in India (Haswell-Elkins et al., 1987), Madagascar (Kightlinger et al., 1995), Nigeria (Holland et al., 1989), Panama (Robertson et al., 1989; Figure 11.17), and Sierra Leone (Webster et al., 1990). Booth et al. (1998), in a study of 1,539

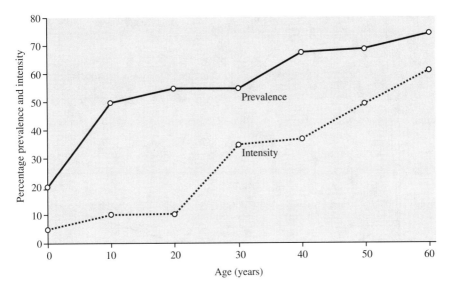

FIGURE 11.11 Hookworm prevalence against age in Zimbabwe. The intensity plot (no scale) shows the relationship with prevalence. From Figure 2 (Crompton, 2000) with permission from Cambridge University Press and Routledge/Taylor & Francis Group, LLC.

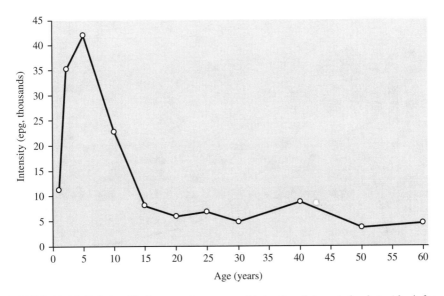

FIGURE 11.12 Relationship between host age and intensity of *Ascaris lumbricoides* infection in rural Jiangxi Province, P R China. From Figure 2 (Peng Weidong et al., 1996) with permission from Cambridge University Press.

school children living on Pemba Island, noted that 58% of the group harbored *Ascaris,* hookworm, and *Trichuris*. These children were obviously at greater risk of morbidity than the others not just because of having the three infections but because

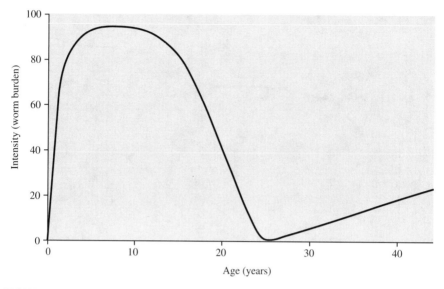

FIGURE 11.13 Relationship between host age and intensity of *Trichuris trichiura* infection in St Lucia. From Figure 1 (Stephenson et al., 2000a) with permission from Cambridge University Press.

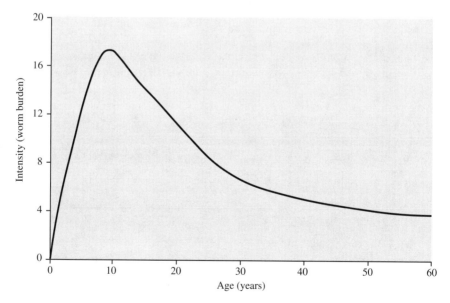

FIGURE 11.14 Relationship between host age and intensity of *Ascaris lumbricoides* infection in rural Myanmar. From Figure 8 (Crompton, 1987) with permission from Elsevier and Routledge/Taylor & Francis Group, LLC.

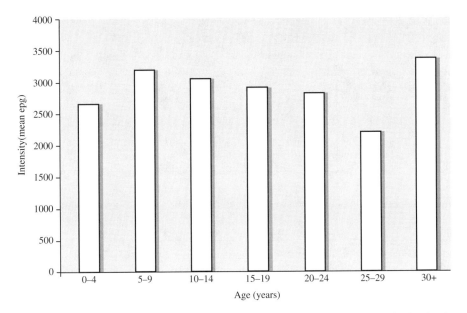

FIGURE 11.15 Relationship between host age and the intensity of hookworm infection (epg) in Thailand. From data published by Preuksaraj et al. (1983).

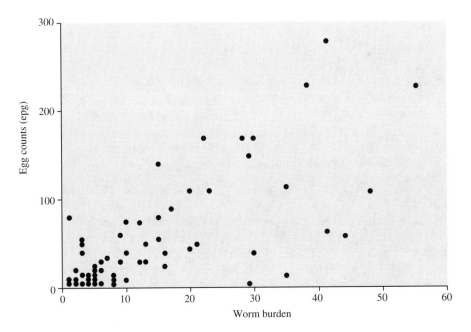

FIGURE 11.16 Relationship between intensity of *Ascaris lumbricoides* infection, measured as epg, and worm burden. From Figure 6 (Crompton, 1994) with permission from Elsevier and Cambridge University Press.

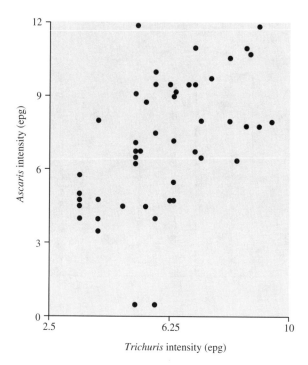

FIGURE 11.17 Intensities (epg) of *Ascaris* and *Trichuris* infections occurring concurrently in individual children. From Figure 10 (Crompton, 1994) with permission from Elsevier and Cambridge University Press.

they generally had higher intensities of infection of each species. *Ascaris* and *Trichuris* infections are also observed to occur in clusters; the bigger the household the more likely that more people will be infected (Williams et al., 1974; Forrester et al., 1988).

11.4.3 Distribution of Numbers of Soil-Transmitted Helminth per Host

Numbers of a species of STH are not usually distributed evenly or at random among individuals in a community. Few hosts harbor most of the worms (Figure 11.18). The worms are found to be aggregated or overdispersed (Anderson and Gordon, 1982) and the negative binomial distribution gives the best fit for the observed distribution (Anderson and May, 1992). In a population of hosts, the worm burdens of the few heavily infected individuals put their hosts at risk of death, discharge most transmission stages into the environment, and play a role in regulating the worm population in the host population.

In practice, health authorities in the south may have insufficient resources to measure intensity. Intuition suggests that where the prevalence of an STH infection is high there will be a considerable number of heavily infected individuals. Guyatt et al. (1990) examined the influence of the frequency distribution of worms per host on the relationship between prevalence and intensity of *A. lumbricoides* infections

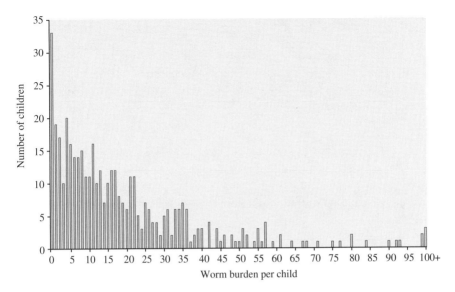

FIGURE 11.18 Numbers of *Ascaris lumbricoides* per child (worm burden) in rural Madagascar. From Figure 2 (Kightlinger et al., 1995) in the *Journal of Parasitology* 81, 159–169, with permission from the American Society of Parasitologists.

with data collected from 13 studies. The mathematics underlying this exercise are complex, but the resulting plot (Figure 11.19) suggests that inability to measure intensity should not deter health authorities from actions to control morbidity.

11.4.4 PREDISPOSITION TO INFECTION WITH SOIL-TRANSMITTED HELMINTHS

A good case can be made for identifying individuals with high worm burdens and keeping them free from STH infection by means of regular anthelminthic chemotherapy. The case is strengthened by the discovery that individuals appear to be predisposed to acquire and maintain worm burdens of particular intensities. When a host population is treated at regular intervals and the worms passed on each occasion are counted, statistical investigation reveals evidence of predisposition of some individuals to a given worm burden. Predisposition holds for *Ascaris* infections in China, India, Iran, Malaysia, Mexico, Myanmar, Nigeria, and Thailand (Croll et al., 1982; Haswell-Elkins et al., 1987; see Crompton, 2001); for *Trichuris* infections in India and the West Indies (Haswell-Elkins et al., 1987; Bundy and Cooper, 1989); and for hookworm infections in India (Schad and Anderson, 1985; Haswell-Elkins et al., 1987). Interestingly, Holland et al. (1989) noted that some individuals appeared to be predisposed to avoid *Ascaris* because they showed no evidence of infection even though they were exposed to as many infective eggs as those who were repeatedly found to be infected. Genetic susceptibility to infection (or not) might seem to be the obvious explanation for predisposition, but recently Moraes and Cairncross (2004) have suggested that variation in exposure to infection due to environmental factors might contribute to the process.

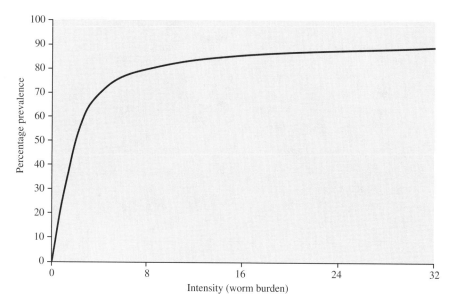

FIGURE 11.19 Prevalence, intensity, and frequency distribution (k = 0.543) of 13 *Ascaris lumbricoides* infections. From Figure 7 (Crompton, 1994) with permission from Elsevier and Cambridge University Press.

11.5 RISK GROUPS, MORBIDITY, AND PATHOLOGY

The groups in society at risk of or experiencing soil-transmitted helminthiasis are those in whom the intensity of infection is or is likely to be greatest. School-aged children comprise the largest group that is most vulnerable to ascariasis and trichuriasis in addition to suffering from hookworm disease (Figure 11.12 and Figure 11.13; Bundy and Guyatt, 1996; Drake et al., 2002). Adolescent girls and women of reproductive age comprise the group most vulnerable to hookworm disease. The intensity of hookworm infection does not decline during adulthood and, since hookworms cause blood loss, girls and women are more at risk of iron loss and anemia than others (WHO, 1996a; Crompton, 2000). Unraveling the contributions to ill health made by STH infections is difficult because of concurrent parasitic infections and other communicable diseases, poor nutrition, pressure on childcare, cultural constraints, poverty, and inequality (Figure 2.1). There is no doubt, however, that the quality of life for the poorest of the poor improves when morbidity due to STH infections is reduced (Table 1.2 and Table 11.2). Since the severity of morbidity is related to the intensity of infection, guidelines have been issued to help health professionals decide as to whether infections should be classified as light, moderate, or heavy (Table 11.3). Concern for identifying risk groups should not divert attention from other groups or from individuals in need of personal attention under the care of a physician (Seltzer et al., 2006).

TABLE 11.2
Effects of Soil-Transmitted Helminthiasis on Human Nutrition, Growth, and Development. From Table 2 (Crompton and Nesheim, 2002). Reprinted with permission from the *Annual Review of Nutrition*, volume 22; copyright 2002 by Annual Reviews (www.annualreviews.org).

Nutrient	Consequences of Parasitism	Effects on Host	Vulnerable Times
Energy	Reduced appetite, lowered energy intake	Inadequate energy intake to support growth, work, fetal development, cognitive development	Early growth period 1–5 years, school ages, reproductive years
Protein	Slight reduction in digestion and absorption; endogenous losses may increase; reduced energy intake results in dietary protein metabolized to supply energy	Poor child growth, low infant birth weight	Early growth years, reproductive ages
Fat	Reduced absorption of dietary fat resulting in lower energy intake and inefficient absorption of vitamin A precursors	Decreased vitamin A availability contributes to deficiency disease	Lactating mothers and young children
Lactose	Intestinal lactase activity reduced	Poor lactose digestion and lactose intolerance, reduced consumption of milk	Young children
Iron	Iron loss from blood in intestine; reduced intake from reduced appetite	Anemia, retarded cognitive development, impaired work output, poor pregnancy outcome	Children and adults
Vitamin A	Reduced absorption and utilization of vitamin A precursors	Contributes to vitamin A deficiency	Children and adults
Other micronutrients	Reduced food intake may result in inadequate intake of micronutrients, especially zinc, folate, and B12	Micronutrient deficiency	Children, pregnant and lactating women

11.5.1 Ascariasis

Evidence for the conclusion that ascariasis is detrimental to the health, growth, and development of children (Table 11.2) has been published in peer-group-reviewed, primary research journals and can be traced by reference to Stephenson (1987), Thein Hlaing (1993), O'Lorcain and Holland (2000), Crompton and Nesheim (2002),

TABLE 11.3

Classes of Intensity for Soil-Transmitted Helminth Infections. From Table 7 (WHO, 2002a) with permission from the World Health Organization (www.who.int). See also Table 4.2.

Helminth	Light-Intensity Infections	Moderate Intensity Infections	Heavy-Intensity Infections
A. lumbricoides	1–4,999 epg[a]	5,000–49,999 epg	≥ 50,000 epg
T. trichiura	1–999 epg	1,000–9,999 epg	≥ 10,000 epg
Hookworms	1–1,999 epg	2,000–3,999 epg	≥ 4,000 epg

[a] epg, eggs per gram of feces.

and Drake et al. (2002). The evidence has been obtained from many investigations in many countries where the disease is endemic. The major burden of disease occurs during the intestinal phase of the infection, which probably affects some 320 million children aged from 5 to 14 years harboring *Ascaris* infections (Drake et al., 2002). Ascariasis also affects some 122 million younger children with equally serious consequences for growth and development. *Ascaris* grows to the same size whether the host is 2 or 12 years old. Children acquire *Ascaris* infections during their first year of life and *Trichuris* follows soon after (Montresor et al., 2003). The conclusion from a recent risk-benefit analysis is that children as young as 12 months can be treated with the appropriate dose of a WHO-recommended anthelminthic drug to expel STHs (see Chapter 14).

The annual mortality rate from ascariasis has been estimated to range from 4,000 to 60,000 (Table 1.2) or 100,000 (Pawlowski and Davis, 1989; Drake et al., 2002). Most deaths result from acute complications that arise during the course of the infection rather than from nutritional interference. Death from intestinal obstruction caused by a knotted bolus of worms is observed most frequently followed by death from biliary duct obstruction (de Silva et al., 1997; Table 11.4). Chai et al. (1991) noted that in Korea middle-aged men and women were susceptible to biliary ascariasis. Reduction in the number of cases presenting with biliary ascariasis since the introduction of control measures has proved to be a valuable indicator of sustained progress in reducing morbidity in Korea (Figure 11.20). Adult *Ascaris* have a tendency to migrate from the intestinal lumen; they have been retrieved from the brain, heart, kidney, placenta, and many other sites in the body. In the North, where Western medicine prevails, surgical intervention may be the preferred means of dealing with complications. Although that option may not always be available in the South there are other approaches to the problem. Zhou et al. (1999) reported that out of 4,167 cases of biliary ascariasis in China full cure was achieved in 4,060 cases by administration of doses of a traditional herbal medicine. Recently, however, a nonsurgical regimen was used successfully in Ecuador where 68 out of 69 patients with biliary ascariasis were relieved through treatment with 800 mg of albendazole given orally (Gonzalez et al., 2001).

TABLE 11.4
Complications of Acute Ascariasis Caused by Adult *Ascaris lumbricoides*
in Relation to Host Age and Gender. From Table 4 (Crompton, 1994) with
permission from Elsevier.

	All Cases	Age		Gender		Deaths	
		Child	Adult	Male	Female	Child	Adult
Biliary system	1124	380	374	92	117	4	23
Gastrointestinal tract	3408	2149	73	572	716	85	9
Hepatic abscess	100	52	6	17	15	13	1
Pancreatitis	67	28	26	11	15	1	2
Miscellaneous complications	94	77	15	50	41	9	3
Totals	4793	2686	494	742	904	112	38

Note: The cases recorded in this table were abstracted from 230 reports published between 1971 and
1992. The cases described were from 54 countries. Often reports were found to be incomplete, lacking
information about the age or gender of the patients.

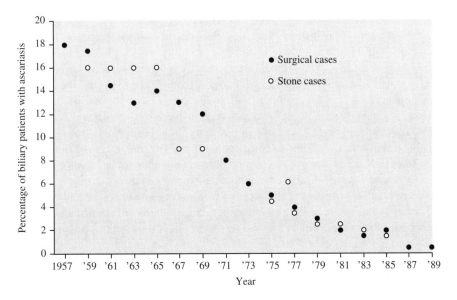

FIGURE 11.20 Benefit of ascariasis control in the Republic of Korea, as shown by the decline
in numbers of ascariasis patients among biliary cases generally. From Figure 2 (Chai et al.
1991) with permission from the Korean Society for Parasitology.

11.5.2 HOOKWORM DISEASE (ANCYLOSTOMIASIS AND NECATORIASIS)

Hookworm infection is inextricably linked with iron deficiency (ID) and iron deficiency anemia (IDA) (Roche and Layrisse, 1966; Crompton and Stephenson, 1990;
Pawlowski et al., 1991; Table 1.2). The detrimental effects of hookworm infections

on rural poor people in the southern states of the U.S. led John D. Rockefeller in 1909 to endow his Sanitary Commission with USD 1 million for dealing with hookworm disease (Ettling, 1990). That decision and that financial commitment from such an astute businessman so long ago is testimony to the public health and economic importance of hookworm disease.

Iron deficiency arises when the complex array of factors that influence iron availability and iron metabolism are disturbed (Table 11.5) and IDA invariably follows. ID and IDA constitute a global nutritional disorder affecting at least 2 billion people (Table 11.6; Figure 11.21). Stoltzfus et al. (2004) concluded that IDA causes about 840,000 deaths annually and is responsible for morbidity amounting to about 35 million DALYs. Ninety percent of the anemic people live in the south (WHO, 2003; SCN, 2004) where hookworm infections are endemic. The contribution of hookworm infections to the DALY estimate for IDA has still to be agreed on. Adults and children, born and unborn, are affected by the disease, resulting from blood loss as these insidious helminths bite into the delicate blood vessels in the intestinal mucosa (Table 11.2; Figure 11.3 and Figure 11.22). Hookworms cause chronic blood loss because they secrete an anticoagulant that allows the feeding site to continue to bleed into the gut lumen after feeding has temporarily stopped (Table 11.7). Occult blood in the stools is an indicator of a likely hookworm infection; the stools are seen to be black and have a glutinous constituency.

In hookworm investigations, IDA is usually detected when the concentration of hemoglobin in the blood falls below certain arbitrary values (Table 11.8). Severe anemia is defined as < 70 g/l, moderate anemia from 70 to 99 g/l, and mild anemia from 100 to 109 g/l (SCN, 2004). These average values apply to adults and do not take into account the age, gender, and physiological state of the individual or confounding factors such as altitude, malaria, urinary schistosomiasis (Chapter 10), trichuriasis (Section 11.5.3), and iron availability (Table 11.5). Nor do measurements of hemoglobin concentrations give any information about the iron status of the body. Iron stores will have been depleted by the time blood hemoglobin begins to decline. Published results have established that adults who are both infected with hookworm and anemic are less productive, have pregnancy problems with serious outcomes for both fetus and infant, and that children experience cognitive difficulties and unsatisfactory educational performance including absenteeism from school (Figure 11.22). Furthermore, iron deficiency complicates the course and outcome of all forms of ill health.

11.5.2.1 Hookworm Disease and Worker Productivity

Iron is a component of the system for transporting oxygen to the body's tissues and also a component of the enzymes that release energy in the process of cellular respiration. Accordingly there can be no doubt that ID and IDA will cause decreased physical fitness and a reduced capacity to carry out sustained periods of work (Haas and Brownlie, 2001; Stoltzfus et al., 2004). During the 1990s the estimated productivity losses in South Asia attributed to iron deficiency cost close to USD 1 billion (see Stephenson et al., 2000b). In many countries of the south the reduced capacity for physical work is as important to the household and to childcare (Chapter 2) as

TABLE 11.5
Factors Influencing the Iron Status of a Hypothetical Adult Human (70 kg Body Weight). Reproduced from Table 15.4 (Crompton and Stephenson, 1990) with permission from Routledge/Taylor & Francis Group, LLC.

Dietary Intake

Depends on energy intake: menstruating women 28 mg or 14 mg if < 10% or > 25% energy derived from animal food

Heme iron (animal food)	11.4 mg/100 g raw liver
Nonheme iron (vegetable food)	0.5 mg/100 g rice
Sociocultural factors, e.g., food fads, religious beliefs, traditions	

Absorption

About 5–10% of dietary intake

Luminal Factors

Enhancers	*Inhibitors*
Animal protein (cysteine)	Bran (?Phytates)
Heme iron	Tea (tannates)
Ascorbic acid	Egg proteins
Lactic acid	Calcium
Human milk	Phosphates
Gastric acid	Pica (clay)

Mucosal Factors

Iron absorption mainly in duodenum and upper jejunum.
Impaired in tropical sprue
Celiac's disease
Protein-losing enteropathy

Bioavailability

Absorption increased as body iron decreases. Up to 42% reabsorption of blood iron from gut during hookworm infection.

Body Iron Distribution

Heme iron (mg)		Ferritin and hemosiderin	
Hemoglobin (red blood cells)	2600	Liver	410
Myoglobin (muscle)	400	Spleen	48
Cytochromes (mitochondria)	17	Kidney	11
Cytochromes (microsomes)	3	Muscle	730
Catalase (RBCs and liver)	5	Bone marrow	300
Transferrin	8	Brain	60
Nonheme iron (stored)		Heme iron	3033
		Nonheme iron	1559
		Total iron	4592

Physiological Losses

0.7 mg daily during menstruation	0.5–1.0 mg daily during lactation
300 mg during birth	0.6 mg daily in men and women (excluding menstruation)

TABLE 11.5 (continued)
Factors Influencing the Iron Status of a Hypothetical Adult Human (70 kg Body Weight). Reproduced from Table 15.4 (Crompton and Stephenson, 1990) with permission from Routledge/Taylor & Francis Group, LLC.

Physiological Requirements
2×10^{11} red blood cells enter circulation daily 300 mg transferred to fetus, mainly during last trimester
Up to 500 mg for increase in RBC mass during pregnancy

Pathological Losses

Infections (e.g., malaria)	Hemorrhoids
Endogenous gut hemorrhages	Hookworm infections (Table 11.7)

TABLE 11.6
Prevalence of Anemia (%) By Population Groups in Regions of the Nonindustrialized South (Figure 1.1). From data published by Stephenson et al. (2000b).

	Children (0 to 4 years)	Children (5 to 14 years)	Pregnant Women	All Women	All Men
Africa	33.1	52.0	46.9	37.9	28.0
Americas	22.9	36.9	39.0	31.0	11.0
Eastern Mediterranean	38.3	30.8	63.9	51.1	32.7
South and SE Asia	52.7	63.9	76.9	60.0	42.4
Western Pacific	14.7	56.9	38.5	33.8	36.0

Note: There are no grounds for expecting the anemia situation to have changed since these data were compiled (see Stoltzfus et al., 2004).

it is to labor productivity in formal employment. Attempts to measure the effects of hookworm-induced ID and IDA on worker productivity have been made on various occasions (see Crompton and Stephenson, 1990). There are difficulties in deciding how much of the impaired productivity is due to malnutrition or other diseases in addition to hookworm infection. Anthelminthic treatment to expel hookworms, however, is followed by better productivity, especially if iron supplements are also provided.

11.5.2.2 Hookworm Disease, Pregnancy, and Lactation

St George (1976), working in Trinidad and Tobago, recorded that women infected with hookworm endured an average duration of labor that was nearly 5 h longer than that of uninfected women. Also, the average birth weight of babies born to infected women was from 227 g to 284 g less than that of babies born to uninfected women. The incidence of premature birth was higher in women with hookworm

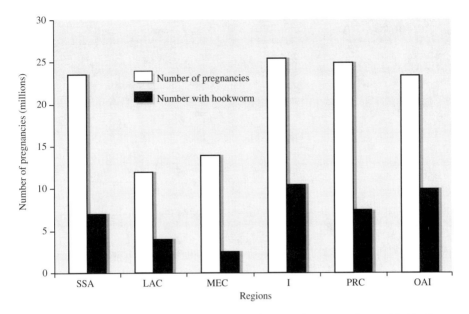

FIGURE 11.21 Numbers of hookworm infections that occur in pregnancies worldwide. From data published in WHO (1996a).

infections. Hookworm infections are often found to occur concurrently with *Ascaris,* *Trichuris* and intestinal protozoan infections. Villar et al. (1989) showed that the risk of intrauterine growth retardation leading to babies of low birth weight increased when women harbored two or more intestinal infections during pregnancy. Of STH infections, however, hookworm disease merits most concern because it leads to ID and IDA. The impact of hookworms on the health of pregnant women, intrauterine growth, preterm delivery, low birth weight, and perinatal mortality should not be ignored (Allen, 1997) and measures to alleviate the problem should be made available (WHO, 1996a; Torlesse et al., 2003).

Studies on hookworm infections during pregnancy have been impeded because of concern over the risk of congenital defects were pregnant women to be given an anthelminthic drug. However, Atukorala et al. (1994) treated a large number of pregnant women from the tea plantations in Sri Lanka with mebendazole. This study showed that mebendazole treatment plus iron-folate supplementation was accompanied by better gains in hemoglobin concentration and iron status than the effect of supplementation alone. There were clearly benefits to be gained from the expulsion of hookworms during pregnancy, but information about birth defects was lacking. Reassurance about using routine anthelminthic treatment during pregnancy in places where hookworm infection is endemic was obtained by de Silva et al. (1999) who compared aspects of pregnancy outcomes between women treated with mebendazole and women who were not treated. The rate of pregnancy defects was not significantly different between the two groups, the proportions of still births and perinatal deaths were significantly lower in the treated group, and the proportion of low-birth-weight babies was lower in the treated than in the untreated group. Diav-Citrin et al. (2003)

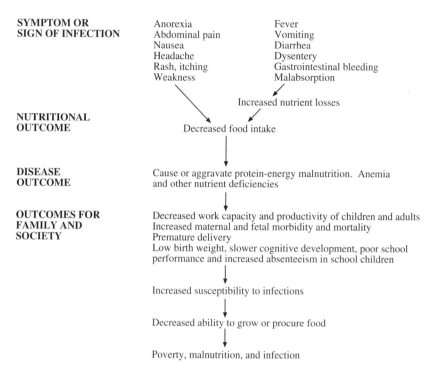

SYMPTOM OR SIGN OF INFECTION

Anorexia   Fever
Abdominal pain   Vomiting
Nausea   Diarrhea
Headache   Dysentery
Rash, itching   Gastrointestinal bleeding
Weakness   Malabsorption

Increased nutrient losses

NUTRITIONAL OUTCOME

Decreased food intake

DISEASE OUTCOME

Cause or aggravate protein-energy malnutrition. Anemia and other nutrient deficiencies

OUTCOMES FOR FAMILY AND SOCIETY

Decreased work capacity and productivity of children and adults
Increased maternal and fetal morbidity and mortality
Premature delivery
Low birth weight, slower cognitive development, poor school performance and increased absenteeism in school children

Increased susceptibility to infections

Decreased ability to grow or procure food

Poverty, malnutrition, and infection

FIGURE 11.22 Conceptual framework of how hookworm disease impairs health and development. From Figure 15.1 (Crompton and Stephenson, 1990) with permission from Routledge/Taylor & Francis Group, LLC.

have also added support to the view that anthelminthic treatment during pregnancy is justified. They investigated 192 pregnancies in which mebendazole had been given to relieve infections with pinworm (*Enterobius vermicularis*). Importantly, although over 70% of the expectant mothers had been given mebendazole during the first trimester, there was no difference in the birth defect rate between the exposed and the unexposed group. Diav-Citrin et al. concluded that mebendazole does not represent a major teratogenic risk in humans when used in doses commonly prescribed for pinworms. The recommended anthelminthic treatment using mebendazole for enterobiasis is a single oral dose of 100 mg repeated after an interval of 2 to 3 weeks (WHO, 2004b).

The case for offering anthelminthic treatment during pregnancy to control hookworm infections has been strengthened by investigations in Sierra Leone and Nepal. Torlesse and Hodges (2000, 2001) explored the effects of albendazole and iron-folate supplementation, alone and in combination, on the course of pregnancies in Sierra Leone. A single dose of albendazole, given after the first trimester, markedly reduced the intensity of *Necator* and *Ascaris* infections to an extent that remained for the duration of the rest of the pregnancy. The mean decline in hemoglobin concentration between the first trimester and end of the third trimester in women given albendazole was 6.6 g/l less than in the untreated women. The corresponding decline in women given both albendazole and the supplements was 13.7 g/l less.

TABLE 11.7
Estimated Blood and Iron Losses during Hookworm Infections. From
Table 3 (Pawlowski et al., 1991) with permission from the World Health
Organization (www.who.int).

	A. duodenale		N. americanus	
	Mean	Range	Mean	Range
Daily blood loss, ml/worm	0.20	0.14–0.26	0.04	0.02–0.07
No of worms causing blood loss of 1 ml/day	5	4–7	25	14–50
Daily egg production per female worm at 1,000 eggs/g of feces[a]		10,000–20,000		5,000–10,000
Estimated hookworm load[b]	11 (5F, 6M)		32 (13F, 19M)	
Blood loss ml/day	2.2	1.54–2.86	1.3	0.82–2.24
Iron loss mg/day[c]	0.76		0.45	

Note: F, female: M, male.

[a] Based on stool weight of 135 g.
[b] At 25,000 eggs/day for *A. duodenale* and 10,000 eggs/day for *N. americanus*.
[c] Based on a hemoglobin level of 100 g/liter of blood.

TABLE 11.8
Hemoglobin and Hematocrit (PCV) Concentration
below Which Anemia Is Present. From Table 13.1
(WHO, 1996a) with permission from the World
Health Organization (www.who.int).

Age/Sex Group	Hemoglobin[a]		Hematocrit
	g/l	mmol/l	
Children under 5 years	110	6.83	0.33
Children 6–14 years	120	7.45	0.36
Nonpregnant women	120	7.45	0.36
Pregnant women	110	6.83	0.33
Men	130	8.07	0.40

[a] Severe anemia in pregnancy is Hb < 70 g/l. Very severe anemia
with risk of congestive heart failure is Hb < 40 g/l.

This work demonstrated that anthelminthic treatment after the first trimester led to
improvements in the iron status of pregnant women and that the cessation of blood
loss from hookworm feeding activities improved the effectiveness of dietary iron-
folate supplementation.

TABLE 11.9

Effect of Anthelminthic Treatment during Pregnancy on Birthweight and 6-Month Infant Mortality. From the table in Christian et al. (2004) with permission from Elsevier.

Albendazole Dosage	Birthweight (g)		Difference in g[a] Mean (95% CL)	6-Month Mortality Number of Deaths/Total, Rate per 1000	Relative Risk[a] (95% CL)
	n	mean (SD)			
No doses	58	2473 (520)	—	25/261, 95.8	1.0
One dose	543	2519 (490)	31 (−94 to 157)	88/866, 101.6	0.86
					(0.49–1.54)
Two doses	2726	2639 (420)	59 (19 to 98)	116/2981, 38.9	0.59
					(0.43–0.82)

Note: CL, confidence limits

[a] Multiple linear regression analysis of differences in birthweight and logistic regression analysis for the relative odds of mortality, were adjusted for nutrient supplement group, maternal parity, tobacco smoking, early pregnancy, weight, height, ethnic group, literacy, gestational duration of pregnancy, and social status.

In Nepal, Christian et al. (2004) enrolled 4,998 women in a study designed to examine prospectively the association between anthelminthic treatment (albendazole) and maternal anemia, birthweight, and infant mortality. Two doses of albendazole were given; the first, 12 weeks after urine-based detection of pregnancy and the second in the third trimester. The prevalence of *Anc. duodenale* in the study area was 74%, that of *A. lumbricoides* 59%, and *T. trichiura* 5.3%. Infants born to women given two doses of albendazole had an average rise in birthweight of 59 g and infant mortality at 6 months was found to have fallen by 41% compared with outcomes for infants from untreated women (Table 11.9). A marked reduction in the severity of ID and IDA was detected in the third trimester in women who had been given the first dose of albendazole. These important results must not be ignored; antenatal anthelminthic treatment to reduce the intensity of STH infections, particularly where the prevalence of hookworm is high, deserves high priority.

11.5.2.3 Hookworm Disease, Cognitive Development, and Educational Achievement

There is extensive evidence to show that ID and IDA are associated with poorer performances in the cognitive development and educational prowess of children (Nokes et al., 1998). Iron deficiency adversely affects psychomotor and mental development scales and behavioral ratings in infants. Lower scores are obtained on cognitive function tests in iron-deficient preschool children, and lower scores are found on cognitive function and educational achievement tests in iron-deficient school-aged children. The mental development and learning skills of children experiencing ID decline as the degree of ID increases. Stoltzfus et al. (2004) conducted a meta-analysis of the relationship between ID and IDA and intelligence in children aged from 2 to 7 years and found that the average deficiency outcome was to reduce IQ from the expected population mean of 100 to a value of < 70. Hookworm infections occur in children of all ages (Preuksaraj et al., 1983; Kightlinger et al., 1995; Montresor et al., 2002) and cause blood loss and iron depletion in children of all ages (Pawlowski et al.,1991). The cognitive development and educational achievement of children must surely be put at risk if they live where hookworm infections are endemic unless control interventions are in place.

11.5.3 TRICHURIASIS

Trichuriasis is now recognized as a cause of severe morbidity in children and a public health problem of global magnitude (Table1.1, Bundy and Cooper, 1989). The disease is due to the intimate contact between the adult worms and the mucosa of the distal section of the host's alimentary tract. The whip-like anterior part of the each worm becomes embedded in the mucosa causing chronic inflammation, lacerations, and hemorrhage. Details of the cellular pathology, in comparison with the pathology of ulcerative colitis and Crohn's disease, have been published by Bundy and Cooper, 1989. The greater the degree of tissue damage, the greater the opportunity for bacterial invasion. Individuals harboring *Trichuris* infections may be asymptomatic, may experience subtle morbidity, or may be more seriously ill with TDS or *Trichuris* dysentery syndrome (Table 11.10).

Authorities now accept that blood loss occurs during *Trichuris* infection with the risk of ID and IDA. For example, Ramdath et al. (1995) compared the blood picture of 264 *Trichuris*-infected children with that of 157 matched, uninfected children. Children with heavy infections (> 10,000 epg) were found to have significantly lower hemoglobin concentrations and lower mean cell volumes than uninfected children (P < 0.05) and 33% of them were diagnosed as anemic.

Trichuriasis is a factor in childhood growth retardation. Children with heavy infections presented with blood in their stools and some degree of iron deficiency and their nutritional status improved following anthelminthic treatment (Gilman et al., 1983). In a group of 260 children in St Lucia, Cooper and Bundy (1986)

TABLE 11.10
Morbidity Associated with *Trichuris trichiura* Infection in Children. From data published by Bundy and Cooper (1989) and Stephenson et al. (2000a).

Signs and Symptoms	Main Outcomes
Abdominal pain, nausea, vomiting	Iron deficiency and anemia
Reduced food intake	Stunting and growth retardation
Digital clubbing	Reduced cognitive development
Chronic diarrhea with blood and mucus	and educational performance
Rectal prolapse	School absenteeism
TDS (*Trichuris* dysentery syndrome)	

detected a strong association between heavy *Trichuris* infection and depressed growth rates. Subsequently, Cooper et al. (1990) linked depressed growth rates to TDS; in 19 children anemia was found. Dramatic catch-up growth was observed in 11 of these children six months after anthelminthic treatment.

Trichuriasis is associated with negative effects on cognitive development, educational achievement, and school attendance. The Partnership for Child Development has been a leading advocate in the movement to deworm primary school children because that would aid cognitive performance (Drake et al., 2002). Stephenson et al. (2000a) reviewed the results of 12 studies, 11 of which were carried out in Jamaica where trichuriasis is common, published since 1992. The conclusion drawn from this body of work is that when the intensity of heavy *Trichuris* infections is reduced by anthelminthic treatment, measures of cognitive function, auditory and retrieval memory, and learning skills improve. School attendance also improved. These are the sort of findings that highlight the importance of providing regular treatment of school-aged children to deal with soil-transmitted helminthiasis.

In a study involving 246 primary school children in Guatemala, Watkins et al. (1996) investigated school performance after deworming with albendazole. At baseline the prevalences of *A. lumbricoides* and *T. trichiura* were 91 and 82% respectively. Tests with the children were carried out at 0 and at 24 weeks and the study design involved a placebo group. Comparison of the treated and the placebo groups failed to reveal any difference after deworming. *Ascaris* infection was cleared by albendazole, but not *Trichuris* infection, so the significance of the results is difficult to interpret. Investigations with S-TH infections invariably give a spectrum of results.

Recently, *Trichuris* has been found to have a seemingly positive, if not bizarre, effect on health. Summers et al. (2003, 2005) have found that in some patients chronic inflammatory bowel disease will respond to a treatment that involves oral doses with eggs of the porcine whipworm *T. suis*. A randomized, double-blind, placebo-controlled trial was carried out with 54 patients suffering from active colitis. Some received regular doses of 2,500 eggs and some the placebo. Improvement in the state of the colitis was detected in 13 of the 30 patients given eggs and in 4 of the 24 patients given placebo. Adverse side effects were not observed in any of the patients in the trial. Understandably, concern has been expressed about the fate of the *T. suis* eggs in the 17 patients whose colitis did not improve (Hsu et al., 2005). For helminthologists, however, the exciting aspect of this discovery is the capacity of a worm infection to modulate the nature of the human host response.

11.5.4 LARVAL MIGRATIONS AND MORBIDITY

Larval migrations are not a feature of trichuriasis. Little is known about the health impact of migrating larval *A. lumbricoides* until their arrival in the lungs. Presumably the larvae invade the liver and induce similar pathology to that observed with *A. suum* in pigs (Crompton, 2001). The pulmonary manifestations of ascariasis range from a mild cough to a Loeffler's syndrome involving severe cough, transient pulmonary infiltrates, dyspnea, and eosinophilia (Little, 1985; Seltzer et al., 2005). In places where *Ascaris* transmission is continuous, pneumonitis is uncommon. When transmission is seasonal, as occurs in Saudi Arabia, outbreaks of pneumonitis coincide with the time when transmission peaks (Gelpi and Mustafa, 1967; Spillman, 1975). The migration of larval *A. lumbricoides* should not be confused with visceral larval migrans, a term for the migrations of larval *Toxocara canis* and the larvae of other species of zoonotic nematodes (Little, 1985).

During skin penetration hookworm larvae may cause a form of dermatitis known as "ground itch" characterized by irritation, erythema, localized edema, and papulovesicular eruption (Pawlowski et al., 1991). Apparently people who are used to hookworm invasion may be less aware of the process than visitors. Skin penetration by zoonotic hookworms (*A. braziliense* and *A. caninum*) is known to cause ground itch. Former U.S. President Jimmy Carter, in an autographical volume about growing up in rural Georgia, describes how he, his friends and others in the community were afflicted by ground itch (Carter, 2001).

Larval *A. braziliense* may move about in the skin giving rise to a lesion called "creeping eruption" or cutaneous larva migrans; a transient form of this can occur with larval *N. americanus* (Little, 1985). Generally, hookworm larvae do not elicit pneumonitis. A condition called "Wakana disease" has been described from people in Japan following oral infection with *A. duodenale*. The symptoms include coughing, nausea, vomiting, and eosinophilia, but are thought to involve larvae in the intestinal mucosa rather than in the lungs (Little, 1985).

11.5.5 Soil-Transmitted Helminths, HIV/AIDS, Malaria, and TB

STH infections may exacerbate malaria and facilitate HIV and TB infections. People living where STH infections are endemic and where control interventions are lacking or inadequate will be constantly exposed to antigens secreted by the worms. We have known for years that helminth infections in humans are accompanied by elevated levels of T-helper cells (Th-2), cytokines, and IgE in addition to the characteristic eosinophilia. Might this modulatory effect of STH infections have favorable consequences for other infections to the detriment of the human host? That proposition has been investigated by Bentwich et al. (1999) who studied Ethiopian immigrants on their arrival in Israel. Many of the immigrants were infected with STHs and some were both STH and HIV positive. The STH-infected people had a dominant Th-2 profile that returned to normal on deworming. Their blood mononuclear cells were highly susceptible to HIV infection associated with increased expression of HIV co-receptors. When immigrants with both infections were dewormed, they responded to retroviral therapy in the same manner as HIV positive patients who had not been infected with STHs. Importantly, in HIV patients, the plasma viral load decreased following deworming. Bundy et al. (2000), in a generally supportive critique of this research, raised the possibility of other factors such as poor nutrition being involved in the STH/HIV synergism, but Bentwich (2000) offered further information that strongly supported the view that STH infections render their hosts more susceptible to HIV infection.

Fincham et al. (2003), after examining 109 research papers, further supported the conclusion that the immunomodulatory effects of STH infections would enhance the establishment of HIV infection and quicker progression to AIDS. They also agreed that the presence of STH infections might interfere with the effectiveness of vaccines developed to combat various infections (Bentwich et al., 1999; Bundy et al., 2000). Perhaps therein lies the explanation for BCG vaccine offering better protection against TB in the north than in the south (see Fincham et al., 2003).

Spiegel et al. (2003) found that in Senegal the risk of presenting with a clinical malaria attack increased significantly in subjects with STH infections as compared with uninfected subjects living under the same conditions ($P = 0.003$). Lack of helminth infections appeared to confer the same degree of protection from *falciparum* malaria as did carriage of the sickle cell trait. A similar observation had been made earlier in Madagascar, but with less control over confounding variables. This important claim needs further investigation. Murray et al. (1978), working on the island of Anjouan in the Comoros, noted that children with heavy *Ascaris* infections (> 50,000 epg) were little troubled by malaria until given anthelminthic treatment, whereupon what was diagnosed as recrudesence of *vivax* malaria occurred.

That STH infections might exacerbate other health conditions is not a recent conclusion. Charles Wardell Stiles, who described and identified *N. americanus* and stimulated the formation of The Rockefeller Sanitary Commission for the Eradication of Hookworm Disease (see Ackert, 1952), wrote when reflecting on all his experience " ... and even a light hookworm infection might be the "last straw" in a case of typhoid fever, tuberculosis, diphtheria, or some other condition" (Stiles, 1930). Hookworm disease is no longer a public health problem in the southern states of the U.S. The provision of sanitation secured eradication.

11.6 ALLERGY AND IMMUNITY

Potent, volatile allergens are released by adult and larval *A. lumbricoides*. Deliberate experimental inhalation of *Ascaris* molecules causes asthmatic attacks. There has been considerable interest in the possibility that *Ascaris* infection induces asthma in some people (Coles, 1985). Do responses to these allergenic molecules cause asthma? Undoubtedly uninfected people can become highly sensitized to *Ascaris* and then experience respiratory spasms, skin rashes, and gastrointestinal disorder. This problem is a health hazard for laboratory workers and medical staff who should be protected when working with the worms or their products.

The human host recognizes *Ascaris* as not-self and there is an excessive production of nonspecific IgE during the course of an infection. Pritchard (1993) proposed that the superfluous IgE, stimulated by the arrival of the worms, might serve as an adaptation for survival in the immunocompetent host by saturating receptors on the surface of mast cells. Some degree of partial or protective immunity may develop in people experiencing regular exposure to infective larvae. That might explain the decline in the intensity of *Ascaris* infections with age (Figure 11.12). The results of much research, however, have still to offer realistic prospects for the development of an affordable ascariasis vaccine that would contribute to public health in the countries of the south.

There is abundant evidence to show that the human immune system is responsive to hookworm infections, but little evidence to indicate that the responses are protective. Some years ago an effective vaccine was developed to protect dogs against

A. caninum. The vaccine, which involved inoculation with live, attenuated L3, was subsequently withdrawn in 1975 because it failed to live up to economic expectations (Soulsby, 1982). There has been progress recently in the development by the Human Hookworm Vaccine Initiative of a recombinant vaccine based on purified antigen (ASP-2) prepared from infective larvae (Hotez et al., 2003). The availability of a vaccine for human use would be a major contribution to the control of hookworm disease provided the costs were appropriate for low-income countries.

11.7 ANTHELMINTHIC TREATMENT

The World Health Organization (WHO, 2002a) recommends four drugs for the treatment and control of soil-transmitted helminthiasis (Table 11.11) that should used according to WHO guidelines (WHO, 2004b). Although each drug is referred to as having broad-spectrum activity, none has been found to be highly effective on a single-dose basis for trichuriasis. There is no need for the use of purgatives or

TABLE 11.11
Drugs Used for the Treatment of Soil-Transmitted Helminthiasis and Strongyloidiasis. From Table 6 (WHO, 2002a) with permission from the World Health Organization (www.who.int).

Drugs and Formulation	Therapeutic Activity[a]	Dosage
Albendazole	Ascariasis +++	400 mg single dose
(tablets 200 and 400 mg	Trichuriasis ++	400 mg single dose
suspension 100 mg/5 ml)	Hookworm disease +++	400 mg single dose
	Strongyloidiasis ++	400 mg daily for 3 days
Ivermectin	Ascariasis +++	200 µg/kg single dose
(tablets 6 mg)	Trichuriasis +	200 µg/kg single dose
	Strongyloidiasis +++	200 µg/kg single dose
Levamisole	Ascariasis +++	2.5 mg/kg single dose
(tablet 40 mg	Trichuriasis +	2.5 mg/kg single dose
syrup 40 mg/5 ml)	Hookworm disease ++	2.5 mg/kg single dose (for heavy infection repeat after 7 days)
Mebendazole	Ascariasis +++	500 mg single dose
(tablets 100 mg and 500	Trichuriasis ++	100 mg twice daily for 3 days *or* 500 mg single dose (less effective)
mg suspension 100 mg/5 ml)	Hookworm disease ++	100 mg twice daily for 3 days *or* 500 mg single dose (less effective)
Pyrantel	Ascariasis +++	10 mg/kg single dose
(tablets 250 mg suspension 50 mg/5 ml	Hookworm disease ++	For heavy hookworm infection 10 mg/kg repeated for 4 days

[a] +++ ≥ 80% cure rate (CR) or ≥ 80% egg reduction rate (ERR).
  ++ 50–80% CR or 60–80% ERR.
  + 10–50% CR or 10–50% ERR.

other preparation before giving the drugs. The information about the drugs listed in Table 11.11 has been abstracted from Gustafsson et al. (1987), Horton (2000, 2003), Janssens (1985), Vanden Bossche (1995), and WHO (1996a,b, 2002b). These publications give details about pharmacokinetics, clinical trials, contraindications, safety issues, and adverse and side effects. Millions of doses of each drug have been given around the world wherever STH infections are endemic. In all cases the benefits to human health of anthelminthic treatment with these four drugs have been far greater than the risks to health.

Albendazole is a benzimidazole carbamate derivative introduced in Australia in 1977 for the treatment of livestock, and first approved for human use in 1982. Its sulfoxide metabolite binds to nematode tubulin preventing the formation of microtubules and disrupting cell division. Apart from killing adult nematodes, albendazole has larvicidal and ovicidal activity. During its development, albendazole was found to be teratogenic in rats, rabbits, cattle, and sheep when given experimentally at high dosage, but under conditions of animal husbandry that would have involved treatment during early pregnancy, there were no recorded cases of teratogenicity (Bogan and Marriner, 1984). A WHO Informal Consultation concluded "that it is probably safe to use selected benzimidazole drugs in women of child-bearing potential, as differences in the doses and pharmacokinetics and metabolism make it likely that there is little if any risk to pregnant and lactating women or to the breast-fed baby" (WHO, 1996a). Another WHO Informal Consultation has concluded that albendazole may be given to children as young as 12 months for the treatment of soil-transmitted helminthiasis (WHO 2002b; Montresor et al., 2003).

Ivermectin is a macrocyclic lactone that originated from the actinomycete *Streptomyces avermitilis*. The drug interferes with the binding of gamma-aminobutyric acid (GABA) at postsynaptic sites at the neuromuscular junction with the effect of paralyzing susceptible nematodes. Its widespread use in the treatment of onchocerciasis (Chapter 9) means that its safety record is well documented. Preparation for its use against lymphatic filariasis has found that the drug is well tolerated and that any side effects are due to inflammatory reactions against dead microfilariae rather than to the drug (Brown et al., 2000). At present there in insufficient information about whether the drug might be used during pregnancy. Pacque et al. (1990) reported on a three-year study in Liberia during which ivermectin (150 µg/kg body weight) was given inadvertently to 200 pregnant women. Two hundred and three children were born to women who had received ivermectin and no statistically significant differences were detected in the course of the pregnancies or the pregnancy outcomes including birth defect rates between those given ivermectin and those not.

Levamisole is a levorotary isomer of tetramisole that originated as an anthelminthic for veterinary use and was introduced for human use in 1965. Levamisole binds to the acetylcholine receptors of nematodes causing paralysis followed by the elimination of the worms. Under different dosages from those in Table 11.11 levamisole has an immunomodulatory effect. Experimentally, levamisole has not been found to have any teratogenicity in rats and livestock given doses of up to 40 mg/kg body weight; the dose recommended for humans is 2.5 mg/kg body weight.

Mebendazole, like albendazole, is a benzimidazole derivative that binds to nematode tubulin and so prevents cell division. Mendazole may have larvicidal and

TABLE 11.12
Single, Oral Dose Combinations of Anthelminthic Drugs for the Treatment of Trichuriasis in Children in Sri Lanka. From data published by Ismail and Jayakody (1999).

Drug Regimen	n	Effectiveness (%)	
		CR (Cure Rate)	ERR (Egg Reduction Rate)
Albendazole (400 mg)	55	44	70
Albendazole (400 mg) with Diethylcarbamazine (6 mg/kg)	47	30	69
Albendazole (400 mg) with Ivermectin (200 µg/kg)	53	79	92

Note: Stools examined by Kato Katz procedure before and 3 weeks after treatment.

ovicidal properties. Mebendazole has been found to have teratogenic effects in rats given doses of 10 mg/kg body weight. The WHO Informal Consultations (see above) have concluded that mebendazole may be used to treat soil-transmitted helminthiasis during pregnancy and in children as young as 12 months.

Pyrantel (embonate or pamoate) is a pyrimidine derivative that began human clinical trials in 1966. The drug binds to acetylcholine receptors of the worms causing paralysis followed by passive elimination from the body. Tests in rats, rabbits, and sheep found no evidence of teratogenic effects. The WHO Informal Consultation concluded that "levamisole (see above) and probably pyrantel are not experimental teratogens or genotoxicants, so these drugs may appear safer to use in women and infants as far as these risks are concerned." In all cases where anthelminthic drugs are being used in a public health setting, every effort should be made to avoid giving them to women during the first trimester of pregnancy.

Anthelminthic treatment to deal effectively with trichuriasis by giving a single dose of an approved drug remains problematic. Ismail and Jayakody (1999) may have found the answer by offering albendazole in combination with ivermectin. Their results are summarized in Table 11.12. This study has important implications for the control of STH infections in programs aimed primarily at the control of lymphatic filariasis (Chapter 8).

11.8 CONTROL STRATEGIES FOR THE USE OF ANTHELMINTHIC DRUGS

Periodic anthelminthic treatment of people in communities where STH infections are endemic has been accepted as the overall public strategy for controlling morbidity due to these infections (Albonico et al., 1998). This approach is possible because the drugs (1) have demonstrably satisfactory safety records, (2) are highly effective with a fair degree of broad spectrum activity (Table 11.11), (3) are given as single oral doses in tablet form, and (4) are now of low cost. The reduction in price has come about because the original proprietary drugs are of sufficient age to have lost the protection of patents and are now available as generic formulations. The approach is also achievable as long as the difficulties can be overcome of getting the drugs to the people who need them, monitoring drug effectiveness, and sustaining the effort. The problem is one of scale.

The fact that generic versions are so much cheaper than drugs protected by patent rights is what has made them accessible for use in community treatment programs. Horton (2003) drew attention to two issues that decision makers should note when planning to use generics in the community. First, the basis for licensing the drugs for human use was a consideration of efficacy and safety data obtained with the proprietary product. Secondly, the generic compound must be known to contain the correct amount of active ingredient and must behave in a manner similar to the original product. There are procedures for establishing drug quality (Horton, 2003) and these should be employed before generics are purchased. Care must be taken to avoid counterfeit drugs.

There are three public health applications for providing treatment to control morbidity due to soil-transmitted helminthiasis in the community: universal; targeted; and selective (see Chapter 4; Anderson 1989; WHO 1996b). **Universal** treatment, sometimes referred to as mass or blanket treatment, concerns population level application of the drug in which everybody is treated irrespective of age, gender, occupation, infection status, or social characteristics. Children younger than 12 months should not be treated and every attempt should be made to withhold treatment from women in the first trimester of pregnancy. In practice, identifying women in the early stage of pregnancy may be difficult and some may be unaware of their condition (Pacque et al., 1990). Such information as is available (WHO 1996a; de Silva et al., 1999) indicates that lack of information about early pregnancy should not prevent universal treatment. Although there is an ethical argument for undertaking diagnosis before treatment, the expense of diagnosis for the health care services of low-income countries and the quality and safety record of anthelminthic drugs (Table 11.11) justifies access to treatment without diagnosis where STH infections are known to be entrenched.

Targeted treatment concerns group-level application of the drug where the group may be defined in terms of risk or other characteristics such as school-aged children, pregnancy, plantation workers, adolescent girls, or preschool children, again irrespective of infection status. In places where STH infections are known to be endemic and have a history of affecting health, there is no need to spend meager resources on diagnostic surveys. If girls and women of childbearing age are targeted, care must be taken to avoid treatment during the first trimester of pregnancy. Similarly, preschool children younger than 12 months should not be treated.

Selective treatment concerns individual application of the drug after determination of infection status. Theory indicates that treating the few individuals who harbor high worm burdens (Figure 11.18) will reduce the risk of life-threatening events, reduce severe morbidity and reduce exposure of others in the individual's circle to the probable risk of infection. The work prior to selective treatment is expensive in terms of time and resources and is intrusive for many people who are put through diagnostic procedures but may not get treatment. An example of the comparison of

TABLE 11.13
Anthelminthic Treatment and Other Interventions for the Control of Soil-Transmitted Helminth Infections. From Table 1 (Olsen, 2003) with permission from Elsevier.

Community Category	Prevalence	Proportion of Heavy Intensity Infections	Recommended Interventions
I Heavy intensity	Any	$\geq 10\%$	(1) Universal treatment (1 per year) (2) More intense treatment of high-risk groups[a] (2 or 3 per year) (3) Health education activities (4) Sanitation improvement
II High prevalence, light intensity	$\geq 50\%$	$< 10\%$	(1) Targeted treatment of high-risk groups (2) Health education activities (3) Sanitation improvement
III Low prevalence, light intensity	$< 50\%$	$< 10\%$	(1) Health education activities (2) Sanitation improvement (3) Case management

[a] High-risk groups: women of childbearing age, preschool children, and school-aged children.

universal, targeted, and selective treatment for controlling morbidity due to ascariasis is to be found in a study carried out in Nigeria by Asaolu et al. (1991). Anthelminthic treatment has the potential to control and reduce morbidity due to soil-transmitted helminthiasis. Sustainable control will need other interventions in support of drug applications. Guidelines appropriate for general control are shown in Table 11.13 and have been discussed by Olsen (2003).

11.9 TREATMENT OF SCHOOL-AGED CHILDREN

There is now an overwhelmingly strong case for targeting anthelmintic treatment at school-aged children to relieve them from the effects of STH infections. If smoking is bad for your health, then worms are bad for your children. The case, based on compelling evidence obtained from round the world since the 1970s, has been championed by the Partnership for Child Development (PCD, 1996; Bundy and Guyatt, 1996; Drake et al., 2002). Technical advice has been published to assist those given the responsibility of implementing the intervention (Montresor et al., 2002).

The PCD has argued that primary schools are the obvious places to treat school-aged children in the south at regular intervals. Unfortunately, various social and cultural influences may prevent treatment from reaching the children who need it. Olsen (2003) has reviewed experience gained from using schools as centers for anthelminthic treatment and has drawn attention to enrollment statistics. In 1995, there were reckoned to be 1.2 billion school-aged children in the world and that about 400 million were likely to be in school on any given school day. So what were the other 800 million doing? Some would be ill, some would live in places without schools, some would be required for household duties, some would be at work, and many might not have been enrolled at their local school (see Olsen, 2003). Ensuring good coverage is a key component if the treatment of school-aged children is to have its desired health and development benefits.

11.10 PREVENTION AND CONTROL

Control leading to the elimination of soil-transmitted helminth infections cannot be achieved until communities live in clean environments. Sanitation provides the solution to this problem and an introduction to some aspects of the technology is offered in Chapter 16. Prevention for any form of helminthiasis needs the support of an educated community so that health messages can be understood. Such messages must be realistic and in tune with local perceptions and traditions (see Chapter 15). Implementation of UNICEF's effort to promote clean water, good sanitation, and hygiene education in schools (UNICEF, 2005) will help to sustain the impact of anthelminthic treatment delivered to children who are able to attend schools in places where soil-transmitted helminthiasis is endemic.

REFERENCES

Ackert JE. 1952. Some influences of the American hookworm. *American Midland Naturalist* **47**, 749–762.

Allen LH. 1997. Pregnancy and iron deficiency: unresolved issues. *Nutrition Reviews* **55**, 91–101.

Albonico M, Crompton DWT and Savioli L. 1998. Control strategies for human intestinal nematode infections. *Advances in Parasitology* **42**, 277–341.

Anderson RM. 1989. Transmission dynamics of *Ascaris lumbricoides* and the impact of chemotherapy. In: *Ascariasis and its Prevention and Control* (eds. DWT Crompton, MC Nesheim and ZS Pawlowski). London, New York and Philadelphia: Taylor and Francis Ltd. pp. 253–273.

Anderson RM and Gordon DM. 1982. Processes influencing the distribution of parasite numbers within host populations with special emphasis on host mortalities. *Parasitology* **85**, 373–398.

Anderson RM and May RM. 1992. *Infectious Diseases of Humans.* Oxford University Press (paperback edition).

Asaolu SO, Holland CV and Crompton DWT. 1991. Community control of *Ascaris lumbricoides* in rural Oyo State, Nigeria: mass, targeted and selective treatment with levamisole. *Parasitology* **103**, 291–298.

Ash LR and Orihel TC. 1990. *Atlas of Human Parasitology,* 3rd edition. Chicago: American Society of Clinical Pathologists Press.

Atukorala TMS, de Silva LDR, Dechering WHJC et al. 1994. Evaluation of the effectiveness of iron-folate supplementation and anthelminthic therapy against anemia in pregnancy — a study in the plantation sector of Sri Lanka. *American Journal of Clinical Nutrition* **60**, 286–292.

Beall J. 2000. Life in the cities. In: *Poverty and Development into the 21ˢᵗ Century* (eds. T Allen and A Thomas). Milton Keynes: The Open University in association with Oxford University Press. pp. 425–442.

Bentwich Z. 2000. Reply to Bundy et al. 2000. *Parasitology Today* **16**, 312.

Bentwich Z, Kalinkovich A, Weisman Z et al. 1999. Can eradication of helminth infections change the face of AIDS and tuberculosis? *Immunology Today* **20**, 485–487.

Bogan JA and Marriner SE. 1984. Phamacodynamic and toxicological aspects of albendazole in man and animals. In: *Albendazole in Helminthiasis* (ed. M Firth). London: The Royal Society of Medicine. pp. 13–18.

Booth M, Bundy DAP, Albonico M. et al. (1998). Associations among multiple geohelminth species infections in school children from Pemba Island. *Parasitology* **116**, 85–93.

Brooker S, Bethony J and Hotez PJ. 2004. Human hookworm infection in the 21ˢᵗ century. *Advances in Parasitology* **58**, 197–288.

Brooker S, Rowlands M, Haller L. et al. 2000. Towards an atlas of human helminth infection in sub-Saharan Africa; the use of Geographical Information Systems (GIS). *Parasitology Today* **16**, 303–307.

Brown KR, Ricci FM and Ottesen EA. 2000. Ivermectin: effectiveness in lymphatic filariasis. *Parasitology* **121**, S133–S146.

Bundy DAP. 1990. Is hookworm just another geohelminth? In: *Hookworm Disease* (eds. GA Schad and KS Warren). London and Philadelphia: Taylor & Francis. pp. 147–164.

Bundy DAP and Cooper ES. 1989. *Trichuris* and trichuriasis in humans. *Advances in Parasitology* **28**, 107–173.

Bundy DAP and Guyatt HL. 1996. Schools for health: focus on health, education and the school-age child. *Parasitology Today* **12**, 1–16.

Bundy D, Sher A and Michael E. 2000. Good worms or bad worms: do worm infections affect the epidemiological patterns of other diseases? *Parasitology Today* **16**, 273–274.

Carter J. 2001. *An Hour Before Daylight: Memories of a Rural Boyhood.* New York and London: Simon & Schuster.

Chai JY, Cho SY, Lee SH and Seo BS. 1991. Reduction in incidence of biliary and other surgical complications of ascariasis according to the decrease in national egg prevalence in Korea. *Korean Journal of Parasitology* **29**, 101–111.

Chan MS, Medley GF, Jamison D and Bundy DAP. 1994. The evaluation of potential global mortality attributable to intestinal nematode infections. *Parasitology* **109**, 373–387.

Cheesbrough M. 1998. *District Laboratory Practice in Tropical Countries Part 1.* March, Cambridgeshire: Tropical Health Technology.

Chitwood BG and Chitwood MB. 1974. *Introduction to Nematology.* Baltimore, London, Tokyo: University Park Press. (Consolidated edition of parts first published in 1937, 1938, and 1941 and offered as a single edition in 1950.)

Christian P, Khatry SK and West KP. 2004. Antenatal anthelmintic treatment, birthweight, and infant survival in rural Nepal. *Lancet* **364**, 981–983.

Coles GC. 1985. Allergy and immunopathology of ascariasis. In: *Ascariasis and its Public Health Significance* (eds. DWT Crompton, MC Nesheim and ZS Pawlowski). London and Philadelphia: Taylor & Francis Ltd. pp. 167–184.

Cooper ES and Bundy DAP. 1986. Trichuriasis in St Lucia. In: *Diarrhoea and Malnutrition in Children* (eds. AS McNeish and JA Walker-Smith). London: Butterworths. pp. 91–96.

Cooper ES and Bundy DAP. 1987. Trichuriasis. In: *Bailliere's Clinical Tropical Medicine and Communicable Diseases. Intestinal Helminth Infections* (ed. ZS Pawlowski). London, Philadelphia, Sydney, Tokyo, Toronto: Bailliere Tindall. pp. 629–643.

Cooper ES, Bundy DAP, MacDonald TT and Golden MH. 1990. Growth suppression in the *Trichuris* dysentery syndrome. *European Journal of Clinical Nutrition* **44**, 285–291.

Croll NA, Anderson RM, Gyorkos TW and Ghadirian E. 1982. The population biology and control of *Ascaris lumbricoides* in a rural community in Iran. *Transactions of the Royal Society of Tropical Medicine and Hygiene* **76**, 187–197.

Crompton DWT. 1987. Human helminthic populations. In: *Bailliere's Clinical Tropical Medicine. Intestinal Helminthic Infections* (ed. ZS Pawlowski). London and Philadelphia: Bailliere Tindall. pp. 489–510.

Crompton DWT. 1989a. Biology of *Ascaris lumbricoides*. In: *Ascariasis and its Prevention and Control* (eds. DWT Crompton, MC Nesheim and ZS Pawlowski). London, New York and Philadelphia: Taylor & Francis. pp. 9–44.

Crompton DWT. 1989b. Prevalence of ascariasis. In: *Ascariasis and its Prevention and Control* (eds. DWT Crompton, MC Nesheim and ZS Pawlowski). London, New York, Philadelphia: Taylor & Francis. pp. 45–69.

Crompton DWT. 1994. Ascaris lumbricoides. In: *Parasitic and Infectious Diseases* (eds. ME Scott and G Smith). San Diego, New York, London: Academic Press. pp. 175–196.

Crompton DWT. 2000. The public health importance of hookworm disease. *Parasitology* **121**, S39–S50.

Crompton DWT. 2001. *Ascaris* and ascariasis. *Advances in Parasitology* **48**, 285–375.

Crompton DWT and Nesheim MC. 2002. Nutritional impact of intestinal helminthiasis during the human life cycle. *Annual Review of Nutrition* **22**, 35–59.

Crompton DWT and Savioli L. 1993. Intestinal parasitic infections and urbanization. *Bulletin of the World Health Organization* **71**, 1–7.

Crompton DWT and Stephenson LS. 1990. Hookworm infection, nutritional status and productivity. In: *Hookworm Disease* (eds. GA Schad and KS Warren). London, New York, Philadelphia: Taylor & Francis. pp. 231–264.

de Silva N, Guyatt, H and Bundy D. 1997. Morbidity and mortality due to *Ascaris*-induced intestinal obstruction. *Transactions of the Royal Society of Tropical Medicine and Hygiene* **91**, 31–36.

de Silva NR, Sirisena JLGJ, Gunasekera DPS et al. 1999. Effect of mebendazole therapy during pregnancy on birth outcome. *Lancet* **353**, 1145–1149.

Diav-Citrin O, Shechtman S, Arnou J. et al. 2003. Pregnancy outcome after gestational exposure to mebendazole: a prospective controlled cohort study. *American Journal of Obstetrics and Gynecology* **188**, 282–285.

Drake L, Maier C, Jukes M. et al. 2002. School-age children: their nutrition and health. *SCN News* **25**, 4–30.

Ettling J. 1990. The role of the Rockefeller Foundation in hookworm research and control. In: *Hookworm Disease* (eds. GA Schad and KS Warren). London, New York, Philadelphia: Taylor & Francis. pp. 3–14.

Fagerholm HP, Nansen P, Roepstorff A et al. 2000. Differentiation of cuticular structures during the growth of the third-stage larva of *Ascaris suum* (Nematoda, Ascaridoidea) after emerging from the egg. *Journal of Parasitology* **86**, 421–427.

Fincham JE, Markus MB and Adams VJ. 2003. Could control of soil-transmitted helminthic infection influence the HIV/AIDS pandemic? *Acta Tropica* **86**, 315–333.

Forrester JE and Scott ME. 1990. Measurement of *Ascaris lumbricoides* infection intensity and the dynamics of expulsion following treatment with mebendazole. *Parasitology* **100**, 303–308.

Forrester JE, Scott ME, Bundy DAP and Golden MHN. 1988. Clustering of *Ascaris lumbricoides* and *Trichuris trichiura* infections within households. *Transactions of the Royal Society of Tropical Medicine and Hygiene* **82**, 282–288.

Gelpi AP and Mustafa A. 1967. Seasonal pneumonitis with eosinophilia: a study of larval ascariasis in Saudi Arabia. *American Journal of Tropical Medicine and Hygiene* **16**, 646–657.

Gilman RH, Chong YH, Davis C. et al. 1983. The adverse consequences of heavy *Trichuris* infection. *Transactions of the Royal Society of Tropical Medicine and Hygiene* **77**, 432–438.

Gonzalez AH, Regalado VC and Van den Ende J. 2001. Non-invasive management of *Ascaris lumbricoides* biliary tract migration: a prospective study in 69 patients from Ecuador. *Tropical Medicine and International Health* **6**, 146–150.

Gustafsson LL, Beerman B and Aden Abdi Y. 1987. *Handbook of Drugs for Tropical Parasitic Infections.* London, New York and Philadelphia: Taylor and Francis Ltd.

Guyatt HL, Bundy DAP, Medley GF and Grenfell BT. 1990. The relationship between the frequency distribution of *Ascaris lumbricoides* and the prevalence and intensity of infection in human communities. *Parasitology* **101**, 139–143.

Haas JD and Brownlie T. 2001. Iron deficiency and reduced work capacity: a critical review of the research to determine a causal relationship. *Journal of Nutrition* **131**, 676S–688S.

Habermann RT and Williams FP. 1957. Diseases seen at autopsy on 708 *Macaca mulatta* (rhesus monkey) and *Macaca philippinensis* (cynomolgos monkey). *American Journal of Veterinary Research* **18**, 419–426.

Haswell-Elkins MR, Elkins DB and Anderson RM. 1987. Evidence for predisposition in humans to infection with *Ascaris*, hookworm, *Enterobius* and *Trichuris* in a South Indian fishing community. *Parasitology* **95**, 323–337.

Henry FJ. 1988. Reinfection with *Ascaris lumbricoides* after chemotherapy: a comparative study in three villages with varying sanitation. *Transactions of the Royal Society of Tropical Medicine and Hygiene* **82**, 460–464.

Hoaglund RE and Schad GA. 1978. *Necator americanus* and *Ancylostoma duodenale*: life history parameters and epidemiological implication of two sympatric hookworms of humans. *Experimental Parasitology* **44**, 36–49.

Holland CV, Asaolu SO, Crompton DWT et al. 1989. The epidemiology of *Ascaris lumbricoides* and other soil-transmitted helminth infections in primary school children from Ile-Ife, Nigeria. *Parasitology* **99**, 275–285.

Horton J. 2000. Albendazole: a review of anthelminthic efficacy and safety in humans. *Parasitology* **121**, S113–132.

Horton J. 2003. The efficacy of anthelminthics: past, present and future. In: *Controlling Disease due to Helminth Infections* (eds. DWT Crompton, A Montresor, MC Nesheim and L Savioli). Geneva: World Health Organization. pp. 143–155.

Hotez PJ, Brooker S, Bethony JM et al. 2004. Hookworm infection. *New England Journal of Medicine* **351**, 799–807.

Hotez, PJ, Zhan, B, Bethony JM et al. 2003. Progress in the development of a recombinant vaccine for human hookworm disease: the Human Hookworm Vaccine Initiative. *International Journal of Parasitology* **33**, 1245–1248.

Hotez PJ, Bundy DAP, Beegle K et al. 2006. Helminth infections. Soil-transmitted helminth infections and schistosomiasis. In: *Disease Control Priorities Project*, 2nd edition (eds. DT Jamison, G Alleyne, J Breman et al.). Washington DC: The World Bank in association with Oxford University Press. Chapter 24.

Hsu S-J, Tseng P-H and Chen P-J. 2005. *Trichuris suis* therapy for ulcerative colitis: nonresponsive patients may need anti-helminth therapy. *Gastroenterology* **129**, 768–769.

Ismail MM and Jayakody RL. 1999. Efficacy of albendazole and its combinations with ivermectin or diethylcarbamazine (DEC) in the treatment of *Trichuris trichiura* infections in Sri Lanka. *Annals of Tropical Medicine and Parasitology* **93**, 501–504.

Janssens PG. 1985. Chemotherapy of gastrointestinal nematodiasis in man. In: *Handbook of Experimental Pharmacology* Vol. 77 (eds. H Vanden Bossche, D Thienpont and PG Janssens). Berlin and New York: Springer-Verlag. pp. 183–406.

Kightlinger LK, Seed JR and Kightlinger MB. 1995. The epidemiology of *Ascaris lumbricoides, Trichuris trichiura*, and hookworm in children in the Ranomafana rainforest, Madagascar. *Journal of Parasitology* **81**, 159–169.

Little MD. 1985. Nematodes of the digestive tract and related species. In: *Animal Agents and Vectors of Human Disease*, 5th edition (eds. PC Beaver and RC Jung). Philadelphia: Lea & Febiger. pp. 127–170.

Montresor A, Crompton DWT, Gyorkos TW and Savioli L. 2002. *Helminth Control in School-age Children. A Guide for Managers of Control Programmes.* Geneva: World Health Organization.

Montresor A, Awasthi S and Crompton DWT. 2003. Use of benzimidazoles in children younger than 24 months for the treatment of soil-transmitted helminthiasis. *Acta Tropica* **86**, 223–232.

Moraes LRS and Cairncross S. 2004. Environmental interventions and the pattern of geohelminth infections in Salvador, Brazil. *Parasitology* **129**, 223–232.

Murray J, Murray A, Murray M and Murray C. 1978. The biological suppression of malaria: an ecological and nutritional relationship of a host and two parasites. *American Journal of Clinical Nutrition* **31**, 1363–1366.

Murrell KD, Eriksen L, Nansen P. et al. 1997. *Ascaris suum*: a revision of its early migratory path and implications for human ascariasis. *Journal of Parasitology* **83**, 255–260.

Nokes C, van den Bosch C and Bundy DAP. 1998. The effects of iron deficiency and anemia on mental and motor performance, educational achievement, and behavior in children. A Report of the International Nutritional Anemia Consultative Group. Washington DC: International Nutritional Anemia Consultative Group.

O'Lorcain P and Holland CV. 2000. The public health importance *of Ascaris lumbricoides. Parasitology* **121**, S51–S71.

Olsen A. 2003. Experience with school-based interventions against soil-transmitted helminths and extension of coverage to non-enrolled children. *Acta Tropica* **86**, 255–266.

Orihel TC. 1970. The helminth parasites of nonhuman parasites and man. *Laboratory Animal Care* **20**, 395–401.

Pacque M, Munoz B, Poetschke G. et al. 1990. Pregnancy outcome after inadvertent ivermectin treatment during community-based distribution. *Lancet* **336**, 1486–1489.

PCD (Partnership for Child Development). 1996. Cost of school-based drug treatment in Tanzania. *Health Policy and Planning* **13**, 384–396.

Pawlowski ZS and Davis A. 1989. Morbidity and mortality in ascariasis. In: *Ascariasis and its Prevention and Control* (eds. DWT Crompton, MC Nesheim and ZS Pawlowski). London, New York, Philadelphia: Taylor & Francis. pp. 71–86.

Pawlowski ZS, Schad GA and Scott GJ. 1991. *Hookworm Infection and Anaemia.* Geneva: World Health Organization.

Peng Weidong, Zhou Xianmin, Cui Xiaomin et al. 1996. *Ascaris,* people and pigs in a rural community of Jiangxi Province, China. *Parasitology* **113**, 545–557.

Peng W, Zhou X and Crompton DWT. 1998. Ascariasis in China. *Advances in Parasitology* **41**, 109–148.

Preuksaraj S, Jeradit C, Sathitayathai A et al. 1983. Studies on prevalence and intensity of intestinal helminthic infections in the rural population of Thailand 1980 to 1981. *Collected Papers on Control of Soil-transmitted Helminthiases* II, pp. 54–65. Tokyo: Asian Parasite Control Organization.

Pritchard DI. 1993. Immunity to helminths: is too much IgE parasite–rather than host protective? *Parasitology Immunology* **15**, 5–9.

Pruss-Ustun A, Kay D, Fewtrell L and Bartram J. 2004. Unsafe water, sanitation and hygiene. In: *Comparative Quantification of Health Risks,* Volume 2 (eds. M Ezzati, AD Lopez, A Rodgers and CJL Murray). Geneva: World Health Organization. pp. 1321–1352.

Ramdath DD, Simeon DT, Wong MS and Grantham-MacGregor SM. 1995. Iron status of schoolchildren with varying intensities of *Trichuris trichiura* infection. *Parasitology* **110**, 347–351.

Robertson LJ, Crompton DWT, Walters DE et al. 1989. Soil-transmitted helminth infections from school children in Cocle Province, Republic of Panama. *Parasitology* **99**, 287–292.

Roche M and Layrisse M. 1966. The nature and causes of hookworm anaemia. *American Journal of Tropical Medicine and Hygiene* **15**, 1029–1102.

St George J. 1976. Intestinal parasitic infestation among parturients in Trinidad and Tobago. *International Surgery* **61**, 222–225.

Salafsky B, Fusco AC and Siddiqui A. 1990. *Necator americanus*: factors influencing skin penetration by larvae. In: *Hookworm Disease* (eds. GA Schad and KS Warren). London, New York and Philadelphia: Taylor & Francis. pp. 329–339.

Schad GA and Anderson RM. 1985. Predisposition to hookworm infection in humans. *Science* **228**, 1537–1540.

Schad GA, Chowdhury AB, Dean CG. et al. 1973. Arrested development in human hookworm infection: an adaptation to a seasonally unfavourable environment. *Science* **180**, 502–505.

Seltzer E, Barry M and Crompton DWT. 2006. Ascariasis. In: *Tropical Infectious Diseases — Principles, Pathogens and Practice,* 2nd edition (eds. RL Güerrant, DA Walker, and PF Wetter). Philadelphia: Churchill Livingstone. pp. 1257–1264.

Soulsby EJL. 1982. *Helminths, Arthropods and Protozoa of Domesticated Animals,* 7th edition. London and Philadelphia: Bailliere Tindall.

Spiegel A et al. 2003. Increased frequency of malaria attacks in subjects co-infected by intestinal worms and *Plasmodium falciparum* malaria. *Transactions of the Royal Society of Tropical Medicine and Hygiene* **97**, 198–199.

Spillman RK. 1975. Pulmonary ascariasis in tropical countries. *American Journal of Tropical Medicine and Hygiene* **24**, 791–800.

Stephenson LS. 1987. *The Impact of Helminth Infections on Human Nutrition.* London, New York, Philadelphia: Taylor & Francis.

Stephenson LS, Holland CV and Cooper ES. 2000a. The public health significance of *Trichuris trichiura*. *Parasitology* **121**, S73–S95.

Stephenson LS, Latham MC and Ottesen EA. 2000b. Global malnutrition. *Parasitology* **121**, S5–S22.

Stiles CW. 1930. Decrease of hookworm disease in the United States. *U.S. Public Health Reports* **45**, 1763–1781.

Stoltzfus RJ, Mullany L and Black RE. 2004. Iron deficiency anaemia. In: *Comparative Quantification of Health Risks,* Volume 1 (eds. E Mazzati, AD Lopez, A Rodgers and CJL Murray). Geneva: World Health Organization. pp. 163–209.

Sturchler D. 1988. *Endemic Areas of Tropical Infections.* Toronto, Lewiston NY, Bern, Stuttgart: Hans Huber Publishers.

SCN (Standing Committee on Nutrition). 2004. *5th Report on the World Nutrition Situation.* Geneva: SCN Secretariat, c/o World Health Organization.

Summers RW, Elliott DE, Qadir K et al. 2003. *Trichuris suis* seems to be a safe and possibly effective in the treatment of inflammatory bowel disease. *Gastroenterology* **98**, 2034–2041.

Summers RW, Elliott DE, Urban JF et al. 2005. *Trichuris suis* therapy for active ulcerative colitis:a randomized controlled trial. *Gastroenterology* **128**, 825–832.

Thein Hlaing. 1993. Ascariasis and childhood malnutrition. *Parasitology* **107**, S125–136.

Thein Hlaing, Than Saw, Htay Htay Aye et al. 1984. Epidemiology and transmission of *Ascaris lumbricoides* in Opko village, rural Burma. *Transactions of the Royal Society of Tropical Medicine and Hygiene* **78**, 497–504

Thienpont D, Rochette F and Vanparijs OFJ. 1979. *Diagnosing Helminthiasis through Coprological Examination.* Beerse, Belgium: Janssen Research Foundation.

Torlesse H and Hodges M. 2000. Anthelminthic treatment and haemoglobin concentrations during pregnancy. *Lancet* **356**, 1083.

Torlesse H and Hodges M. 2001. Anthelminthic therapy and reduced decline in haemoglobin concentrations during pregnancy (Sierra Leone). *Transactions of the Royal Society of Tropical Medicine and Hygiene* **95**, 195–201.

Torlesse H, Crompton DWT and Savioli L. 2003. Anthelminthic treatment during pregnancy. In: *Controlling Disease due to Helminth Infections* (eds. DWT Crompton, A Montresor, MC Nesheim and L Savioli). Geneva: World Health Organization. pp. 135–142.

UNICEF. 2005. Water, sanitation and education. www.unicef.org/wes/index_schools.

Vanden Bossche H. 1995. Principles of anthelminthic chemotherapy. In: *Enteric Infection 2. Intestinal helminths* (eds. MJG Farthing, GT Keusch and D Wakelin). London: Chapman and Hall. pp. 267–286.

Villar J, Klebanoff M and Kestler E. 1989. The effect on fetal growth of protozoan and helminthic infection during pregnancy. *Obstetrics and Gynecology* **76**, 915–9920.

Watkins WE, Cruz JR and Pollitt E. 1996. The effects of deworming on indicators of school performance in Guatemala. *Transactions of the Royal Society of Tropical Medicine and Hygiene* **90**, 156–161.

Webster J, Hodges ME, Crompton DWT and Walters DE. 1990. Intestinal parasitic infections in children from Freetown, Sierra Leone. *Journal of the Sierra Leone Medical and Dental Association* **5**, 144–155.

Williams D, Burke G and Owen Hendly J. 1974. Ascariasis: a family disease. *Journal of Pediatrics* **84**, 853–854.

WHO. 1980. *Manual of Basic Techniques for a Health Laboratory.* Geneva: World Health Organization.

WHO. 1994. *Bench Aids for the Diagnosis of Intestinal Parasites.* Geneva: World Health Organization.

WHO. 1996a. Report of the WHO informal consultation on hookworm infection and anaemia in girls and women. Geneva: World Health Organization (WHO/CTD/SIP/96.1).

WHO. 1996b. Report of the WHO Informal Consultation on the use of chemotherapy for the control of morbidity due to soil-transmitted nematodes in humans. Geneva: World Health Organization (WHO/CTD/SIP/96.2).

WHO. 2002a. Prevention and control of schistosomiasis and soil-transmitted helminthiasis. Report of an Expert Committee. *Technical Report Series* 912, Geneva: World Health Organization.

WHO. 2002b. Report of the WHO Informal Consultation on the use of praziquantel during pregnancy/lactation and albendazole/mebendazole in children under 24 months. Geneva: World Health Organization. (WHO/CDS/CPE/PVC/2002.4).

WHO. 2003. *Battling iron deficiency anaemia.* www.who.int/nut/ida.htm.

WHO. 2004a. Intensified control of neglected diseases. Report of an international workshop. Geneva: World Health Organization (WHO/CDS/CPE/CEE/2004.45).

WHO. 2004b. *WHO Model Formulary 2004.* Geneva: World Health Organization.

Xu Long-Qi, Yu Sen-Hai and Xu Shu-Hui. (eds.). 1999. *Distribution and Pathogenic Impact of Human Parasites in China.* Shanghai: People's Medical Publishing House.

Yu Sen-hai and Shen Wei-xia. 1990. Hookworm infection and disease in China. In: *Hookworm Disease* (eds. GA Schad and KS Warren). London, New York and Philadelphia: Taylor & Francis. pp. 44–54.

Zaman V and Keong LA. 1989. *Handbook of Medical Parasitology,* 2nd edition. Singapore: KC Ang Publishing Pte Ltd.

Zhou Xianmin, Peng Weidong, Crompton DWT and Xiong Jiangqin. 1999. Treatment of biliary ascariasis in China. *Transactions of the Royal Society of Tropical Medicine and Hygiene* **93**, 561–564.

12 Strongyloidiasis

The diseases of poor people when related to infection with helminths are invariably classified as the neglected or forgotten diseases (Molyneux, 2004). That view may prevail in the north and it may apply particularly to the north's response to strongyloidiasis, which often appears to be the most neglected of all. Strongyloidiasis is an assemblage of syndromes that usually accompany infection with *Strongyloides stercoralis*, but can also occur from infection with *S. fuelleborni* and *S. fuelleborni kellyi*. Millions of people living in tropical and subtropical countries are at risk of infection and disease (Crompton, 1999; Grove, 1989a, 1996; Sturchler, 1988). Strongyloidiasis is yet another drain on the health and energy of people who live where sanitation is inadequate. The report of a WHO Expert Committee, published in 1987 under the title *"Prevention and Control of Intestinal Parasitic Infections"* and extending to 86 pages, devoted only seven lines to the prevention and control of strongyloidiasis and concluded "further work is needed before any realistic and specific control measures can be suggested" (WHO, 1987).

The persistence of *Strongyloides* depends on the interaction between people and contaminated soil. For some arbitrary reason *Strongyloides* is often excluded from the quartet of common soil-transmitted helminths (Chapter 11). In their seminal work on the dynamics and control of the infectious diseases of humans, Anderson and May (1992) did not cover strongyloidiasis. Perhaps its control has seemed intractable because the four broad-spectrum anthelminthic drugs recommended for dealing with soil-transmitted helminthiasis in the community are not particularly effective against *Strongyloides* infections. Our ignorance still prevents us from designing a realistic control strategy, almost 20 years after the WHO report. The omission of strongyloidiasis from the public health agenda needs to be put right. Strongyloidiasis deserves more than lip service in reports.

12.1 SPECIES, HOST RANGE, AND LIFE HISTORY

Thirteen species of *Strongyloides* have been reported to infect humans (Chapter 1, List; Coombs and Crompton, 1991). Members of this genus of nematodes are remarkable for being able to exploit a parasitic and a free-living life style, a direct (homogonic) and indirect (heterogonic) developmental pattern, sexual and asexual (parthenogenetic) reproduction, multiplication within the definitive host, and autoinfection. The helminth's ability to switch between developmental states is controlled by chemosensory neurons that respond to environmental cues. Under experimental conditions, Ashton et al. (1998) used laser microbeams to knock out different neurons and so produced larvae that adopted either the homogonic or the heterogonic developmental pattern in relation to the presence or absence of particular neurons. All adult *Strongyloides* are small worms measuring a few millimeters in length. *Strongyloides stercoralis* is of the greatest public health concern, but for some communities *S. fuelleborni* and *S. fuelleborni kellyi* also pose a risk to health.

12.1.1 *STRONGYLOIDES STERCORALIS*

The morphology of the free-living and parasitic stages of *S. stercoralis* has been described by Georgi (1982), Schad (1989), and Muller (2002). Patterns in the worm's versatile life history and development are shown in Figure 12.1 and the morphology of some of the developmental stages is shown in Figure 12.2. The following summary of the life history follows that given by Schad (1989). Infection of the human host occurs when third-stage larvae (L3) penetrate skin. Some of these larvae remain in the skin while others enter blood vessels and are then assumed to reach the lungs and settle in the alveoli before migrating to the small intestine via the trachea. Two more molts must occur before maturity is attained in the small intestine. The population of *S. stercoralis* that becomes established in the mucosa of the small intestine consists of parthenogenetic females. These worms deposit eggs in the mucosal tissues that hatch to release first-stage larvae (L1). Most L1 pass out of the host in the feces, but some may remain and attain maturity in the gut thereby giving rise to a long-lasting infection through autoinfection. Under certain circumstances, hyperinfection occurs when numerous infective larvae (L3) invade the body after development in the host's gut (Figure 12.1).

The L1 that reach the external environment feed, molt twice to form infective L3 (homogonic development) or give rise to a free-living generation (heterogonic development) that flourishes in the soil (Figure 12.1). The free-living generation culminates with sexually mature male and female worms whose offspring develop to form infective, skin-penetrating L3. Superficially *Strongyloides* L3 have the

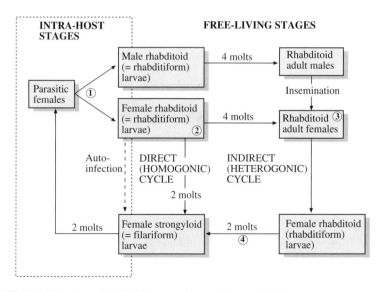

FIGURE 12.1 The alternative life-history pathways of *Strongyloides stercoralis*. From Figure 4 (Schad,1989) with permission from Routledge/Taylor & Francis Group, LLC.

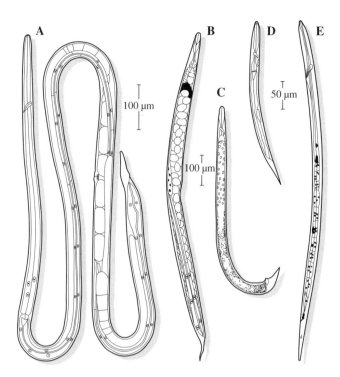

FIGURE 12.2 *Strongyloides stercoralis*. A. Parasitic female. B. Free-living female. C. Free-living male. D. First-stage (rhabditoid) larva from fresh feces. E. Third-stage (filariform) larva from surface of fecal specimen 24 hours after collection. From illustrations in Little (1985), *Journal of Parasitology* 52, 69–84, with permission from the American Society of Parasitologists.

appearance of hookworm L3 except that they are not enclosed in the cuticular sheaths formed when L2 molt to become L3. Schad pointed out that the conditions that stimulate the L1 to develop directly to become infective L3 or indirectly to become infective L3 after a free-living cycle are not well understood. Favorable conditions in the host and unfavorable conditions outside the host are thought to promote homogonic development, while reverse circumstances are thought to promote heterogonic development. The evidence for this tentative conclusion comes from experimental work with species of *Strongyloides* infecting pigs, rabbits, and sheep. From a public health perspective, however, possession of such developmental plasticity may impede control activities.

Autoinfection is another consequence of developmental versatility. Female worms resulting from L1 that have not left the gut may remain there and deposit eggs. If the resulting larvae develop to infectivity (L3) without being passed out of the host, autoinfection can occur. Apparently autoinfection usually takes place when L3 invade the colonic mucosa (internal autoinfection) or the tissues in the perianal region (external autoinfection) after expulsion from the gut. Attaining infectivity before expulsion in the feces means that *S. stercoralis* may be transmitted during anal intercourse (Schad, 1989). Autoinfection explains how some individuals have remained infected with *S. stercoralis* for years after arriving in an area where the infection is not endemic. The condition is well known with regard to former prisoners of war who were found to be infected years after having left the Far East and returned home to a *Strongyloides*-free environment (Pelletier, 1984).

Strongyloides stercoralis is not restricted to the human host. Natural infections of *S. stercoralis* have been found in domesticated cats and dogs, chimpanzees, gibbons, orangutans, and various species of monkey (Genta and Grove, 1989). Zoonotic transmission of *S. stercoralis* is possible, however, and has been reported to occur from asymptomatic dogs to animal workers (Georgi and Sprinkle, 1974). Ackert and Payne (1922) demonstrated that *S. stercoralis* L3 remained viable after passing through the alimentary tract of pigs and Ackert (1922) found that *S. stercoralis* L1 were able to develop to infective L3 after passing through chickens. There is little information to indicate whether nonhuman hosts or the activities of domesticated animals have the potential to upset control activities.

12.1.2 *Strongyloides fuelleborni*

Strongyloides fuelleborni relies mainly on nonhuman primates for its hosts. Ashford and Barnish (1989) have recommended that *S. fuelleborni* is the name to be used for *Strongyloides* infecting wild and captive Old World nonhuman primates and humans from the same locations. *Strongyloides fuelleborni* is probably a widespread infection among rural people in sub-Saharan Africa. Its life history pattern is basically the same as that of *S. stercoralis* (Figure 12.1), but the eggs released by the parthenogenetic female worms do not hatch in the gut and instead are carried out of the host in the feces. For some as yet unknown reason the eggs of *S. fuelleborni* are seen to be passed in clumps or strings. The L1 escape from their surrounding eggshells in the host's environment and then embark on either the homogonic or heterogonic developmental cycle. Infection is assumed to occur when infective L3 penetrate skin. There is circumstantial evidence to indicate that infections observed in neonatal monkeys had become established by either the transplacental or transmammary route (Wong, 1989). In a study in rural Zaire, Brown, and Girardeau (1977) found that 26 out of 76 infants aged less than 200 days were infected with *S. fuelleborni*. This prevalence rate (34%), although less than that of 44% in the general population, seemed too high to be explained by the children having contact with contaminated soil. Three larval *S. fuelleborni* were then recovered from a sample of human breast milk. Autoinfection ought not to occur with *S. fuelleborni* if the eggs always pass out of the host before hatching. Similarly, hyperinfection ought not to occur because there will be no completion of the homogonic cycle in the gut of the host.

12.1.3 *Strongyloides fuelleborni kellyi*

Kelly and Voge (1973) published a brief paper describing a form of *Strongyloides* infection in people living in New Guinea. The nematode clearly resembled *S. fuelleborni* because eggs not larvae were passed in host feces, but since wild nonhuman primates are not found in New Guinea a firm identification was not made. Further research revealed that the infection was distributed widely throughout Papua New

Guinea (Ashford et al., 1992) and a new subspecies has been recognized and named *Strongyloides fuelleborni kellyi* by Viney et al. (1991).

12.2 IDENTIFICATION AND DIAGNOSIS

For all but dedicated experts, the identification of species of *Strongyloides* is a daunting and difficult task. Live worms are required for processing and examination. Similarly, diagnosis is tricky and depends on a combination of clues from parasitological, immunological, and clinical investigations. A basic framework for reliable diagnosis (Figure 12.3) has been devised by Grove (1989b). Detailed clinical indicators of strongyloidiasis are discussed in Section 12.4.

A key for the identification of genera in the superfamily Rhabditoidea, to which *Strongyloides* belongs, has been constructed by Anderson and Bain (1982) and information about how to identify species of *Strongyloides* has been published by Speare (1989). In most public health situations, however, a pragmatic approach to identification and parasitological diagnosis will have to be adopted because of constraints on resources and time. Infection with *Strongyloides* spp. should be anticipated if a parasitological survey is to be carried out where the species are known to be endemic and technical staff should be trained accordingly. Two important points should be considered with reference to the *Bench Aids for the Diagnosis for Intestinal Parasites* (WHO, 1994) and Appendix 3.

First, although the presence of larvae in stool samples is generally assumed to be diagnostic for *S. stercoralis*, the sample should be as fresh as possible to avoid

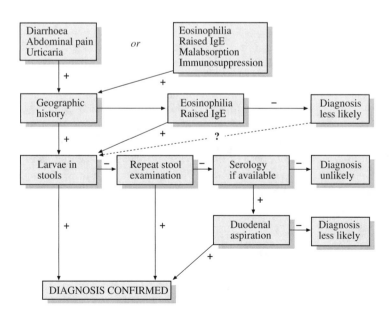

FIGURE 12.3 A flow chart for the diagnosis of strongyloidiasis. From Figure 9 (Grove, 1989b) with permission from Routledge/Taylor & Francis Group, LLC.

contamination with any larvae of free-living nematodes (Figure 12.2). *Strongyloides* larvae are not always released in large numbers. Grove (1989b) describes how an investigator was unable to find larvae in the stools of 20 patients who were known to be infected because larvae were present in their duodenal aspirates. Concentration techniques may need to be used before stool samples are examined. Stool samples may need to be collected over several days. Secondly, care should be taken to confirm that thin-shelled eggs found in stool samples belong to *S. fuelleborni* and not to one of the common species of hookworm. Mixed infections may occur. Eggs of *S. fuelleborni* measure about 50×34 μm. Those of *S. fuelleborni kellyi* are slightly longer and a little narrower and are found in clumps or strings. Hookworm eggs measure 60 to 75×36 to 40 μm.

Immunodiagnosis can be used provided that adequate laboratory facilities and equipment are available. Once work has been done to prepare reagents, enzyme-linked immunoabsorbent assays (ELISAs) offer a sensitive means of detecting IgG antibodies produced against *Strongyloides*. Grove (1989b) contains a protocol for an ELISA with a sensitivity of 93% and a specificity of 95% when a reading of 0.2 optical density units was taken to be the cut-off point. Such techniques are well suited for diagnosis in a hospital setting.

12.3 DISTRIBUTION AND ABUNDANCE

Although millions of people may be at risk of infection with *Strongyloides* spp. (Table 12.1) an accurate estimate of the numbers who are infected and unwell remains elusive. Individuals who are infected or even quite seriously ill may not be passing larvae in their stools at the time of a survey. Sturchler (1988), in a projection dated 23rd August, 1987, and based on stool examinations conducted since the 1970s, indicated that infection with *S. stercoralis* had a global distribution encompassing 110 countries that extended from Belgium in the north to the Republic of South Africa in the south. Infection with *Strongyloides* spp. in many of the 110 countries was considered to be sporadic rather than endemic. Sturchler proposed that a population prevalence of 1 to 10% indicated that the infection was endemic and a prevalence of > 10% revealed the infection to be hyperendemic. Ukoli (1984), in contrast to Sturchler (1988) and Grove (1989a), seemed to play down the public health significance of infection with *Strongyloides* spp. He stated that its prevalence is less than 20% in most parts of tropical Africa and even when present it is generally not considered of great medical importance.

The distribution of *S. fuelleborni* in Africa was elucidated by Pampiglione and Ricciardi (1972) who collected stool samples from 4,577 individuals at 45 centers in West, Central, and East Africa over a period of 4 years. Evidence for the infection was obtained from Benin, Burundi, Cameroon, Central African Republic, Congo, Democratic Republic of the Congo, Ethiopia, Mozambique, Sierra Leone, Somalia, Tanzania, Togo, and Uganda. Prevalence values ranged from 0.2 to 100%; no significant difference was found between males and females, but rates were somewhat higher in children (aged 2 to 12 years) than in adults. All the centers were situated where people would be expected to share their environment with nonhuman primates.

TABLE 12.1
Frequencies of Symptoms in Four Series of Patients With Chronic Strongyloidiasis (Expressed As a Percentage). From Table 1 (Grove, 1989c) with permission from Routledge/Taylor & Francis Group LLC.

	Hinman (1937)* n = 66	Jones (1950)* n = 100	Milder et al. (1981)* n = 56	Pelletier (1984) n = 52
Abdominal pain	67	79	39	67
Anorexia	—	28	—	27
Nausea ± vomiting	11	54	32	27
Diarrhea	15	36	41	42
Constipation	2	26	20	41
Pruritus ani	—	—	4	85
Weight loss	11	53	20	—
Urticaria	—	22	4	100
Larva currens	—	—	2	92
Chest pain	—	6	—	50
Cough	—	21	5	25
Dyspnea	—	18	—	52
Wheeze	—	—	2	40
Malaise/weakness	14	16	—	—
Fever	6	13	—	—
Nervousness	—	27	—	—
Headache	—	18	—	—
Vertigo	—	14	—	—

Note: —, not mentioned

* See Grove (1989c).

This distribution clearly overlaps with that plotted for *S. stercoralis* by Sturchler (1988).

People in sub-Saharan Africa probably suffer most from strongyloidiasis. Sturchler (1988) calculated the median prevalence of infection to be 3% of the continent's population based on pooling results from 400,000 stool samples. If the figure of 3% is applied to the current population of sub-Saharan Africa (see UNICEF, 2004) about 20 million people there will be infected with *Strongyloides*. From a review of data undertaken by Pawlowski (1989) the distribution of *Strongyloides* infections appears to be patchy and not to show any age- or gender-specific patterns. More seems to be known about the detail of the infection in nonendemic countries such as those in Europe to which infected people have returned after spells abroad. In genuinely endemic locations, the infection persists wherever sanitation and public and personal hygiene are inadequate to reduce transmission.

The distribution of *S. fuelleborni kellyi* in Papua New Guinea is now well known. Transmission occurs in the eastern part of the country and particularly around Kanabea, Gulf Province (Ashford and Barnish, 1987). High prevalence values have been measured in young children and the infection is observed to decline in older children and adults. Infection has been detected in children as young as 18 and 21 days. How did such young children become infected?

12.4 PATHOLOGY, MORBIDITY, AND MORTALITY

Strongyloidiasis due to *S. stercoralis* is considered by Genta and Caymmi Gomes (1989) to be a systemic infection resulting from the helminth's ability to invade all parts of the host's body. Detailed analysis of the pathogenesis of *S. stercoralis* infections and the pathology this elicits has been provided by Galliard (1967) and by Genta and Caymmi Gomes (1989). The strength of the host's immunocompetence undoubtedly influences the course of the disease.

Grove (1996) recognizes two syndromes resulting from infection with *S. stercoralis*. These are (1) chronic uncomplicated strongyloidiasis and (2) severe complicated strongyloidiasis, with conditions intermediate and overlapping between the two. Danescu (1976) proposed that the intensity of an intestinal infection of *Strongyloides* could be based on counting the numbers of larvae produced per unit time from a stool sample of known mass. The general relationship between the intensity of a *Strongyloides* infection and the seriousness of the accompanying morbidity has been found to apply; as the number of worms in the host increases, the morbidity worsens (Rawlins et al., 1983; Figure 12.4). Many infected people are probably asymptomatic while many others with the chronic form of the disease will experience numerous symptoms and be ill.

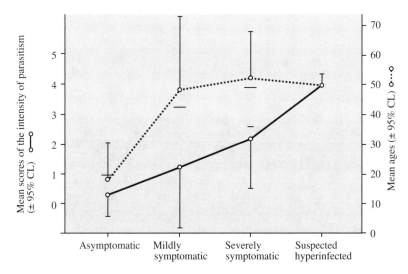

FIGURE 12.4 Variations of morbidity with intensity of infection and ages of patients. Intensity of parasitism was scored as 1, < 100; 2, 100 to 349; 3, 350 to 499; 5, > 1000 larvae per ml stool. From Figure 1 (Grove, 1989c) with permission from Routledge/Taylor & Francis Group, LLC.

12.4.1 Chronic Uncomplicated Strongyloidiasis

Results from four investigations of the symptoms of patients judged to have chronic uncomplicated strongyloidiasis are summarized in Table 12.1. Grove (1996) concluded that the triad of abdominal pain, diarrhea, and urticaria are strongly suggestive of strongyloidiasis, but that the key diagnostic feature is the retrieval and identification of the parasite. The two types of skin lesion listed in Table 12.1 are interesting. Urticaria lasting for a few days occurs on various parts of the body, but most commonly on the buttocks. The problem recurs often weeks or months later and probably has an allergic basis. Larva currens (Table 12.1) was first described by Fuelleborn and was named by Arthur and Shelley (1958). The lesion appears to be specific to infection with *Strongyloides* and should not be confused with cutaneous larva migrans (creeping eruption) that is a feature of infection with the larvae of zoonotic hookworms and other parasites (Beaver, 1956). Larva currens, which is extremely itchy and is more common in people infected in Asia than elsewhere, is caused by the "rapid" movement of larval *Strongyloides* that may travel in the subcutaneous tissue as much as 100 mm in an hour. In contrast, the nematodes responsible for cutaneous larva migrans may take 24 hours to move a few millimeters (see Grove, 1989c).

12.4.2 Severe Complicated Strongyloidiasis

This outcome of infection with *Strongyloides* is also referred to as disseminated strongyloidiasis, overwhelming strongyloidiasis, hyperinfective strongyloidiasis, and massive strongyloidiasis and it develops when the tissues are invaded by innumerable worms. Grove (1989c) considered that the condition is most likely to arise when asymptomatic carriers or patients experiencing chronic uncomplicated strongyloidiasis undergo some impairment of the immune system. His literature investigation revealed that out of 180 patients who had developed severe complicated strongyloidiasis, 127 had been prescribed immune suppressant drugs to alleviate their primary health problem. Since then, Heyworth (1996) has advised that the administration of immunosuppressant corticosteroid drugs to patients with chronic *S. stercoralis* infection can trigger a life-threatening form of strongyloidiasis. HIV/AIDS patients would appear to be at obvious risk of hyperinfective strongyloidiasis given the fact that the virus causes devastating immunosuppression. Interestingly, Grove (1989c) found only one case, reported in 1987, of hyperinfective strongyloidiasis in an AIDS patient. In strong support of this observation is the finding that in individuals with normal immune function, as would usually be the case if corticosteroid drugs had been prescribed, *S. stercoralis* undergoes direct development and is unlikely to invade the body. When immune function has been damaged, as occurs when HIV is established, *S. stercoralis* undergoes indirect development and the risk of the hyperinfective syndrome is avoided (Viney et al., 2004).

The gastrointestinal tract, the lungs, and the central nervous system are most commonly affected during severe complicated strongyloidiasis. The most frequent manifestations are steatorrhea, protein-losing enteropathy, hypoalbuminemia, weight loss, edema, and pulmonary and neurological complications (Grove, 1996). Secondary bacterial invasion may lead to meningitis and septicemia. The literature concerning severe complicated strongyloidiasis regularly refers to autopsies and *post mortem* examinations. Whether strongyloidiasis is a direct cause of death or not is unclear, but the severe form of the disease may well ensure that concurrent health problems become life threatening.

What may be classed as a bizarre form of severe complicated strongyloidiasis is swollen belly (or baby) syndrome (SBS) that was discovered in Papua New Guinea (see Ashford and Barnish, 1987). SBS has proved to be a disease with a high fatality rate that strikes infants aged about 2 months. The infants present in a highly distressed state with a swollen abdomen and respiratory difficulties. Ashford and Barnish reported that by the mid-1980s the SBS case rate was about eight per 100 live births and that SBS was responsible for about 30% of infant deaths in the region (Section 12.3). There is no doubt that *S. f. kellyi* was involved in the etiology of SBS with infected infants passing more than 100,000 eggs per ml of loose stool.

12.5 ANTHELMINTHIC TREATMENT

Infection with *Strongyloides* spp. has proved to be difficult to treat successfully. WHO (2004) recommends that all infected patients should be treated with albendazole or ivermectin. The emphasis on infected patients implies that accurate diagnosis will have been carried out. But diagnosis of *Strongyloides* is a difficult task for public health services in a country where the infection is endemic (Section 12.2). Albendazole given as an oral dose of 400 mg once or twice daily for 3 consecutive days can clear up to 80% of infections and is well tolerated by those who receive it. According to WHO (2004) ivermectin is the drug of choice because it can be given as a single oral dose of 200 μm/kg body weight on 2 consecutive days. Neither of these treatment regimes is ideal for a community program intended to reduce morbidity.

Marti et al. (1996) compared the effectiveness of ivermectin and albendazole for the treatment of *S. stercoralis* infection in children in rural Zanzibar. The usual group of soil-transmitted helminth infections was established in the children. Ivermectin was given as a single oral dose of 200 μm/kg body weight and albendazole was given as a single oral dose of 400 mg on 3 consecutive days. The effectiveness of the treatment was assessed by each subject bringing stool samples (two per subject on consecutive days) three weeks after receiving the drug. Stools were examined by the Kato Katz, Baermann, and fecal culture methods. The cure rate for ivermectin against *S. stercoralis* was found to be 83% and that for albendazole to be 45%. Ivermectin was effective against *Ascaris lumbricoides* (100%), much less effective against *Trichuris trichiura* (11%), and ineffective against hookworms. Albendazole behaved as expected for a broad spectrum anthelminthic drug against the soil-transmitted helminth infections.

Operational research that seeks to facilitate the use of anthelminthic drugs in the community must pay attention to the frequency and severity of adverse effects following treatment. Adverse effects, however minor, must be identified in advance so that the community can be reassured about their transient impact, otherwise compliance will crash dive. Marti et al. (1996) recruited a Zanzibari medical assistant to investigate adverse effects following treatment. For ivermectin, 45 out of 152 and for albendazole 53 out of 149 subjects experienced some discomfort soon after treatment. Headache and loose stools were the commonest adverse effects, but none lasted beyond 3 days following treatment. These findings indicate that a single oral dose of ivermectin (200 µg/kg body weight) is the best available treatment for chronic uncomplicated strongyloidiasis.

12.6 PREVENTION AND CONTROL

Measures to reduce the impact of strongyloidiasis in the community are essentially similar to those available for soil-transmitted helminthiasis (Pawlowski, 1989). In theory, short-term control of morbidity might be achieved through anthelminthic treatment with long-term control being maintained through access to sanitation. Infections of *S. stercoralis,* however, remain difficult to treat with a single, oral dose of anthelminthic drug and extending the provision of sanitation often seems remote. Pawlowski also discusses measures to interfere with larval development such as growing plants with larvicidal properties and altering the pH of soil with chemical fertilizers. Much more operational research is needed to prepare for the prevention and control of strongyloidiasis.

REFERENCES

Ackert JE. 1922. Investigations on the control of hookworm disease IV. The relation of the domestic chicken to the spread of hookworm disease. *American Journal of Hygiene* **2**, 26–38.

Ackert JE and Payne FK. 1922. Investigations on the control of hookworm disease V. The domestic pig and hookworm dissemination. *American Journal of Hygiene* **2**, 39–50.

Anderson RC and Bain O. 1982. Keys to the genera of the superfamilies Rhabditoidea, Dioctophymatoidea, Trichinelloidea and Muspiceoidea. In: *CIH Keys to the Nematode Parasites of Vertebrates*, No. 9 (eds. RC Anderson, AG Chabaud and S Willmott). Farnham Royal, Bucks: Commonwealth Agricultural Bureaux.

Anderson RM and May RM. 1992. *Infectious Diseases of Humans.* Oxford University Press. (paperback edition).

Arthur RP and Shelley WB. 1958. Larva currens. A distinctive variant of cutaneous larva migrans due to *Strongyloides stercoralis. Archives of Dermatology* **78**, 186–190.

Ashford RW and Barnish G. 1987. Strongyloidiasis in Papua New Guinea. In: *Bailliere's Clinical Tropical Medicine and Communicable Diseases. Intestinal Helminthic Infections* (ed. ZS Pawlowski). London and Philadelphia: Balliere Tindall. pp. 765–773.

Ashford RW and Barnish G. 1989. *Strongyloides fuelleborni* and similar parasites in animals and man. In: *Strongyloidiasis, a Major Roundworm Infection of Man* (ed. DI Grove). London, New York and Philadelphia: Taylor & Francis. pp. 271–286.

Ashford RW, Barnish G and Viney ME. 1992. *Strongyloides fuelleborni kellyi*: infection and *disease* in Papua New Guinea. *Parasitology Today* **8**, 314–318.

Ashton FT, Bhopale VM, Holt D et al. 1998. Developmental switching in the parasitic nematode *Strongyloides stercoralis* is controlled by the ASF and ASI amphidial neurons. *Journal of Parasitology* **84**, 691–695.

Beaver PC. 1956. Parasitological reviews. Larva migrans. *Experimental Parasitology* **5**, 587–621.

Brown RW and Girardeau MHE. 1977. Transmammary passage of *Strongyloides* sp. larvae in the human host. *American Journal of Tropical Medicine and Hygiene* **26**, 215–219.

Coombs I and Crompton DWT. 1991. *A Guide to Human Helminths*. London, New York and Philadelphia: Taylor & Francis.

Crompton DWT. 1999. How much human helminthiasis is there in the world? *Journal of Parasitology* **85**, 397–403.

Danescu P. 1976. Observations concerning the parasite load, duration of infection and clinical manifestations in strongyloidiasis. *Transactions of the Royal Society of Tropical Medicine and Hygiene* **70**, 162–163.

Galliard H. 1967. Pathogenesis of *Strongyloides*. *Helminthological Abstracts* **36**, 247–260.

Genta RM and Caymmi Gomes M. 1989. Pathology. In: *Strongyloidiasis, a Major Roundworm Infection of Man* (ed. DI Grove). London, New York and Philadelphia: Taylor & Francis. pp. 105–132.

Genta RM and Grove DI. 1989. *Strongyloides stercoralis* infections in animals. In: *Strongyloidiasis, a Major Roundworm Infection of Man* (ed. DI Grove). London, New York and Philadelphia: Taylor & Francis. pp. 251–269.

Georgi JR. 1982. Strongyloidiasis. In: *Handbook Series in Zoonoses*. Section C: *Parasitic Zoonoses*, Volume II (ed. MG Schultz). Boca Raton, Florida: CRC Press Inc. pp. 257–267.

Georgi JR and Sprinkle CL. 1974. A case of human strongyloidosis apparently contracted from asymptomatic colony dogs. *American Journal of Tropical Medicine and Hygiene* **23**, 899–901.

Grove DI. (ed.).1989a. *Strongyloidiasis, a Major Roundworm Infection of Man*. London, New York and Philadelphia: Taylor and Francis.

Grove DI. 1989b. Diagnosis. In: *Strongyloidiasis, a Major Roundworm Infection of Man* (ed. DI Grove). London, New York and Philadelphia: Taylor & Francis. pp. 175–197.

Grove DI. 1989c. Clinical manifestations. In: *Strongyloidiasis, a Major Roundworm Infection of Man* (ed. DI Grove). London, New York and Philadelphia: Taylor & Francis. pp. 155–175.

Grove DI. 1996. Human strongyloidiasis. *Advances in Parasitology* **38**. 251–309.

Heyworth MF. 1996. Parasitic diseases in immunocompromised hosts. Cryptosporidiosis, isosporiasis, and strongyloidiasis. *Gastroenterological Clinics in North America* **25**, 691–707.

Kelly A and Voge M. 1973. Report of a nematode found in humans at Kiunga, Western District. *Papua New Guinea Medical Journal* **16**, 59.

Little MD. 1985. Nematodes of the digestive tract and related species. In: *Animal Agents and Vectors of Human Disease*, 5th edition (eds. PC Beaver and RC Jung). Philadelphia: Lea & Febiger. pp. 127–170.

Marti H, Haji HJ, Savioli L et al. 1996. A comparative trial of a single-dose ivermectin versus three days of albendazole for treatment of *Strongyloides stercoralis* and other soil-transmitted helminth infections in children. *American Journal of Tropical Medicine and Hygiene* **55**, 477–481.

Molyneux DH. 2004. "Neglected" diseases but unrecognised successes — challenges and opportunities for infectious disease control. *Lancet* **364**, 380–383.

Muller R. 2002. *Worms and Disease*, 2nd edition. Wallingford, Oxfordshire: CAB International.

Pampiglione S and Ricciardi ML. 1972. Geographic distribution of *Strongyloides fuelleborni* in humans in tropical Africa. *Parassitologia* **14**, 329–338.

Pawlowski ZS. 1989. Epidemiology, prevention and control. In: *Strongyloidiasis, a Major Roundworm Infection of Man* (ed. DI Grove). London, New York and Philadelphia: Taylor & Francis. pp. 233–249.

Pelletier LL. 1984. Chronic strongyloidiasis in World War II Far East ex-prisoners of war. *American Journal of Tropical Medicine and Hygiene* **33**, 55–61.

Rawlins SC, Terry SI and Chen WN. 1983. Some laboratory, epidemiological and clinical features of *Strongyloides stercoralis* in a focus of low endemicity. *West Indies Medical Journal* **32**, 212–218.

Schad GA. 1989. Morphology and life history of *Strongyloides stercoralis*. In: *Strongyloidiasis, a Major Roundworm Infection of Man* (ed. DI Grove). London, New York and Philadelphia: Taylor & Francis. pp. 85–104.

Speare R. 1989. Identification of species of *Strongyloides*. In: *Strongyloidiasis, a Major Roundworm Infection of Man* (ed. DI Grove). London, New York and Philadelphia: Taylor & Francis. pp. 11–83.

Sturchler D. 1988. *Endemic Areas of Tropical Infections,* 2nd edition. Toronto, Lewiston NY, Bern and Stuttgart: Hans Huber Publishers.

Ukoli FMA. 1984. *Introduction to Parasitology in Tropical Africa.* Chichester and New York: John Wiley and Sons Limited.

UNICEF. 2004. *The State of the World's Children 2005.* New York: The United Nations Children's Fund.

Viney ME, Ashford RW and Barnish G. 1991. A taxonomic study of *Strongyloides* Grassi, 1879 (Nematoda) with special reference to *Strongyloides fuelleborni* von Linstow, in man in Papua New Guinea and the description of a new sub species. *Systematic Parasitology* **18**, 95–109.

Viney ME, Brown M, Omoding NE et al. 2004. Why does HIV infection not lead to disseminated strongyloidiasis? *Journal of Infectious Diseases* **190**, 2175–2180.

Wong MM. 1989. Personal communication to Ashford and Barnish (1989 p. 276).

WHO. 1987. Prevention and Control of Intestinal Parasitic Infections. Report of a WHO Expert Committee. *Technical Report Series* 749. Geneva: World Health Organization.

WHO. 1994. *Bench Aids for the Diagnosis of Intestinal Parasites.* Geneva: World Health Organization.

WHO. 2004. *WHO Model Formulary 2004.* Geneva: World Health Organization.

13 Trichinellosis (Trichinosis)

Trichinellosis is a chronic disease that affects about 11 million people around the world (Table 1.1). The problem extends from Argentina to Zimbabwe and accompanies infection with nematode worms of the genus *Trichinella*. Morbidity results from the activities of the minute adult worms in the intestinal mucosa and the colonization of skeletal muscle by first-stage larvae (L1), also called trichinae (Figure 13.1). Once the worm's life cycle had been elucidated and the significance of eating contaminated pork and pork products appreciated, the disease was justifiably given major public health significance in Europe in the 19th century. Considerable prevention and control of trichinellosis in many communities of the North has been achieved by the public health interventions of meat inspection, food hygiene, and improved pig husbandry. Nevertheless, outbreaks of trichinellosis continue to occur in the North and elsewhere around the world and the disease must remain as a standing item on the public health agenda. Vigilant public health professionals should always be assigned to the task of surveillance for outbreaks of trichinellosis.

Trichinella appears to have exceedingly low host specificity and will develop in a wide range of domesticated and wild species of mammal. The infection thrives in carnivorous and scavenging animals. People will always be at risk from this zoonotic infection if their food, food habits, social gatherings, and festive meals either regularly or occasionally include hosts of *Trichinella*. At first, trichinellosis was assumed to be caused by infection with a single species identified as *T. spiralis*. Now other species and strains of *Trichinella* are known to be agents of the disease (List, Chapter 1; Table 13.1) and more may be described since molecular investigations have begun to characterize distinct genotypes within some of the species (Pozio, 2005).

13.1 TRANSMISSION OF SPECIES OF *TRICHINELLA* TO HUMANS

Some information about species of *Trichinella* known to infect humans is introduced in Table 13.1. Much more information can be traced through the references cited in the table. Species and strains of *Trichinella* are adapted to infect a wide range of wild animals. When established naturally in wild animals such as hyenas, their development is referred to as having a sylvatic cycle. When established in domesticated animals such as pigs, their development is said to have a domestic cycle. Humans have steadily intruded into the extensive sylvatic cycle (Figure 13.2) through hunting, animal husbandry, social behavior, and cultural practices associated with food. Animal husbandry set up the domestic cycle that has plagued human health

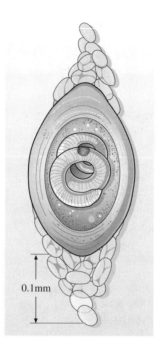

0.1mm

FIGURE 13.1 Cyst of a first-stage larva of *Trichinella spiralis*. From an illustration by Bristowe and Rainey (1854) and reproduced as Figure 72 by Cobbold (1864).

for the thousands of years since wild boar were first domesticated to provide pigs and pork (Clutton-Brock, 1987). *Trichinella* enters the body in raw and undercooked meat contaminated with infective L1. In addition to pork and horse meat, common sources of infection include meals prepared from seals, walruses, bush pigs, wart hogs, bears, and dogs, and bizarre dishes such as barbequed badger and cougar jerky. In countries lacking statutory requirements for carcass inspection before meat and meat products are released for human consumption, people should be advised of the risk of eating undercooked meat from whatever source.

Meat originating from herbivores is no guarantee of protection from infection. People eating horsemeat would not expect to become infected with *Trichinella*, but that has happened frequently since the 1970s in France and Italy (Ancelle, 1998). Perhaps the parasite had been introduced into the food chain by unscrupulous owners adding meat to horse fodder in an attempt to fatten up the animals before sale as happened in Serbia (Murrell et al., 2004). More research is needed to discover how horses become infected; 3,327 cases of trichinellosis, attributable to contaminated horsemeat, have occurred in France and Italy since 1975 (Pozio and Zarlenga, 2005).

There is some evidence to suggest that migrating L1 may pass the placenta. A pregnant woman was one of 336 people infected with *T. britovi* during an outbreak in Slovakia. The woman was infected during the 10th week of her pregnancy and at week 22 the pregnancy was terminated at her request. Immunocytochemical examination revealed that the fetus had also been infected (Dubinsky et al., 2001).

TABLE 13.1

Species of *Trichinella* Infecting Humans. From Capo and Despommier (1996), Pozio and La Rosa (2000), and Owen et al. (2005).

Species (Genotype)	L1	Resistance of L1 to Freezing [a]	Distribution	Common Sources of Infection	Cases of Human Infection
T. spiralis (T1)	E [b]	None	Cosmopolitan (temperate)	Pig (pork) Wild boar Horse	U.S. (Steele, 1982) Italy (Pozio et al., 1993) France, Italy (Ancelle, 1998)
T. nativa (T2, T6)	E	High	Arctic, subarctic, holoarctic	Dog Walrus	NE China (Cui and Wang, 2001) Quebec, Canada (Proulx et al., 2002)
T. nelsoni (T7)	E	None	East Africa	Bush pig	Kenya (Campbell, 1991)
T. britovi (T3, T8, T9)	E	None	Palaearctic (temperate)	Wild boar	Umbria, Italy (Piergili-Fioretti et al., 2005)
T. murrelli (T5)	E	None	Continental U.S., Canada, France	Horse	France (see Pozio and Zarlenga, 2005)
T. pseudospiralis (T4)	NE [c]	None	Palaearctic, Nearctic	Wild boar Wild pig	France (Ranque et al., 2000) Thailand (Jongwutiwes et al., 1998)
T. papuae	NE	None	Oceania	Wild pig	Papua New Guinea (Owen et al., 2005)

[a] Recommended control intervention for killing infective L1 of all species of *Trichinella* with the exception of freeze-resistant *T. nativa*, is $-15\,^{\circ}$C for 30 days (WHO, 2000). Discussion in Pozio and Zarlenga (2005) indicates that more studies are needed on the cold tolerance of L1 stages.

[b] E, encapsulated L1 that are restricted to development in the mammals with body temperatures ranging from 37 to 40°C (Pozio, 2000).

[c] NE, nonencapsulated L1. *Trichinella pseudospiralis* can complete development over a body temperature range of 37 to 42°C and so are infective to mammals and birds (Pozio, 2000).

13.2 DEVELOPMENT AND LIFE HISTORY

Trichinella is remarkable not only for its extensive host range (Figure 13.2) but also for the fact the worm completes its development and life cycle without ever leaving its host. In the natural world *Trichinella* is passed between hosts through hunting or feeding on carrion. The negligible degree of host specificity shown by *Trichinella*

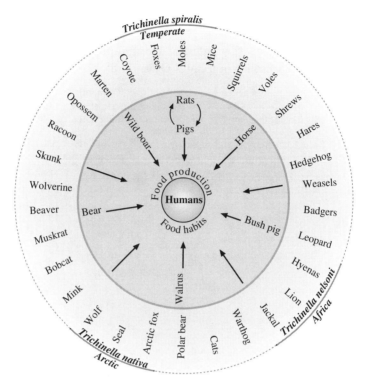

FIGURE 13.2 Human exposure to infection with *Trichinella*. Based on information in Zimmerman (1971), Muller (2002), and Pozio (2005).

has facilitated establishing experimental infections in laboratory animals giving a wealth of information about how *Trichinella* develops and the events in its life history. The following account is taken mainly from Capo and Despommier (1996) and Despommier (1998) and is assumed to apply to the worm's development in humans:

- Infective larvae (L1) are swallowed in raw or undercooked meat. The larvae, which are enclosed in a nurse cell, may be encapsulated (Figure 13.1) or not, depending on the species (Table 13.1).
- The host's digestive processes release the minute larvae from the meat.
- The larvae colonize the mucosa of the small intestine, undergo four molts and attain sexual maturity about 30 hours after the meat was eaten. Adult worms are small; males measure about 1.5 mm in length and about 40 μm in diameter and females about 3 mm and 60 μm.
- Mating occurs and after about 6 days after the meat was eaten female worms begin to discharge the next generation of L1. *Trichinella* is ovoviviparous. There are no obvious egg shells and some authors describe the females as giving birth to their progeny. During a brief reproductive life span, each female worm probably produces about 500 to 1,500 larvae.

- The larvae leave the intestinal mucosa, penetrate the blood and lymphatic systems and so migrate throughout the host's body.
- Those that successfully penetrate striated muscle cells become coiled and are seen to be enveloped in a nurse cell which takes about 20 days to form.
- Larvae may survive in muscle for months or even years. In due course, calcification of dead larvae occurs.

The larva-nurse cell complex is another remarkable feature of the biology of *Trichinella*; and a most informative and well illustrated account of the process of nurse cell formation is available at www.trichinella.org/bio_nursecell.htm. Having entered a striated muscle cell, the larva stimulates the cell to transform into a nurse cell which bears no resemblance to a muscle cell. The fully formed nurse cell is served by a blood supply, presumably to provide nutrients and remove waste products, although oxygen is not needed since L1 stages obtain energy anaerobically. In species that are seen to be encapsulated (Figure 13.1; Table 13.1) the nurse cell initiates the secretion of collagen on its outer surface. Nurse cells probably shield their occupants from host immune responses. L1 are infective to the next host about 16 days after entering striated muscle cells.

13.3 MUSCLE DISTRIBUTION OF *TRICHINELLA* LARVAE

There is much evidence to show that some striated muscles harbor more L1 than others. This finding is important for diagnosis, for meat inspection, and for understanding transmission. Kapel et al. (2005) have recently investigated the distribution of *Trichinella* larvae from domestic and sylvatic hosts in experimentally infected domesticated and wild animals. Samples of muscle tissue, collected at post mortem 5, 10, 20, and 40 weeks after infection from 17 sites in the body, were minced and digested for a set time at 45°C in an aqueous solution containing pepsin and HCl. Larval burdens were determined and the results are shown by rank in Table 13.2. Predilection sites, independent of larval burden, were found to be striated muscle fibers in the tongue, diaphragm, and masseter of pigs, wild boar, and horses, and in the tongue, diaphragm, and forelimbs of foxes. Minor differences to this pattern occur according to the larval genotype.

13.4 DIAGNOSIS

13.4.1 ANIMAL HOSTS

Virchow, the German pathologist who contributed much to public health and the control of trichinellosis, wrote "If one has slaughter houses, then nothing is simpler than setting up microscopes there, and preventing the sale of any pork unless there is an official permit certifying the purity of the said animal." Methods for the detection of *Trichinella* spp. in slaughtered animals and the current state of meat hygiene regulations in the north have been reviewed by Nockler et al. (2000). Detection depends on the reliability of sampling procedures; time constraints prevent the thorough examination of whole carcasses of animals as big as pigs. Trichinoscopy

TABLE 13.2
Rank of Predeliction Sites of Encapsulating and Nonencapsulating
Trichinella spp. in Muscle Tissues of Experimentally Infected Animals.
From Table 1 (Kapel et al., 2005) with permission from Elsevier.[a,b]

	Pig/Wild Boar		Horses		Foxes	
	T. spiralis, *T. nativa* *T. britovi,* *T. nelsoni*	*T. pseudo.* (Russia) (Australia) (U.S.)	*T. spiralis,* *T. britovi*	*T. pseudo.* (Russia)	*T. spiralis,* *T. nativa,* *T. britovi,* *T. murrelli,* *Trichinella T6*	*T. pseudo.* (Russia) (Australia) (U.S.)
Tongue base	1	2	1	3	1	9
Diaphragm	2	1	3	1	3	1
Masseter	6	3	2	2	6	7
Tongue tip	3	5	4	4	—	—
Neck	4	4	5	5	7	5
Abdomen	7	8	6	8	—	—
Tenderloin	10	7	7	6	—	—
Throat	5	6	—	—	—	—
Shoulder	—	—	9	7	—	—
Intercostals	9	10	10	11	—	—
Upper jaw	11	12	—	—	—	—
Upper forelimb	13	9	11	9	4	2
Lower forelimb	14	13	5	14	2	8
Upper hindlimb	—	—	14	10	5	4
Lower hindlimb	12	11	12	15	8	6
Rump	8	14	13	13	—	—
Filet	15	15	15	12	9	3

[a] The study involved 102 pigs, 36 wild boar, 30 horses, and 108 foxes and encapsulated and nonencapsulated *Trichinella*.

[b] Rank: 1, highest mean infestation; 15, lowest mean infestation; —, muscle not sampled.

and pooled sample digestion are two useful methods for the detection of *Trichinella* in animals. In trichinoscopy small samples of muscle (2 × 10 mm) are taken from predilection sites (Table 13.2 and Table 13.3), compressed between glass plates until translucent, and then examined with a light microscope for the presence of larvae. In pooled sample digestion, as much as 100 g of muscle from predilection sites in several animals are digested in 2 l of an aqueous solution of 1% pepsin and 1% HCl for 30 min at 46 to 48° C with continuous mixing from a magnetic stirrer. At the end of the digestion, the brew is passed through a sieve to remove undigested matter and the larvae per unit volume are counted under the microscope. If more than diagnosis is required, the numbers of larvae can be expressed as larvae per gram of muscle (LPG). Beck et al. (2005) compared diagnosis by trichinoscopy with diagnosis by digestion and concluded that digestion should be the method of choice on account of its greater sensitivity. Gajadhar and Forbes (2002) have devised a quality

TABLE 13.3
Predilection Sites for *Trichinella* Larvae in Different Animal Species.
From Table 1 (Nockler et al., 2000) with permission from Elsevier.

Animal Species	Predilection Sites[a,b]	Aim of Examination
Domestic pig	Diaphragm, tongue, masseter	Meat inspection (Domestic meat)
Horse	Tongue, masseter	Meat inspection (Domestic meat)
Wild boar	Forearm, diaphragm	Meat inspection (Game meat)
Bear	Diaphragm, masseter, tongue	Meat inspection (Game meat)
Walrus	Tongue	Diagnosis of infection
Fox	Diaphragm, forearm muscles	Epidemiological studies (reservoir animals)
Raccoon	Diaphragm, forearm muscles	Epidemiological studies (reservoir animals)

[a] 0.5 g samples are recommended for trichinoscopy.
[b] 5 g samples are recommended for pooled sample digestion.

assurance system for diagnostic parasitology in public health based on experience gained with trichinellosis.

Serodiagnostic methods are also available for the detection of *Trichinella* in animal hosts and their merits have been reviewed and compared by Nockler et al. (2000). From all the available tests, the most useful is enzyme-linked immunosorbent assay (ELISA) that can be automated, deal with many samples, and detect the presence of a single larva in 100 g of tissue. A problem with serodiagnosis, however, is the period during which larvae have become established in muscles before antibodies can be detected, thereby giving false or misleading results. Also, serodiagnosis may be less reliable for detecting long-established infections. After much comparative research, Gamble (1998) advised that serodiagnosis was of value for epidemiological work with herds of pigs (or other animals), but that muscle digestion was more reliable and the best means of protecting the public from infection with *Trichinella*.

13.4.2 HUMAN HOSTS

The range and sophistication of the diagnostic methods available for dealing with trichinellosis will be determined by the location of the outbreak. Muscle biopsies, histology, DNA amplification by PCR, and immunodiagnosis will be available in the North where modern hospitals exist with technical staff and adequate resources. Patients in the North who present with the symptoms of infection can expect an accurate diagnosis to be made (Capo and Despommier, 1996).

13.5 EPIDEMIOLOGY

Species of *Trichinella* are ubiquitous, presumably because so many mammal species (Figure 13.2) sustain the worms in the sylvatic cycle and because birds and reptiles also serve as hosts (Pozio and Zarlenga, 2005). *Trichinella*-free zones probably do

TABLE 13.4
Information about the Distribution of Species of *Trichinella* Known to Infect Humans. From Table 1 (Bruschi and Murrell, 2002) with permission from the BMJ Publishing Group.

Trichinella Species	Muscle Capsule	Infectivity for Pigs	Freeze Resistance	Molecular Markers (PCR)	Reported Sources	General Geographic Distribution
T. spiralis	+	++++	–	173 bp	Pork, game, horse meat	Cosmopolitan
T. britovi	+	+	±	127, 252 bp	Pork, game, horse meat	Temperate Europe/Asia
T. pseudospiralis	–	+	–	300, 360 bp	Pork, game	Cosmopolitan
T. papuae	–	+	–	240 bp	Pork, game	Papua New Guinea
T. nativa	+	–	+++	127 bp	Game	Arctic/ subarctic
T. nelsoni	+	+	–	155, 404 bp	Game	Africa (south of Sahara)
T. murrelli	+	–	–	127, 361 bp	Game, horse meat	North America

Note: PCR, polymerase chain reaction; bp, base pairs.

not exist. We might expect knowledge of the distribution and host range to increase as more vertebrate species are examined. Perhaps helminthologists collecting material from vertebrates that have had little parasitological attention should also sample muscles as a matter of routine even though *Trichinella* was not the reason for their survey. We might expect *Trichinella* to spread even further through the illegal importation of meat and meat products from countries that cannot afford to support veterinary services or a public health system that requires meat inspection.

A glimpse into the global distribution of seven species of *Trichinella* known to infect humans is given in Table 13.4. National prevalence values for *Trichinella* infections are difficult to obtain because so many outbreaks are on a small scale in terms of the numbers of people involved, often because they relate to a particular batch of meat or an unusual event. For example, since 1962, the Ministry of Public Health in Thailand has reported 118 outbreaks of trichinellosis involving 5,400 patients, 95 of whom died (Khamboonruang, 1991). These outbreaks occurred mainly in rural communities where people attending traditional ceremonies had eaten a special dish consisting of spiced, finely chopped wild pig meat served nearly raw. In China, over 500 outbreaks of trichinellosis have been documented affecting 20,000 people resulting in over 200 deaths (Takahashi et al., 2000). The vast number of domestic pigs belonging to rural households in China is likely to put many people at risk of infection. Herraez Garcia et al. (2003) describe an outbreak of trichinellosis in 52 people who ate contaminated pork and Ancelle (1998) has collated details of 12 discrete outbreaks relating to infected horsemeat in France and Italy. Outbreak after outbreak can be traced through the published literature.

13.6 ALLERGY AND IMMUNITY

Allergic (hypersensitivity) and immune responses to *Trichinella* have been studied extensively in experimental infections in rats and mice. The fruits of this work may seem somewhat remote from public health issues but they help to explain aspects of the morbidity observed during trichinellosis in humans, provide the basis for immunodiagnostic tests, and account for the pronounced eosinophilia characteristic of *Trichinella* infections (Despommier, 1998; Bruschi and Murrell, 2002).

13.7 PATHOGENESIS AND MORBIDITY

The course of trichinellosis has two phases, an enteral (intestinal) phase and a parenteral (tissue) phase (Capo and Despommier, 1996). Many people on first becoming infected with *Trichinella* may be asymptomatic if relatively few L1 have been ingested. In heavier infections mild diarrhea and nausea develop, and in still heavier infections abdominal pain, vomiting, and low-grade fever last for a few days. These symptoms, which accompany establishment of the worms in the mucosa, occur during other intestinal disorders such as food poisoning or even cholera and early trichinellosis may be misdiagnosed.

Symptoms during the first part of the parenteral phase relate to the migrating L1 and their establishment in striated muscle. Mild to moderate infections produce general myalgia (muscle pain), peri-orbital and facial edema, respiratory problems, headache, conjunctivitis, skin rash, fever, weight loss, visual disturbances, and swallowing difficulties. These symptoms, related in part to hypersensitivity and inflammation, are some of the many attributed to the first part of the parenteral phase. They do not always occur in every case, but they are always more severe as the larval burden increases. The course of the disease and the rate at which the L1 become encapsulated vary in relation to the species of *Trichinella* that has become established. Neurological disorders including vertigo, convulsions, meningitis, and encephalitis may develop if larvae have reached the brain. Endocarditis and myocarditis (inflammation and damage to heart valves and muscle) sometimes develop. Misdiagnosis remains a problem because other diseases may generate similar symptoms. Data reported in Section 13.5 show a death rate of 295 out of 25,400 cases of trichinellosis. According to Capo and Despommier (1996) death is most likely to occur in heavily infected patients during the acute parenteral phase. Capo and Despommier have also devised a most informative chart (Figure 3 on page 50 of their review) that enables health workers to follow the course of the development of fluctuating morbidity during trichinellosis and highlights circumstances that might lead to death. In countries of the North, trichinellosis is an expensive disease. Ancelle (1998) reckoned that in France in 1985 the average cost of a case for medical services was FRF 10,000 and FRF 27 million nationally.

13.8 TREATMENT

After diagnosis, anthelminthic treatment to expel adult worms should begin promptly to prevent the further production of larvae. WHO (2004) recommends the following effective drugs and dosages: albendazole (400 mg daily for 3 days); mebendazole (200 mg daily for 5 days); or pyrantel (10 mg/kg body weight daily for 5 days). Prednisolone, a representative corticosteroid compound, can be given (40 to 60 mg daily) to alleviate allergic and inflammatory symptoms. The manufacturer's instructions and information about contraindications should be checked at the time of treatment.

13.9 PREVENTION AND CONTROL

The dosing regimes (above) for the treatment of trichinellosis do not lend themselves to the control of morbidity in the community by means of universal, targeted, or selective anthelminthic treatment. The proven public health interventions for the protection of the public from trichinellosis are high standards of animal husbandry, meat inspection by highly trained staff, and good hygiene practices in the food industry, with all measures being supported by legislation (see Kapel, 2005). Freezing meat and meat products is not such a secure measure as was previously thought (WHO, 2000) since the discovery that *Trichinella* larvae in horsemeat can survive for up to 4 weeks at –18°C (Kapel, 2005). Educational drives to warn the public of the risks of eating undercooked meat from whatever source should be strengthened in places with a record of outbreaks of trichinellosis.

REFERENCES

Ancelle T. 1998. History of trichinellosis outbreaks linked to horse meat consumption 1975 to 1998. *Eurosurveillance* **3**, 86–89.

Beck R, Mihaljevic Z and Marinculic A. 2005. Comparison of trichinelloscopy with a digestion method for the detection of *Trichinella* larvae in muscle tissue from naturally infected pigs with low level infections. *Veterinary Parasitology* **132**, 97–100.

Bruschi F and Murrell KD. 2002. New aspects of human trichinellosis: the impact of new *Trichinella* species. *Postgraduate Medical Journal* **78**, 15–22.

Campbell WC. 1991. *Trichinella* in Africa and the *nelsoni* affair. In: *Parasitic Helminths and Zoonoses in Africa* (eds. ANL Macpherson and PS Craig). London: Unwin Hyman. pp. 83–100.

Capo V and Despommier DD. 1996. Clinical aspects of infection with *Trichinella* spp. *Clinical Microbiology Reviews* **9**, 47–54.

Clutton-Brock J. 1987. *A Natural History of Domesticated Mammals*. British Museum (Natural History) and Cambridge University Press.

Cobbold TS. 1864. *Entozoa: An Introduction to the Study of Helminthology with reference, more particularly, to the Internal Parasites of Man.* London: Groombridge and Sons.

Cui J and Wang ZQ. 2001. Outbreaks of human trichinellosis caused by consumption of dog meat in China. *Parasite* **8**, S74–S77.

Despommier DD. 1998. *Trichinella* and *Toxocara.* In: *Topley and Wilson's Microbiology and Microbial Infections,* 9th edition, Volume 5, *Parasitology* (eds. FEG Cox, JP Kreir and D Wakelin). London, Sydney, Auckland: Hodder Headline Group. pp. 597–607.

Dubinsky P, Boor A, Kincekova J et al. 2001. Congenital trichinellosis? Case report. *Parasite* **8**, S180–S182.

Gajadhar AA and Forbes LB. 2002. An internationally recognized quality assurance system for diagnostic parasitology in animal health and food safety, with example data from trichinellosis. *Veterinary Parasitology* **103**, 133–140.

Gamble HR. 1998. Sensitivity of artificial digestion and enzyme immunoassay methods of inspection for trichinae in pigs. *Journal of Food Protection* **61**, 339–343.

Herraez Garcia J, Leon Garcia LA, Lanusse Senderos C et al. 2003. Outbreak of trichinellosis in the region of la Vera (Carceres, Spain) caused by *Trichinella britovi. Anales de Medicina Interna* **20**, 63–66.

Jongwutiwes S, Chantachum N, Kraivichian P et al. 1998. First outbreak of trichinellosis caused by *Trichinella pseudospiralis. Clinical Infectious Diseases* **26**, 111–115.

Kapel CMO. 2005. Changes in the EU legislation on *Trichinella* inspection — new challenges in the epidemiology. *Veterinary Parasitology* **132**, 189–194.

Kapel CMO, Webster P and Gamble HR. 2005. Muscle distribution of sylvatic and domestic *Trichinella* larvae in production animals and wildlife. *Veterinary Parasitology* **132**, 101–105.

Khamboonruang C. 1991. The present status of trichinellosis in Thailand. *Southeast Asian Journal of Tropical Medicine and Public Health* **22**, 312–315.

Muller R. 2002. *Worms and Human Disease,* 2nd edition. Wallingford, Oxon: CABI Publishing.

Murrell KD, Djordjevic M, Cuperlovic K et al. 2004. Epidemiology of *Trichinella* infection in the horse: the risk from animal feeding practices. *Veterinary Parasitology* **123**, 223–233.

Nockler K, Pozio E, Voigt WP and Heidrich J. 2000. Detection of *Trichinella* infection in food animals. *Veterinary Parasitology* **93**, 335–350.

Owen IL, Gomez Morales MA, Pezzotti P and Pozio E. 2005. *Trichinella* infection in a hunting population of Papua New Guinea suggests ancient relationship between *Trichinella* and human beings. *Transactions of the Royal Society of Tropical Medicine and Hygiene* **99**, 618–624.

Piergili-Fioretti D, Castagna B, Frongillo RF and Bruschi F. 2005. Re-evaluation of patients involved in a trichinellosis outbreak caused by *Trichinella britovi* 15 years after infection. *Veterinary Parasitology* **132**, 119–123.

Pozio E. 2000. The domestic, synanthropic and sylvatic cycles of *trichinella* and flow between them. *Veterinary Parasitology* **93**, 241–262.

Pozio E. 2005. The broad spectrum of *Trichinella* hosts: from cold- to warm-blooded animals. *Veterinary Parasitology* **132**, 3–11.

Pozio E and La Rosa G. 2000. *Trichinella murrelli* n.sp: etiological agent of sylvatic trichinellosis in temperate areas of North America. *Journal of Parasitology* **86**, 134–139.

Pozio E and Zarlenga DS. 2005. Recent advances on the taxonomy, systematics and epidemiology of *Trichinella. International Journal for Parasitology* **35**, 1191–1204.

Pozio E, Varese P, Morales MA et al. 1993. Comparison of human trichinellosis caused by *Trichinella spiralis* and by *Trichinella britovi. American Journal of Tropical Medicine and Hygiene* **48**, 568–575.

Proulx JF, MacLean JD, Gyorkos TW et al. 2002. Novel prevention programme for trichinellosis in inuit communities. *Clinical Infectious Diseases* **34**, 1508–1514.

Ranque S, Fugere B, Pozio E et al. 2000. *Trichinella pseudospiralis* outbreak in France. *Emerging Infectious Diseases* **6**, 543–547.

Steele JH. 1982. Trichinosis. In: *Handbook Series in Zoonoses*. Section C: *Parasitic Zoonoses* Volume II (ed. MG Schultz). Boca Raton, Florida: CRC Press Inc. pp. 293–324.

Takahashi Y, Mingyuan L and Waikagul J. 2000. Epidemiology of trichinellosis in Asia and the Pacific Rim. *Veterinary Parasitology* **93**, 227–239.

WHO. 2000. *Foodborne Disease: a Focus for Health Education* (prepared by Y Motarjemi). Geneva: World Health Organization.

WHO. 2004. *WHO Model Formulary 2004*. Geneva: World Health Organization.

Zimmerman WJ. 1971. Trichinosis In: *Parasitic Diseases of Wild Animals* (eds. JW Davis and RC Anderson). Ames, Iowa: The Iowa State University Press. pp. 127–139.

Part III

Control Interventions

To those of us who have been interested in helminth infections for over three decades, the availability of broad spectrum, single dose, low toxicity drugs in the last ten years seems a veritable miracle.

Kenneth S. Warren, 1989

Anthelminthic treatment is an intervention that is already bringing relief from helminthiasis to many people and has the potential to bring relief to millions more. We subscribe to the proposal that efforts to control helminthiasis should be incorporated into a multidisease approach together with malaria, HIV/AIDS, and TB. Helminthiasis falls into the group of "neglected" diseases — the time for intervention is long overdue.

Molyneux DH and Nantulya VM. 2004. Linking disease control programmes in rural Africa: a pro-poor strategy to reach Abuja targets and millennium development goals. *BMJ* **328**, 1129–1132.

Montresor A, Crompton DWT, Gyorkos TW, and Savioli L. 2002. *Helminth Control in School-age Children. A Guide for Managers of Control Programmes.* Geneva: World Health Organization.

Warren KS. 1990. An integrated system for the control of the major human helminth parasites. *Acta Leidensia* **59**, 433–442.

World Health Organization. 2004. *Waterborne Zoonoses. Identification, Causes and Control* (eds. JA Cotruvo, A Dufour, G Rees, J Bartram, R Carr, DO Cliver, GF Craun, R Fayer and VPJ Gannon). London: IWA Publishing on behalf of the World Health Organization.

New York, October 26, 1909

The Rockefeller Commission
For the Extermination of the Hookworm Disease

Today this organization was made, with a gift of a million dollars by John D. Rock-efeller. The one greatest single cause of anaemia and stagnation in the South will by this fund be ultimately removed, and 2,000,000 inefficient people will be made well.

From The Early Life of Walter H. Page 1855–1913 by Burton J Hendrick (1928)

14 Use of Anthelminthic Drugs in the Public Health Arena*

The uneven distribution of the world's economic resources has ensured that the research-based pharmaceutical industry and the market it serves belong to the North (WHO, 1988). Medicines sold annually for billions of U.S. dollars are manufactured to alleviate cardiovascular, psychotropic, and inflammatory illnesses for affluent people in the North. The economy of the South is too weak to offer a viable market to support the research and development needed to provide new medicines to combat the infectious diseases that plague the population there. Sadly, despite the vast number of people in need of anthelminthic treatment (Table 1.1), there is no commercial incentive to develop new drugs for a market that cannot pay. The process of drug development is long, complex, and extremely expensive (McIntosh et al., 1989). The research-based pharmaceutical industry is obliged to reward shareholders, must pay wages, and must invest for its future survival. The handful of ageing essential anthelminthic drugs now available in generic form for the treatment of helminthiasis in humans (WHO, 2004) owes much to the demand created by the North's lucrative veterinary market.

We seek to interrogate the best use of deworming drugs (Table 14.1) in communities where helminthiasis is endemic regardless of whether control programs are in progress, being planned, or waiting to join the health agenda (WHO, 2003). "Best use" challenges public health professionals to engage with several penetrating questions:

- Does the risk of infection and disease to members of the community justify the use of anthelminthic drugs in community-based control programs?
- Are some groups in the community more at risk of disease than others?
- What benefits will accrue to the health and development of the community if an appropriate control program is implemented?
- Do available drugs have a history of vigilant safety recording, proven quality, demonstrable effectiveness, and stability under basic storage conditions?
- Can reliable drug procurement be guaranteed for the duration of the program?
- Does political will exist to support the deworming program?

---

* The use of WHO-recommended anthelminthic drugs brings many health benefits, some temporary side effects and extremely rare adverse events. Before rounds of treatment begin, public health professionals responsible for intervention with anthelminthic drugs in community programs must ensure that provision has been made to deal with adverse events. Advice about all aspects of anthelminthic use in the community is available from the Department of Control of Neglected Tropical Diseases, World Health Organization, 1211 Geneva 27, Switzerland (www.who.int/neglected_diseases/en/).

TABLE 14.1
WHO-Recommended Anthelminthic Drugs for Helminth Control Programs in the Community Based on Single Doses Administered by Mouth. From data in WHO (2004).[1-4]

Infections (Chapter)	Drug	Dose
Cestodiasis (5)		
Hymenolepis nana	Niclosamide	Adult and child over 6 years, 2 g as single dose on first day then 1 g daily for 6 days; child of 2–6 years, 1 g on first day then 500 mg daily for 6 days; child under 2 years, 500 mg daily on first day then 250 mg daily for 6 days
	Praziquantel	Adult and child over 4 years, 15–25 mg/kg as a single dose
Taenia saginata[5,6] and *T. solium*	Niclosamide	Adult and child over 6 years, 2 g as a single dose after a light breakfast, followed by a purgative after 2 hours. Child of 2–6 years, 1 g; child under 2 years, 500 mg
	Praziquantel	Adult and child over 4 years, 5–10 mg/kg as a single dose
Enterobiasis		
Enterobius vermicularis	Albendazole	Adult and child over 2 years, 400 mg Child 12 months – 2 years, 200 mg
	Mebendazole	Adult and child over 2 years, 100 mg as a single dose, repeated after interval of 2–3 weeks
	Pyrantel	Adult and child, 10 mg/kg as a single dose; second dose after 2–4 weeks
Food-borne trematodiasis (7)		
Clonorchis sinensis	Praziquantel	Adult and child over 4 years, 40 mg/kg
Fasciolopsis buski	Praziquantel	Adult and child over 4 years, 25 mg/kg
Heterophyes heterophyes[7]	Praziquantel	Adult and child over 4 years, 25 mg/kg
Opisthorchis felineus	Praziquantel	Adult and child over 4 years, 40 mg/kg
Opisthorchis viverrini	Praziquantel	Adult and child over 4 years, 40 mg/kg
Paragonimus westermani	Praziquantel	Adult and child over 4 years, 40 mg/kg
Fasciola gigantica	Triclabendazole	Adult and child over 4 years, 10 mg/kg
Fasciola hepatica	Triclabendazole	Adult and child over 4 years, 10 mg/kg
Lymphatic filariasis (8)		
Brugia malayi *Brugia timori* *Wuchereria bancrofti*	Albendazole + Ivermectin or Albendazole + Diethylcarbamazine	Adult and child over 5 years, (> 15 kg) Albendazole 400 mg + Ivermectin 200 µg/kg Albendazole 400 mg + Diethylcarbamazine 6 mg/kg
Onchocerciasis (9)		
Onchocerca volvulus	Ivermectin	Adult and child over 5 years (> 15 kg), 150 µg/kg once a year

TABLE 14.1 (continued)
WHO-Recommended Anthelminthic Drugs for Helminth Control Programs in the Community Based on Single Doses Administered by Mouth. From data in WHO (2004).[1-4]

Infections (Chapter)	Drug	Dose
Schistosomiasis (10)		
All species of *Schistosoma*	Praziquantel	Adult and child over 4 years, 40–60 mg/kg
Soil-transmitted helminthiasis (11)[8]		
Ascaris lumbricoides	Albendazole	Adult and child over 2 years, 400 mg as a single
Ancylostoma duodenale		dose. Child of 12 months to 2 years, 200 mg as
Necator americanus		a single dose
Trichuris trichiura	Albendazole	Adult and child over 2 years, 400 mg as a single
		dose (moderate infections), or 400 mg daily for
		3 days (severe infections)
		Child of 12 months to 2 years, 200 mg as a single
		dose (moderate infections), or 200 mg initially
		then 100 mg twice daily for 3 days (severe
		infections)
Ascaris lumbricoides	Levamisole	Adult and child, 2.5 mg/kg as a single dose; in
Ancylostoma duodenale		severe hookworm infections, a second dose may
Necator americanus		be given after 7 days
Ascaris lumbricoides	Mebendazole	Adult and child over 1 year, 500 mg as a single
Ancylostoma duodenale		dose
Necator americanus		
Trichuris trichiura		
Ascaris lumbricoides	Pyrantel	Adult and child, 100 mg/kg as a single dose
Ancylostoma duodenale		
Necator americanus		
Strongyloidiasis (12)		
Strongyloides stercoralis	Ivermectin	All patients, 200 µg/kg

[1] With the exception of intervention for soil-transmitted helminth infections, the drugs would usually be given as an annual treatment. Intervals between doses depend on epidemiological data.

[2] Advice about adverse reactions, contraindications, and precautions set out in the *WHO Model Formulary 2004* should be followed. Circumstances may indicate that pregnant and lactating women may be in need of treatment. If so, every effort should be made to avoid treatment during the first trimester (see Chapter 4, Chapter 10, Chapter 11, and Chapter 14 for further discussion) unless decided otherwise for an individual patient by a qualified medical practitioner.

[3] Helminth infections not included in this list (echinococcosis, loasis, strongyloidiasis, trichinellosis, and many listed at the end of Chapter 1) are not readily treated in community control programs. These infections should not be ignored and access to appropriate drugs should be available for people in need.

[4] Anthelminthic drugs must be of good quality from a reputable manufacturer.

[5] For *T. saginata*, half the dose of Niclosamide may be taken after breakfast and the remainder an hour later followed by the purgative 2 hours after that.

[6] Presumably, *Taenia asiatica* may be treated in the same manner as *T. saginata*, but this recommendation is not yet included in the WHO Model Formulary 2004.

[7] Representing species of minute intestinal flukes.

[8] Albendazole, levamisole, mebendazole, and pyrantel have varying degrees of activity against these four species of soil-transmitted helminth.

- How can the community be empowered to support and comply with a control program based on the distribution of anthelminthic drugs?
- Which strategy for drug distribution will be most effective for achieving the program's objectives?
- When is the optimum time to offer drug treatment in the community?
- Will diagnosis be necessary before administration of the drugs?
- If diagnosis is required, are resources available to do it properly?
- Would the perceived benefits of the program override the decision not to undertake diagnosis if resources for diagnosis are not forthcoming?
- How, where, and by whom will the drugs be distributed in the community assuming the drugs are to be given as oral doses?
- Is there any need to calculate doses based on individual body weight?
- How many doses will be needed per person annually to achieve the program's objectives?
- Can the necessary coverage be attained and sustained for the duration of the program?
- How will the impact, progress, and sustainability of the program be measured?
- Can a workable system of pharmacovigilance be established to detect and report suspected side effects or adverse events?
- Can steps be taken to protect members of the community from known contraindications?
- What policies are to be adopted regarding infants and pregnant and lactating women?
- Will there be access to medical care in the event of a crisis in relation to drug administration?
- What measures will be established to screen for the emergence of drug resistance?
- Should actions be prepared and implemented to offset the emergence of drug resistance?
- Are enough trained personnel available to deliver and administer the proposed program?
- Should the program be integrated into existing public health measures, be used to introduce new health services, or become part of a development initiative?
- Are there social factors and cultural traditions that may impair or enhance the program's progress?
- What will the program cost and will this be at the expense of other programs?
- Where will sustainable funding be obtained for the duration of the program and beyond?

Responses to these questions depend on national health priorities, local conditions and on the nature of the helminthiasis and its public health significance. Responses will often overlap. For example, coverage may suffer if it coincides with a rainy season when roads become impassable for vehicles, the time of coffee picking, or some traditional festival. There are, however, some principles that seem to apply to all community-based helminth control programs centered on the use of anthelminthic drugs. These principles emerge in relation to (1) epidemiology, (2) the strategy chosen for drug administration, (3) drug effectiveness and cure rates, (4) the threat of drug resistance, and (5) costs. Throughout this chapter we are concerned mainly with drugs recommended by the World Health Organization for the treatment and control of helminthiasis. We recognize that weighing scales may not always be available for calculating doses based on body weight (Table 14.1). The use of calibrated dose poles may be helpful for treating onchocerciasis (Alexander et al. 1995) and schistosomiasis (Montresor et al. 2005).

14.1   EPIDEMIOLOGY

Before anthelminthic drugs are used to control an identified helminth infection, epidemiological information must be obtained about (1) the distribution and transmission of the infection in the community, (2) the amount and severity of the morbidity that accompanies the infection, and (3) relevant characteristics of the community. Evaluation of this data is essential for planning, implementing, and monitoring the progress of the program. Without data, risk groups cannot be identified, targets cannot be set, drug needs cannot be calculated, drug dosing regimens and coverage cannot be determined, infrastructural resources such as transport and equipment cannot be estimated, and so on. Every aspect of helminth control depends on a secure epidemiological foundation maintained by careful surveillance throughout the program.

Every health worker assigned to helminth control must understand that the objective is to attain a sustainable reduction in the intensity of infection or worm burden in each infected person and thereby in the community. Intensity determines the severity of morbidity, the rate of transmission, and the persistence of stable helminth populations in the infected community (Bundy and Michael, 2001). The allocation of technical time and resources to measure and track intensity during the course of a control program facilitates the detection of changes in morbidity, drug effectiveness, transmission, and helminth populations.

14.1.1   THEORETICAL ASPECTS

The theoretical basis of helminth population biology has established that in most cases a helminth's population in a community of hosts will decline and be destroyed if the helminth's basic reproductive rate (R_0) falls below 1 and remains there (Anderson and May, 1992; Smith and Scott, 1994; see Chapter 4). In life, anthelminthic drugs offer health professionals a powerful tool for reducing the real reproductive rate (R_e) of the helminth of concern provided that the best possible epidemiological information has been obtained about the distribution and abundance of the worms and the environment and activities of the host community (Smith and Scott, 1994).

14.1.2   PRACTICAL EPIDEMIOLOGY IN RELATION TO THE USE
OF ANTHELMINTHIC DRUGS

The information on which a helminth control program is to proceed is derived from contemporary surveys, records kept by health services, published literature, and local knowledge. Numbers from health services may exaggerate the real situation because the people from whom they were obtained probably suspected their infection status. Numbers from the literature may be out of date or of inadequate origin.

Many publications offer guidance about epidemiological surveying. Deciding on the sample size is difficult. Random sampling, in which every individual in the community to be surveyed should have an equal chance of being sampled, may not be possible in a country where resources for health care are scarce and where sophisticated, contemporary demographic data is lacking. Advice about the determination of sample sizes is available from the World Health Organization (WHO, 1986) and other sources (Chitsulo et al., 1995; Montresor et al., 2002; De et al., 2003). As usual in the realm of public health, there is no universal paradigm. Perhaps the best advice is to sample as many individuals as possible while trying to avoid collecting from more than one member of the same family. The distribution, abundance, and significance of intermediate and reservoir hosts should not be overlooked. Helminth control would never have started had perfect sampling been an obligatory requirement.

14.2 SAFETY AND THERAPEUTIC EFFICACY AND EFFECTIVENESS

For each anthelminthic drug there is an efficacy value defined as its therapeutic effect when used under ideal, standard conditions in isolation. The effectiveness of the drug is defined as its therapeutic effect when used under operational conditions such as those encountered in a public health program (WHO, 2002). Effectiveness can be described in terms of the cure rate (CR), egg reduction rate (ERR), or another appropriate measurement such as decline in microfilaremia. Cure rate is the number (usually expressed as a percentage) of previously infected people judged to be uninfected following anthelminthic treatment. Egg reduction rate is the percentage decline in egg counts at a set time after anthelminthic treatment based on the examination of a stool or urine sample (WHO, 1999). Generally, an egg count is made from a stool sample collected on a given day and no account is taken of the subject's daily fecal production (Table 16.4). The possibility of such changes should not be ignored when egg counts are made at different times from the same person. In the case of lymphatic filariasis or onchocerciasis, effectiveness is measured in terms of the microfilaremia in tissue samples.

That an anthelminthic drug should be safe is more important than its therapeutic activity. Helminth control based on deworming would collapse if effectiveness overshadowed safety concerns. What is meant by safe in this context? Davis (1985) stated that in the case of large-scale drug usage, it had always been his opinion that safety is paramount, that efficacy came second, and that it did not worry him whether a drug was 60 or 75% effective. He went on to state that what really counted was the population tolerance and good toxicology of the drug so that it could be handled by paramedical personnel. Expert staff will be required to decide which drug should be used for deworming in a particular program and their choice should note that delivery of the drug to the community will depend on individuals, such as school teachers, traditional healers, and a range of volunteers with far less training than paramedical personnel (see WHO, 1995).

Decision about whether a drug is safe enough for use in a public health setting requires an informed judgment about the benefit/risk ratio (WHO, 1995). Such judgment involves comparison of the scale of the direct and indirect benefits stemming from the drug's use with the extent and severity of side effects and the potential for misuse of the drug. The direct benefit from anthelminthic treatment is the reduction of morbidity outlined in previous chapters. Indirect benefit includes increased productivity, poverty reduction, improved school attendance, and so on. Do side effects outweigh the benefits gained at relatively low cost? Registration of a new drug for human use includes scrutiny of records of side effects and adverse events during clinical trials. These results, however encouraging, should not have the same influence as the vast experience of the drug's use among many more people without the strict supervision of a trial. In the case of deworming drugs (Table 14.1), with the exception of triclabendazole, for which there may still be insufficient information about its use in humans, millions of doses have been taken and the side effects have been minimal.

The history of albendazole exemplifies how an anthelminthic drug may be chosen with confidence for an appropriate deworming program. Information about its clinical trials, efficacy tests, operational research studies, and safety evaluation can be traced by reference to Firth (1983), Horton (2000, 2003), Dayan (2003), and Urbani and Albonico (2003). Albendazole is a broad spectrum anthelminthic compound with activity against several species of intestinal helminth infections in humans (Table 14.2). Its safety record is impressive. In the case of intestinal helminth infections treated with albendazole, the overall frequency of minor gastrointestinal symptoms was just greater than 1% in a sample of 22,810 subjects (Horton, 2000). Generally, side effects occurring for any of the WHO-recommended deworming drugs given at the correct dosage are considered to be negligible and self-limiting (Urbani and Albonico, 2003). Establishing cause and effect is difficult in communities where numerous other factors might be responsible for such transient effects as epigastric pain, headache, and nausea. Also, side effects that were not being experienced may be suggested if interviews are conducted with the purpose of identifying them. Although side effects may be rare and negligible, a control program involving anthelminthic treatment should include systems for monitoring side effects with the support of medical care for affected individuals.

14.3 TREATMENT STRATEGIES

We implore governments to give high priority to providing access to essential anthelminthic drugs in all health systems in all endemic areas regardless of whether a national drug policy exists or not (WHO, 1988). In this context, access implies that sufficient quantities of quality drugs will always be available for those in need of treatment. The exceptionally low cost of generic anthelminthic drugs, the goodwill of several research-based pharmaceutical companies, and the generosity of donors and the endeavors of partnerships have produced an affordable tool for the control of morbidity due to helminth infections in even the lowest-income countries. We

TABLE 14.2
Cure Rates (CR) and Egg Reduction Rates (ERR) in Common Intestinal Helminthiases Found in Humans after Recommended Doses of Albendazole: Summary of Studies Published Up to March 1998. From Table 1 (Horton, 2000) with permission from Cambridge University Press.

	No of Studies	Cure Rate	Efficacy CR (%)	CR Range (%)	ERR (%)
Hookworms	68	4871/6272	77.7	100–33.3	87.8
Ancylostoma duodenale	23	538/586	91.8	100–75.0	—
Necator americanus	30	2606/3547	75.0	100–36.6	—
Trichuris trichiura	57	2050/4301	47.7	100–4.9*	75.4
Ascaris lumbricoides	64	4848/5127	94.6	100–66.9*	98.6
Enterobius vermicularis	27	883/903	97.8	100–40.0*	NA
Strongyloides stercoralis	19	298/479	62.2	100–16.7*	NA
Hymenolepis nana	11	190/377	68.5	100–28.5*	NA
Taenia spp.	7	111/131	84.7	100–75.7*	NA

Note: NA = not applicable.

* Small sample numbers.

are concerned here to discuss the application of anthelminthic drugs that may be chosen for use in community-based control programs where resources for drug application are likely to limit intervention to no more than a few, well-spaced days annually (Table 14.1). Three strategies are available for drug application, namely universal (blanket or mass), targeted, and selective according to needs and resources (Anderson, 1989; WHO, 1996b; see Chapter 4).

To what extent should diagnosis be undertaken before the drug is offered in the case of universal and targeted applications? In deciding whether to provide diagnosis or not, public health managers must confront the following questions:

• Is it best practice to allow people to take a drug without knowledge of their infection status or need?
• Given available resources, can diagnosis be achieved and is it affordable?
• Will the cost of diagnosis be less than the cost of giving the drug to uninfected people?
• Does the drug's safety record suggest that only those diagnosed as having infection and morbidity should be exposed to any risk of side effects or adverse events?
• Should a process be organized for dealing with contraindications?
• Can pregnant and lactating women be included in the program?
• How will the ethical aspects of the proposed use of the drug be explained to the community?

The response to each of these questions must accommodate the type of helm-inthiasis, its epidemiology, and the properties of the anthelminthic drug, including its safety record. The epidemiological survey, on which the selection of treatment strategies will have been based, will have given some diagnostic information, but does this meet the concerns of those who argue that treatment without diagnosis is unethical? Anthelminthic drugs have little or no prophylactic use, although a single weekly dose of diethylcarbamazine is a recommended prophylaxis for nonresidents in areas where infection with *Loa loa* is endemic (WHO, 2004). In a community-based program, uninfected people will inevitably take the drug. The sincere case that there is an ethical obligation to carry out diagnosis before treatment has to be balanced against the sickness and suffering that will persist if treatment is denied because individual diagnosis was beyond the resources of the proposed control program. In a resource-limited setting, a strong case for the use of anthelminthic drugs without prior diagnosis can be made because the drugs have been shown to have an impressive safety record. In any event, program managers should have a system in place for dealing with serious adverse events no matter how remote the risk appears to be.

14.4 REINFECTION

In areas where helminth infections are endemic the inevitability of reinfection either directly or indirectly could undermine plans to embark on morbidity control. Seeking to reduce morbidity is fundamentally different from aiming to disrupt transmission leading to elimination of a helminth infection. Estimations of reinfection rates facilitate the calculation of R_o and guide decisions about the duration of treatment intervals (Anderson and Medley, 1985). Reinfection should not detract from mea-sures to relieve suffering.

14.5 DRUG RESISTANCE

The specter of drug resistance haunts the use of anthelminthic drugs to control morbidity due to helminthiasis. Drug resistance is defined as a genetically transmitted loss of susceptibility to a drug in a worm population that was previously sensitive to the appropriate therapeutic dose. The fact that drug resistance is heritable means that the responsible genes exist in worm populations. Productivity in the livestock industry of the North has been made more difficult and expensive by the development of species and strains of nematodes that have become resistant in various ways to anthelminthic drugs (Taylor, 1992). The widespread use and misuse of drugs has served to select populations of animal-parasitic nematodes resistant to benzimida-zoles, ivermectin, levamisole, and pyrantel, all anthelminthic drugs recommended for the treatment of helminthiasis in humans (WHO, 2004). Furthermore, drug-resistant animal-parasitic worms have a global distribution and are not confined to a single region or continent. It should be noted that in some veterinary control programs animals may be given at least 10 rounds of treatment annually, thereby creating selection pressure much greater than would be exerted in a control program for humans (WHO, 2002).

14.5.1 THEORETICAL ASPECTS

Insight into understanding how drug resistance develops can be gained from research into the development of insecticide resistance (see Comins, 1977). Imagine a hypothetical worm population that is closed to other worms of the same species so that there is no gene flow. In such worms, assume that two alleles (pairs of contrasted genetic characters such as susceptibility and resistance) are involved at a single locus in a diploid worm for the development of drug resistance. Then assume that universal anthelminthic treatment is applied to the worms' hosts in a homogeneous manner to provide continual selection pressure. Theory indicates that under these conditions it will be only a matter of time before a drug-resistant population of worms will be established in the hosts. The generation time of the worms is a crucial factor in the process. A species with a short generation time will respond to the selection pressure faster than a species with a longer generation time. Antibiotic resistance can emerge rapidly in bacteria where generation times as short as 20 minutes may prevail. If drug resistance in worms is indeed similar to insecticide resistance, then the time for it to develop will probably fall within the range of from 5 to 100 generations (WHO, 1996a,b, 1999).

14.5.2 THE REALITY OF DRUG RESISTANCE

Resistance to benzimidazoles. The molecular basis of resistance to benzimidazoles in *Haemonchus contortus,* a damaging parasite of sheep, was demonstrated by Roos et al. (1990). The primary action of benzimidazoles is to bind to tubulin, the essential protein component of microtubules (WHO, 1996a; Mansour, 2002). Since cells cannot divide without healthy microtubules, pharmacologists had to explain why benzimidazoles did not stop cell division in the host. Fortunately benzimidazoles have a vastly greater binding affinity for nematode tubulin than for mammalian tubulin so drugs of this class should be safe for human use.

Two β-tubulin genes, termed isotype-1 and isotype-2, have been identified in animal-parasitic nematodes (see Mansour, 2002). Experiments have shown that benzimidazole resistance involves two molecular steps at both of the genetic loci that determine the worm's response to benzimidazoles. In effect, susceptibility to benzimidazoles is controlled at the isotype-1 locus and resistance at the isotype-2 locus. Once resistance has been established, benzimidazoles no longer interfere with β-tubulin so that worm growth and development proceed normally. It is not unreasonable to suppose that such genes form part of the genome of human-parasitic worms. Drug resistance should be suspected if the established effectiveness of an anthelminthic drug is found not to have occurred after it has been used according to the manufacturer's instructions.

Resistance to imidazothiazoles and tetrahydropyrimidines. Levamisole (tetramisole) and pyrantel are representatives of the imidazothiazoles and tetrahydropyrimidines, respectively. They interfere with neurotransmission and act by targeting

the nicotinic receptors of the acetylcholine-gated membrane ion channels of susceptible intestinal helminths (Mansour, 2002). This causes the muscles of the worms
to become paralyzed and the worms are then swept out of the alimentary tract.
Resistance induced in *Caenorhabditis elegans,* that most amenable experimental
animal and atypical nematode, involves changes in the worm's nicotinic receptors
rendering them less ready to bind with the drug (Lewis et al., 1980). Resistance to
levamisole in animal-parasitic nematodes such as *Oesophagostomum dentatum* from
pigs develops after about 10 generations (see Mansour, 2002).

   Resistance to ivermectin. Ivermectin affects the motility of nematode muscles
and especially those of the pharynx, which nematodes use to pump up liquid food
such as blood. Nematodes that are susceptible to ivermectin starve to death (Mansour,
2002). Ivermectin-resistant populations of nematodes that infect sheep are now well
established. The mechanism of resistance has yet to be determined, but Sangster
et al. (1999) have found that ivermectin-resistant isolates of *H. contortus* can tolerate
177-fold higher concentrations of ivermectin than nonresistant isolates.

   Resistance to praziquantel. Praziquantel resistance has been found in an isolate
of *S. mansoni* maintained in laboratory mice. The isolate was obtained from patients
in Egypt who continued to pass eggs of *S. mansoni* after several doses of the drug.
The susceptibility of the flukes to praziquantel was then tested by dosing the mice
and limited resistance was demonstrated. Another laboratory strain of *S. mansoni*
maintained in mice has been exposed to repeated doses of praziquantel. Flukes
subjected to this artificial selection pressure were found to be somewhat resistant to
praziquantel compared with flukes of the same strain meeting the drug for the first
time (Cioli, 2000). While this work shows that praziquantel-resistant genes are
present in some populations of *S. mansoni*, the conditions used to select the resistant
flukes are unlikely to occur in a public health setting and in any case, selection in
this manner might render the schistosomes unable to survive in the real world. It is
interesting to note that schistosomes can be selected for drug resistance under a
range of laboratory techniques (Marshall, 1987) so the threat of drug resistance must
not be ignored.

14.5.3 IS THERE EVIDENCE OF DRUG RESISTANCE IN HELMINTHS INFECTING HUMANS?

De Clercq et al.(1997) reported the presence of a population of *Necator americanus*
that was judged to be resistant to mebendazole in Mali, and Reynoldson et al. (1997)
concluded that they had detected a population of *Ancylostoma duodenale* resistant
to pyrantel in Australia. An anxious, sceptical, and critical response followed these
publications as advocates of deworming sought to allay fears that the control of
morbidity due to soil-transmitted helminthiasis was about to collapse. In fact relatively few subjects had been enrolled in the studies, there was no obvious evidence
of selection pressure through previous exposure to the drugs, there was no demon-
stration of genetically transmitted drug resistance, and other explanations for reduced
effectiveness were not offered. Subsequent work in Mali by Sacko et al. (1999)
found mebendazole treatment to produce a 51% cure rate and a 68% ERR against
N. americanus. The ERR in school children from Pemba Island, Zanzibar, who were

given mebendazole for hookworm infection was found to change from 82 to 52% during exposure to 15 rounds of treatment (Albonico et al., 2003). These results do not eliminate concern over the possible presence of mebendazole resistant hookworms in Mali and Zanzibar.

Since its discovery in 1972, praziquantel has transformed the treatment and control of all forms of schistosomiasis (Marshall, 1987; Engels and Savioli, 2005). In the 1990s anxiety was aroused when treatment with praziquantel for schistosomiasis mansoni in a focus in Senegal was not followed by the expected high cure rates. The subsequent investigation eliminated the initial suspicion of drug resistance and showed that epidemiological circumstances had misled health workers about the effectiveness of the drug. In this case from Senegal, the infected people were living in an area of intense transmission so that even if the drug treatment had killed over 90% of the blood flukes, eggs would have continued to have been discharged. Also the intense transmission would probably have ensured the presence of many immature flukes in the patients. Apparently praziquantel may not be particularly effective against immature flukes and some survivors might have matured and begun to release eggs by the time the impact of the drug was being monitored. This proposition was tested by giving two doses of praziquantel 2 to 3 weeks apart in a regimen to knock out recently matured flukes. The expected effectiveness was then achieved (Renganathan and Cioli, 1998).

The emergence of drug-resistant worms will become more likely as anthelminthic treatment increases in communities. Public health planners cannot afford to adopt a panglossian view for the future. Preparation for dealing with the threat requires (1) sensitive techniques for detecting genuine drug resistance, (2) procedures for delaying the threat, (3) advocacy to convince the research-based pharmaceutical industry that novel anthelminthic compounds are needed, and (4) philanthropic initiatives to pay for the drugs once they become available for human use.

14.5.4 Detection of Drug Resistance

Drug resistance should be suspected if expected measures of cure or effectiveness are not achieved (Table 14.2). Other explanations must be eliminated before the presence of drug resistance is accepted. Consider the case of an apparent fall in the effectiveness of an anthelminthic drug in a program designed to control soil-transmitted helminthiasis. Three lines of enquiry will need to be explored. First, how was the effectiveness (ERR) measured? Second, if the results of the effectiveness measurement are trustworthy, was the drug of good quality and was it administered properly? Third, are the unexpected results due to biological, epidemiological, or social factors in the host-parasite relationship? Elimination of these possibilities indicates that the poor effectiveness might have been caused by drug resistance.

If it be accepted that the fall in the ERR is genuinely less than was expected, attention should be paid to the anthelminthic drug and the circumstances of its use in the community. Assurance about drug quality must be sought before a control program begins. This is most important when generic drugs are used. There is no intention to imply that generic products are of poor quality, but when health is involved, managers must know that the drug has the correct formulation for patient safety and therapeutic effectiveness. Manufacturing and distributing counterfeit drugs represent a criminal abomination and anyone so charged should be examined in court. Sometimes the drugs may not have been distributed in the community as widely as required. Perhaps one of the times when drug distribution was due was missed allowing new worms to mature. Perhaps the conditions under which the drugs had been stored led to their deterioration. Perhaps drugs had been stolen and replaced by defective drugs. Unusual transmission dynamics may cause a drug to appear to have lost effectiveness as happened in an attempt to control schistosomiasis in Senegal (Section 14.5.3). Some highly respected members of the community may have covertly campaigned against the control program and compliance may have fallen in response. No matter how bizarre, every possible explanation for the decline in a drug's effectiveness must be considered during an investigation into suspected drug resistance.

Detection of a genuine decline in effectiveness depends on obtaining an accurate measurement of intensity before the drug is used. In the case of soil-transmitted helminthiasis this will probably entail the Kato Katz procedure (WHO, 1999; Appendix 3) applied to a number of stool samples obtained from a representative cohort in the community. Technical staff must (1) have had proper training in the application of the method, (2) have access to suitable equipment, and (3) must be well managed with a routine that avoids boredom and fatigue. A system of quality control should be established to check the accuracy of the egg counts. Exactly the same procedures should be used throughout the program so that a valid comparison can be made of the results with those from before drug administration began. Repeating the procedures means that stool samples should be collected at the same time of day on each occasion, the interval between collection and examination should be the same, and so on (see Montresor et al., 2002).

Veterinary practice relies on the Faecal Egg Count Reduction Test (FECRT) and the Egg Hatch Assay (EHA) for monitoring drug resistance in intestinal nematodes (see Coles et al., 1992). The FECRT is a standardized method for measuring the ERR, its merit being that it eliminates the possibility of differences due to technical variations. The EHA provides data on the viability of the eggs of intestinal nematodes that have been exposed to benzimidazoles. The test is carried out *in vitro* and benzimidazole resistance is indicated when 50% or more of the eggs hatch in a solution containing thiabendazole at a concentration of 0.1 µg/ml. Albonico et al. (2004, 2005) used the EHA to screen for the presence of benzimidazole resistance in soil-transmitted helminths infections in a community in Zanzibar. They concluded that the EHA can readily be performed as part of a public health program and that it may be most useful in the case of hookworms whose eggs hatch more rapidly than those of other species of soil-transmitted helminth.

If the FECRT and EHA suggest that drug resistance has arrived, public health managers will find themselves shutting the stable door after the horse has bolted. A more sensitive test is needed because when the FECRT and EHA are used under veterinary conditions at least 25% of the population of worms will already be resistant (Martin et al., 1989). Prospects for detecting the presence of drug resistance in a much smaller proportion of a population of worms are likely to depend on an application of the Polymerase Chain Reaction (PCR) once more is known about the molecular basis of drug resistance.

14.5.5 MEASURES TO DELAY AND AVOID THE EMERGENCE OF DRUG RESISTANCE

The simplified theory and a wealth of veterinary experience offer various measures to delay or even avoid the problem of drug resistance in a control program in which reduction of morbidity is the prime objective (Section 14.5).

Treatment of a proportion of the people in a community rather than everybody will ensure that the genes of surviving worms that have not experienced anthelminthic selection pressure will serve to dilute the impact of the genes carrying resistance. That is one of the strengths of the targeted treatment of school-aged children to reduce morbidity due to schistosomiasis and soil-transmitted helminthiasis while withholding treatment from the rest of the community (Bundy and Michael, 2001).

Treatment at intervals greater than the worm's generation time will tend to act against the emergence of drug resistance. This tends to be standard practice in all helminth control programs.

Changing or alternating anthelminthic drugs during the course of a control program should reduce the selection pressure especially when the program is planned to run for years. Changing anthelminthic drugs during the course of a program is not guaranteed to deal with the threat of drug resistance. Veterinary experience shows that side-resistance and cross-resistance may occur (Prichard et al., 1980). Side-resistance exists where the resistance to a compound results from selection pressure from another compound with a similar mode of action. There would unlikely be any benefit from switching from one benzimidazole to another (see Taylor, 1992).

Combination therapy, in which two drugs are used simultaneously, is another approach to reducing selection pressure while controlling the helminth infection (Albonico et al., 2003; Barnes et al., 1995). The benefits of combination therapy as a means of reducing the risk of resistance developing during large-scale deworming programs have been emphasized by Horton (2003). Considerable experience of combination therapy is being gained through the use in combination of ivermectin and albendazole (WHO, 2000a) or diethylcarbamazine and albendazole (WHO, 2000b) in the Global Alliance to Eliminate Lymphatic Filariasis (Chapter 8). The combination of DEC and albendazole is recommended for universal application in communities where *O. volvulus* and *L. loa* are judged to be absent. DEC should not be used in places where *O. volvulus* and *L. loa* are endemic because severe, even-life threatening, adverse effects may occur (WHO, 2004). In practice this means that DEC should no longer be used for the control of LF in Africa. The combination of ivermectin and albendazole is recommended for universal application in communities where LF occurs concurrently with infections of *Onchocerca volvulus* or *Loa loa*.

Janssens (1985) reviewed a range of drug combinations that have been tested against soil-transmitted helminth infections. Most of the drugs discussed by Janssens are not currently recommended for human use by the World Health Organization (WHO, 2004) and much of the information cited appears not to be readily available in the public domain. Nevertheless, finding and testing a series of drug combinations may be the best approach for dealing with drug resistance. Horton (2003) drew attention to the previously used combination of pyrantel and oxantel. Apparently oxibendazole, a benzimidazole derivative used in veterinary practice, is under consideration for use against soil-transmitted helminthiasis in humans, and nitazoxanide produced 30 years ago has now been licensed for human use in Latin and South America (Horton, 2003). Nitazoxanide is an inhibitor of pyruvate ferredoxin oxido-reductase and is used for the treatment of cryptosporidiosis. The drug has activity against *Taenia saginata* and *Hymenolepis nana* when given as a single oral dose of 30 to 45 mg/kg body weight. Nitazoxanide has been shown to have therapeutic activity against soil-transmitted helminths (Gilles and Hoffman, 2002).

Perhaps nitanoxanide, oxantel, and oxibendazole, if used with other anthelminthic drugs, will offer hope for safe and effective combination therapy.

14.5.6 POROUS BORDERS

Human migration is a relentless feature of life in our global village. Economic migrants, refugees, asylum seekers, displaced persons, military personnel, business travellers, student backpackers, holiday makers, and the victims of people traffickers are forever on the move. And their worms go with them. Worms like germs do not need a passport. *Strongyloides stercoralis* has demonstrated a remarkable capacity for traveling and surviving (Chapter 12). Urbanization, in which people leave the countryside to seek prosperity in the cities, dominates demography in low-income countries. And the worms go too. Government officials should be encouraged to consider the possibility of drug-resistant worms being transported from one endemic country to another or from one location to another in the same country. The distribution of helminthiasis is now well known and can be plotted for various purposes by means of GIS (Brooker et al., 2002). Years of careful helminth control could be wrecked if drug-resistant worms were to begin to travel.

14.6 COSTS AND COST EFFECTIVENESS OF DEWORMING

Advocates of helminth control glibly claim that deworming is a most cost-effective public health activity. Is this wishful thinking or is there evidence in support of their claim? So what are the costs? What does effective mean and what is the relationship between cost and effectiveness? Evans and Guyatt (1995) described two forms of cost-effective analysis that can be used when planning deworming campaigns for soil-transmitted helminthiasis. The first seeks to identify the most efficient procedure for delivering the deworming drug to the community. The second is concerned with comparing all the expenditure in delivering a helminth control program with all the expenditure for other public health programs and then comparing the benefits to the individuals and communities. The second is of much importance in setting public health priorities (see Chapter 3).

Simple examples of the first type of cost-effective analysis deal with deworming projects in communities in Kenya, Nigeria, and Bangladesh where declines in prevalence and intensity of infection were the outcome indicators. Stephenson et al. (1983), who organized a deworming project in Kenya lasting for four years and

TABLE 14.3
Comparison of the Costs (Naira) of the Universal, Targeted, and Selective Use of Levamisole to Control Ascariasis during 1989 to 1990 in Nigerian Villages. From Table III (Holland et al., 1996, *Journal of Parasitology* 82, 527–530) with permission from the American Society of Parasitologists.

Effectiveness Measures	Selective (Alakowe)	Targeted (Iyanfoworogi)	Mass (Akeredolu)
Total cost	12,490	3,956	4,701
Reduction (eggs per gram)	2,496	6,478	10,417
Number treated	36	194	455
Cost per 1,000 egg reduction per gram	5,004	611	451
Cost per person treated	347	20.4	10.3

providing three doses of levamisole annually to preschool and school children through local primary schools, estimated that the total cost would be USD 225 per village. During the program, the mean worm burden (*Ascaris lumbricoides*) fell from 10.1 to 4.2 per child where stool collections were made. Holland et al. (1996) calculated the expenditure (manpower, materials, drugs, and transport) involved in providing universal (mass), targeted, and selective treatment with levamisole to communities in rural Nigeria and expressed the results as cost per reduction of 1,000 epg and cost per person treated (Table 14.3). Clearly in the Nigerian project the universal strategy was the most cost effective. Mascie-Taylor et al. (1999) concluded that under the conditions of their study in Bangladesh the most cost-effective strategy for reducing the prevalence and intensity of STHs was universal treatment with albendazole. Cost effectiveness depends on the aims of the health provision. Would a single dose of an anthelminthic drug have been most cost effective if childhood weight gain had been under review?

A helpful management tool for investigating the cost effectiveness of deworming has been modeled by Guyatt et al. (1993). Their model explores how universal drug application would be expected to reduce intensity, prevalence, and morbidity in populations infected with *Ascaris lumbricoides* and living in endemic areas of high ($R_0 = 5$) and low ($R_0 = 2$) transmission rates. The universal drug application is implemented for five years at intervals of 4 months, 6 months, 1 year, and 2 years in both areas and the impact of the treatment is followed for 10 years. Predictions from the model are displayed in Figure 14.1. In the high-transmission area, the model indicates that the greater the frequency of treatment the greater will be the reduction in intensity, prevalence, and disease (morbidity). All the variables are shown to return to pretreatment values within 10 years once the treatment has stopped after 5 years. Interestingly, the model predicts that treatment will have least effect on prevalence that returns to the pretreatment level faster than intensity and morbidity (Figure 14.1). In the low-transmission area, more rounds of treatment provide more health benefits

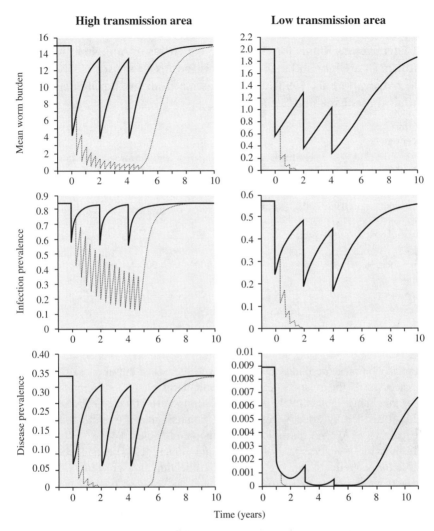

FIGURE 14.1 The effect of different frequencies of treatment on the *Ascaris* population throughout the 10-year time period. For clarity, only the effect of treatment every 4 months and every 2 years is illustrated; the effects of treatment every 6 months and every year lie between these two curves. Note the difference in the scale of the y-axis in the high and low transmission areas. From Figure 1 (Guyatt et al., 1993) with permission from The Royal Society of Tropical Medicine and Hygiene.

for the community and none of the variables reaches its pretreatment level regardless of the intervals between rounds of treatment (Figure 14.1).

The results further emphasize the importance of concentrating on efforts to reduce the intensity of helminth infections in community control programs. Reducing intensity and thereby morbidity is the appropriate measure of effectiveness. The most effective measure revealed by this study is to treat the community every

TABLE 14.4

Cost-Effectiveness Ratios for Different Frequencies of Anthelminthic Treatment in a High- and a Low-Transmission Area for *Ascaris*. From Table 2 (Guyatt et al., 1993) with permission from The Royal Society of Tropical Medicine and Hygiene.

Cost-Effectiveness Ratios for Each Effectiveness Measure	High Transmission Area				Low Transmission Area			
	2 Years	1 Year	6 Months	4 Months	2 Years	1 Year	6 Months	4 Months
Cost per unit reduction in mean worm burden	0.06	0.07	0.08	0.10	0.21	0.24	0.33	0.47
Cost per infection case prevented	5.65	5.44	4.34	3.30	1.30	1.18	1.21	1.69
Cost per disease case prevented	2.82	3.05	3.48	4.23	27.88	40.97	72.01	104.72

Note: Costs in USD, U.S. Dollars

4 months. The measure that is most effective in terms of the impact of deworming is not necessarily the most cost effective, particularly when a severe budget restraint will be inevitable. Guyatt et al.(1993) proposed the use of a cost-effectiveness ratio (CE; cost divided by effectiveness). CEs for the simulated treatment schedules in the two areas of *Ascaris* transmission are shown in Table 14.4.

In their more general analysis, Evans and Guyatt (1995) examined the consequences for cost effectiveness of diagnosis. Collecting and screening stool samples may increase the costs of a program over and above those of drug delivery by a factor of from 2 to 6. Furthermore, compliance may be compromised if people are unwilling to accept the intrusiveness of providing stool samples. Poor community compliance spells failure for helminth control activities.

These studies are examples of the cost effectiveness of vertical control programs. Guyatt (2003) has pointed out that despite the dramatic fall in the price of anthelminthic drugs, countries where helminthiasis is endemic would not be able to finance vertical control programs. For example, if Kenya were to provide albendazole treatment annually for all its school-aged children the costs would vary from USD 3 to 1.5 million; the lower figure being for drug alone. USD 3 million represented 4% of Kenya's annual health budget at the time. Vertical deworming is hard to justify when low-income countries have so many demands on resources for health care.

Guyatt (2002, 2003) published a costed menu for the targeted treatment of school children in a community with schistosomiasis and soil-transmitted helminthiasis (Table 14.5). The approach was generated by the Partnership for Child Development (PCD) working in Ghana and Tanzania in collaboration with the Ministries of Health and of Education. Other services in place may offer hitch-hiking opportunities for

TABLE 14.5
Cost Menu (USD) for Deworming Children Attending an African School in a Region Where Schistosomiasis and Soil-Transmitted Helminthiasis Are Endemic. From Table 4 (Guyatt, 2002) with permission from the World Health Organization (www.who.int) and Table 2 (Guyatt, 2003) with permission from Elsevier.

Item	Base Unit	Base Unit Cost (C)	Quantity per Child or School (Q)	Unit Cost per Child or School (CQ)
Per Child				
Drug				
Albendazole	per tablet	0.02	1.1[a]	0.022
Mebendazole	per tablet	0.02	1.1[a]	0.022
Praziquantel	per tablet	0.02	2.5[a]	0.175
Insurance, freight, clearance	per tablet	0.007	3.6	0.025
Per School				
Equipment				
Dose poles	—	—	—	3.5
Health education				
Adaptation and duplication of materials	—	—	—	10
Training				
Stationery	—	—	—	3.5
Trainers' *per diem* allowances	per person	25	0.1	2.5
Trainees' allowances	per person	2.5	3.0	7.5
Drug distribution				
Officers' training allowances	—	—	—	0.25
Instruction forms	per form	0.05	4.0	0.2
Treatment forms	per form	0.05	10.0	0.5
School officer collection allowance	—	—	—	5.0
Senior officer collection allowance	—	—	—	0.2

Note: Costs in USD, U.S. Dollars

[a] Including approximately 10% wastage

deworming including school feeding programs, mother/child clinics, family planning sessions, vitamin A distribution, and so on. Deworming is no Old Man of the Sea entwined around some poor Sinbad. Deworming enhances and strengthens health services. Studies in the Democratic Republic of the Congo (formerly Zaire) demonstrated how ascariasis control enhanced health awareness and community participation in health services (see Jancloes, 1989). In villages with ascariasis control, immunization coverage increased, tuberculosis screening increased, and the residents became convinced that abdominal pain was a symptom of ascariasis. Despite such seemingly low costs (Table 14.6), improvements in health, and synergistic interactions with other services, Guyatt (2003) concluded that low-income countries must have donor support if helminthiasis control is to progress and be sustained.

TABLE 14.6

Cost Menu for Deworming Communities (Universal Drug Application) in a Region Where Soil-Transmitted Helminthiasis[1] Is Endemic. From Table 1 (Guyatt et al., 1993) with permission from The Royal Society of Tropical Medicine and Hygiene.

Category	Item	Units	Unit Cost in 1988 USD[2]
Labor	Nurse	Per person	0.13
	Health education officer	Per person	0.177
Capital	Vehicle	Per 50,000 persons	10740
Consumables	Drugs	Per person	0.250
	Treatment cups	Per person	0.062
	Stationery	Per person	0.011
	Miscellaneous	Per person	0.008
	Petrol	Per person	0.034
Maintenance	Vehicle maintenance	Per vehicle per year	43
	Vehicle insurance	Per vehicle per year	871
Wastage	10% of consumables	Per person	0.00365

Note: USD, U.S. Dollars

[1] Guyatt et al. (1993) concentrated on ascariasis but, since broad-spectrum anthelminthic drugs are now widely used, the menu can be applied to soil-transmitted helminthiasis.

[2] Costs are based on those incurred during a control program in Montserrat.

REFERENCES

Albonico M, Bickle Q, Ramsan M et al. 2003. Efficacy of mebendazole and levamisole alone or in combination against intestinal nematode infections after repeated targeted mebendazole treatment in Zanzibar. *Bulletin of the World Health Organization* **81**, 343–352.

Albonico M, Engels D and Savioli L. 2004. Monitoring drug efficacy and early detection of drug resistance in human soil-transmitted nematodes: a pressing public health agenda for helminth control. *International Journal for Parasitology* **34**, 1205–1210.

Albonico M, Wright V, Ramsam M et al. 2005. Development of the egg hatch assay (EHA) for detection of anthelminthic drug resistance by human hookworms. *International Journal of Parasitology* **35**, 803–811.

Alexander ND, Cousens SN, Yahaya H et al. 1993. Ivermectin dose assessment without weighing scales. *Bulletin of the World Health Organization* **71**, 361–366.

Anderson RM. 1989. Transmission dynamics of *Ascaris lumbricoides* and the impact of chemotherapy. In: *Ascariasis and its Prevention and Control* (eds. DWT Crompton, MC Nesheim and ZS Pawlowski). London and Philadelphia: Taylor and Francis. pp. 253–273.

Anderson RM and May RM. 1992. *Infectious Diseases of Humans.* Oxford University Press (paperback edition).

Anderson RM and Medley GF. 1985. Community control of helminth infections in man by mass and selective chemotherapy. *Parasitology* **90**, 629–660.

Barnes BH, Dobson RJ and Barger IA. 1995. Worm control and anthelminthic resistance: adventures with a model. *Parasitology Today* **11**, 56–63.

Brooker S, Beasley NMR, Ndinaromtan M et al. 2002. Use of remote sensing and a geographical information system in a national helminth control programme in Chad. *Bulletin of the World Health Organization* **80**, 783–789.

Bundy DAP and Michael E. 2001. Epidemiology and control of helminthiasis of humans. In: *Perspectives on Helminthology* (eds. N Chowdhury and I Tada). Enfield (NH), U.S. and Plymouth, U.K. Springer Publishers Inc. pp. 179–223.

Cioli D. 2000. Praziquantel: is there real resistance and are there alternatives? *Current Opinion in Infectious Diseases* **13**, 659–663.

Chitsulo L, Lengeler C and Jenkins J. 1995. The schistosomiasis manual. Geneva: World Health Organization. TDR/SER/MSR/95.2.

Coles GC, Bauer C, Borgsteede FH et al. 1992. World Association for the Advancement of Veterinary Parasitology (WAAVP) methods for the detection of anthelminthic resistance in nematodes of veterinary importance. *Veterinary Parasitology* **44**, 35–44.

Comins HN. 1977. The management of pesticide resistance. *Journal of Theoretical Biology* **65**, 399–420.

Davis A.1985. Discussion. In: *Ascariasis and its Public Health Significance* (eds. DWT Crompton, MC Nesheim and ZS Pawlowski). London and Philadelphia: Taylor and Francis. p 282.

Dayan AD. 2003. Albendazole, mebendazole and praziquantel. Review of non-clinical toxicity and pharmacokinetitics. *Acta Tropica* **86**, 141–159.

De NV, Murrell KD, Cong LD et al. 2003. The food-borne trematode zoonoses of Vietnam. *Southeast Asian Journal of Tropical Medicine and Public Health* **34** (Suppl. 1), 12–34.

De Clercq D, Sako M, Behnke J et al. (1997). Failure of mebendazole in treatment of human hookworm infections in the southern region of Mali. *American Journal of Tropical Medicine and Hygiene* **57**, 25–30.

Engels D and Savioli L. 2005. Public health strategies for schistosomiasis control. In: *World Class Parasites*: Volume 10: *Schistosomiasis* (eds. WE Secor and DG Colley). New York: Springer Science & Business Media, Inc. pp. 207–222.

Evans DB and Guyatt H L. 1995. The cost effectiveness of mass drug therapy for intestinal helminths. *PharmacoEconomics* **8**, 14–22.

Firth M. 1983. *Albendazole in Helminthiasis*. London: The Royal Society of Medicine and Academic Press Ltd.

Gilles HM and Hoffman PS. 2002. Treatment of intestinal parasitic infections: a review of nitanoxanide. *Trends in Parasitology* **18**, 95–97.

Guyatt H. 2002. Communication to the WHO Expert Committee on Prevention and Control of Schistosomiasis and Soil-transmitted Helminthiasis. *Technical Report Series* 912, p 21. Geneva: World Health Organization.

Guyatt H. 2003. The cost of delivering and sustaining a control programme for schistosomiasis and soil-transmitted helminthiasis. *Acta Tropica* **86**, 267–274.

Guyatt HL, Bundy DAP and Evans D. 1993. A population dynamic approach to the cost-effective analysis of mass anthelminthic treatment: effects of treatment frequency on *Ascaris* infection. *Transactions of the Royal Society of Tropical Medicine and Hygiene* **87**, 570–575.

Holland CV, O'Shea E, Asaolu SO et al. 1996. A cost-effective analysis of anthelminthic intervention for community control of soil-transmitted helminth infection: levamisole and *Ascaris lumbricoides*. *Journal of Parasitology* **82**, 527–530.

Horton J. 2000. Albendazole: a review of anthelmintic efficacy and safety in humans. *Parasitology* **121**, S113–S132.

Horton J. 2003. The efficacy of anthelminthics: past, present, and future. In: *Controlling Disease due to Helminth Infections.* (eds. DWT Crompton, A Montresor, MC Nesheim and L Savioli). Geneva: World Health Organization. pp. 143–155.

Jancloes M. 1989. The case for control: forging a partnership with decision-makers. In: *Ascariasis and its Prevention and Control* (eds. DWT Crompton, MC Nesheim and ZS Pawlowski). London, New York and Philadelphia: Taylor and Francis. pp. 379–393.

Janssens PG. 1985. Chemotherapy of gastrointestinal nematodiasis in man. In: *Handbook of Experimental Pharmacology* vol. 77 (eds. H Vanden Bossche, D Thienpont and PG Janssens). Berlin, Heidelberg, New York and Tokyo: Springer Verlag. pp. 183–406.

Lewis JA, Wu CH, Berg H and Levine JH. 1980. The genetics of levamisole resistance in the nematode *Caenorhabditis elegans. Genetics* **95**, 905–928.

McIntosh DAD, Bax RP and Lewis DJ. 1989. The role of the pharmaceutical industry in the prevention and control of ascariasis. In: *Ascariasis and its Prevention and Control* (eds. DWT Crompton, MC Nesheim and ZS Pawlowski). London, New York and Philadelphia: Taylor and Francis. pp. 275–288.

Mansour TE. 2002. *Chemotherapeutic Targets in Parasites.* Cambridge University Press.

Marshall I. 1987. Experimental chemotherapy. In: *The Biology of Schistosomes.* (eds. D Rollinson and AJG Simpson). New York and London: Academic Press. pp. 401–430.

Martin PJ, Anderson N and Jarret RG. 1989. Detecting benzimidazole resistance with faecal egg count reduction tests and in vitro assays. *Australian Veterinary Journal* **66**, 236–240.

Mascie-Taylor CGN, Alam M, Montanari RM. et al. 1999. A study of the cost effectiveness of selective health interventions for the control of intestinal parasites in rural Bangladesh. *Journal of Parasitology* **85**, 6–11.

Montresor A, Crompton DWT, Gyorkos TW and Savioli L. 2002. *Helminth Control in School-age Children. A Guide for Managers of Control Programmes.* Geneva: World Health Organization.

Montresor A, Odermott P, Muth S et al. 2005. The WHO dose pole for the administration of praziquantel is also accurate in non-African populations. *Transactions of The Royal Society of Tropical Medicine and Hygiene* 99, 78–81.

Prichard RK, Hall CA, Kelly JD et al. 1980. The problem of anthelminthic resistance in nematodes. *Australian Veterinary Journal* **56**, 239–251.

Renganathan E and Cioli D.1998. An international initiative on praziquantel use. *Parasitology Today* **14**, 390–391.

Reynoldson JA, Behnke JM, Pallant LJ et al. 1997. Failure of pyrantel in treatment of human hookworm infections (*Ancylostoma duodenale*) in the Kimberly region of North West Australia. *Acta Tropica* **68**, 301–312.

Roos MH, Boersema JH, Borgst D, Borgsteede FHM et al. 1990. Molecular analysis of selection for benzimidazole resistance in the sheep parasite *Haemonchus contortus. Molecular and Biochemical Parasitology* **43**, 77–88.

Sacko M, De Clercq D, Behnke J et al. 1999. Comparison of the efficacy of mebendazole, albendazole and pyrantel in the treatment of human hookworm infections in the Southern region of Mali, West Africa. *Transactions of the Royal Society of Tropical Medicine and Hygiene* **93**, 195–203.

Sangster NC, Bannan SC, Weiss AS et al. 1999. *Haemonchus contortus*: Sequence heterogeneity of internucleotide binding domains from P-glycoproteins. *Experimental Parasitology* **91**, 250–257.

Smith G and Scott ME. 1994. Model behaviour and the basic reproduction ratio. In: *Parasitic and Infectious Diseases* (eds. ME Scott and G Smith). New York and London: Academic Press. pp. 21–28.

Stephenson LS, Crompton DWT, Latham MC et al. 1983. Evaluation of a four year project to control *Ascaris* infection in children in two Kenyan villages. *Journal of Tropical Pediatrics* **29**, 175–184.

Taylor MA. 1992. Anthelmintic resistance in helminth parasites of domestic animals. *Agricultural Zoology Reviews* **5**, 1–50.

Urbani C. and Albonico M. 2003. Anthelminthic drug safety and drug administration in the control of soil-transmitted helminthiasis in community campaigns. *Acta Tropica* **86**, 215–221.

WHO. 1986. Sample size determination: a user's manual. Geneva: World Health Organization. WHO/HST/ESM/86.1.

WHO. 1988. *The World Drug Situation.* Geneva: World Health Organization.

WHO. 1995. The use of essential drugs. *Technical Report Series* 850, Geneva: World Health Organization.

WHO. 1996a. Report of the WHO Informal Consultation on hookworm infection and anaemia in girls and women. Geneva: World Health Organization. WHO/CTD/SIP/96.1.

WHO. 1996b. Report of the WHO Informal Consultation on the use of chemotherapy for the control of morbidity due to soil-transmitted nematodes in humans. Geneva: World Health Organization. WHO/CTD/SIP/96.

WHO. 1999. Report of the WHO Informal Consultation on monitoring of drug efficacy in the control of schistosomiasis and intestinal nematodes. Geneva: World Health Organization. WHO/CDS/CPC/SIP/99.1.

WHO. 2000a. Preparing and implementing a national plan to eliminate lymphatic filariasis (in countries where onchocerciasis is not co-endemic). Geneva: World Health Organization. WHO/CDS/CPE/CEE/2000.15.

WHO. 2000b. Preparing and implementing a national plan to eliminate lymphatic filariasis (in countries where onchocerciasis is co-endemic). Geneva: World Health Organization. WHO/CDS/CPE/CEE/2000.16.

WHO. 2002. Prevention and control of schistosomiasis and soil-transmitted helminthiasis. *Technical Report Series* 912. Geneva: World Health Organization.

WHO. 2003. *Global Defence against the Infectious Disease Threat.* Geneva: World Health Organization.

WHO. 2004. *WHO Model Formulary.* Geneva: World Health Organization.

15 Health Awareness and Helminthiasis

Ignorance in the community is an obstacle that must be overcome if public health interventions for the sustainable control of helminthiasis are to succeed. Strengthening health awareness should be an integral component of all public health programs. In the context of helminthiasis, health awareness involves communication and understanding between health workers and the community about the nature and consequences of the infections that afflict members of the community and their families. People need to know about (1) the pathogens that share their environment, (2) modes of transmission, (3) risk groups, (4) how to get relief, (5) how to avoid infection, and (6) the benefits of control. Prospects of achieving this agenda of awareness will depend on opportunities for sharing information, knowledge of local resources, sensitivity to cultures and tradition, and the level of education of the people living where control interventions are needed.

Education would be expected to be the key to improved health awareness. Education enables people to communicate, to understand, and to make appropriate decisions. Education stimulates people to find out for themselves. Knowledge may be acquired through teaching and learning, but knowing what to do with the knowledge is the reward of education. Education depends on literacy, the ability to read and write. Regrettably, many of the people suffering from helminthiasis or at risk of infection have the least educational provision and the lowest literacy rates (Table 15.1). The data in this table reveal a marked gender disparity in literacy rates in the populations of the south that are most affected by helminthiasis. Health care for the family and others in the household depends mainly on women. The fact that so many girls still await the opportunity to learn to read and write is likely to have serious effects on measures intended to improve health and well being. Some of the circumstances that hold girls back from completing their primary and secondary education have been highlighted by UNICEF (2003).

15.1 EDUCATION AND THE RISK OF HELMINTH INFECTION

Education in general, without any special attention to health matters, will tend to reduce the risk of helminth infection. Presumably people with more education are better able to recognize and respond to health messages. Low levels of education, however, predominate among poor people so it is difficult to isolate the effect of education or lack of it from other confounding variables. Perhaps inadequate education is another proxy for poverty or inequality. Traub et al. (2004) investigated education status and other potential risk factors associated with soil-transmitted helminth infections in staff and workers on tea estates in Assam. Staff members, the

TABLE 15.1
The State of Education in Regions Where Helminthiasis Persists. From data in UNICEF (2004).

Region	Adult Literacy Rate (%) 2000		Primary School Enrollment Ratio (%) 1997–2000 (net)	
	Male	Female	Male	Female
Sub-Saharan African	69	53	63	58
Mid East and North Africa	74	52	83	75
South Asia	66	42	80	65
East Asia and Pacific	93	81	93	92
Latin America and Caribbean	90	88	96	94
Developing countries	81	67	84	77
Least developed countries	62	42	67	61

Note: Adult literacy rate — percentage of persons aged 15 and over who can read and write.

Primary school enrolment ratio — number of children enrolled in primary school who belong to the age group that officially corresponds to primary schooling, divided by the total population of the same age group.

minority group on an estate, had completed a minimum of secondary schooling, and lived in housing with separate rooms, space between neighboring houses, and latrines connected to septic tanks. Workers, who were mainly tea pluckers, were the majority group on an estate. They were largely illiterate and they lived in overcrowded housing with poor standards of hygiene. The prevalences and intensities of *Ascaris lumbricoides, Trichuris trichiura,* and hookworm infections were found to be greater in the workers than in the staff members. But could educational status be teased out from other risk factors prevailing among the group so obviously vulnerable to infection? After adjusting for other variables, rigorous statistical analysis revealed that individuals were 1.3 times ($P = 0.0581$), 1.4 times ($P = 0.0121$, and 1.6 times ($P = 0.0093$) less likely to be infected with *Ascaris, Trichuris,* and hookworm, respectively, with increasing levels of education.

Evidence indicates that the risk of helminth infection in children is related to the level of education attained by their mothers is suggested by observations on infection with *Ascaris lumbricoides* (Table 15.2). Similar observations could be made about the risk of catching other helminth infections. Improving educational opportunities for girls must improve the health and well being of children.

15.2 ASSESSMENT OF HEALTH AWARENESS

The design and implementation of interventions for the control of helminthiasis should be sensitive to the degree of health awareness in the community (Curtale

TABLE 15.2
Level of Mother's (Caregiver's) Education As a Risk Factor for *Ascaris lumbricoides* Infection in Children[a]

Location	Study Population	Sample Size	Method	Observation	Reference
Panama	Preschool children	212	Caregiver interview	Children with *Ascaris* infection had caregivers with low education ($P = 0.003$).	Holland et al. (1988)
Malaysia	Primary school children	205	Interview	Low level of mother's education was a risk factor for *Ascaris* infection in children > 6 years.	Norhayati et al. (1998)
Malawi	Children aged 3–14 years	553	Interview	Urban mothers with 4–8 years education was a risk factor for *Ascaris* infection in children.	Phiri et al. (2000)
India	Primary school children aged 5–9 years	204	Questionnaire to mothers	Children with high intensity of *Ascaris* infection had mothers with low education.	Naish et al. (2004)
Turkey	School children aged 7–14 years	639	Questionnaire	43% *Ascaris* prevalence in children whose mothers had less than primary school education, 29% when mothers had primary school or more education ($P = 0.035$).	Okyay et al. (2004)

[a] *Ascaris* infections are relatively easy to investigate because family compliance is good and diagnosis is reliable based on the ease of identifying eggs in stool samples examined by the Kato Katz procedure (Appendix 3).

et al., 1998). Health awareness is often investigated through KAP surveys (knowledge, attitudes, and practice), which depend on members of the community giving reliable answers to an appropriately prepared questionnaire (Motarjemi, 2000). Interviews conducted by local field assistants are regularly used to study health awareness. Caution should be exercised over the interpretation of answers since respondents may want to answer in line with the investigator's expectations, may feel they could be disadvantaged if they do not provide desired answers, and may well not do in practice what they have claimed to do. Wherever possible, some form of independent check should be carried out to test the reliability of data gained from KAP surveys and interviews.

In an investigation of hand washing and latrine use, Cairncross et al. (2005) employed "pocket voting" as a means of exploring health awareness and hygiene practice. Pocket voting requires the respondents to put their answers to a few key questions into pockets attached to a portable frame. This has advantages for what might be a culturally delicate issue such as gender and latrine use. Also an independent observer can compare the responses to a question such as one about latrine use to the actual state of the latrines at the time while voting is taking place. Other investigators have used multiple choice questions as a means of getting information about health awareness (Da Silva et al., 2002).

Summaries of information obtained from studies of health awareness are shown in Table 15.3. The findings are diverse and demonstrate that the managers of public health programs must investigate the level of health awareness in each target community. Surveys of the literature tend to suggest that people living in areas where helminthiasis is endemic often have knowledge of disease and the need for treatment, but inadequate knowledge of infectious agents. This opinion is supported by the

TABLE 15.3
Summaries from Surveys Concerning Community Health Awareness of Helminthiasis

Helminthiasis	Location	Respondents	Summary Comments	References
Lymphatic filariasis	Haiti	Residents	Hydrocele was thought to be caused by trauma (60%) or trapped gas (30%) or ceremonial powder (23%).	Eberhard et al. (1996)
Onchocerciasis	Guatemala	Heads of households	100% heard of disease; 95% identified surgery as treatment; 50% knew of insect bites; 39% knew of worm involvement.	Richards et al. (1991)
	Nigeria	Women	KAP survey showed minimal knowledge of onchocerciasis as a disease entity.	Ogbuokiri (1995)
Schistosomiasis mansoni	Brazil	Residents	69.6% know how infection is acquired; 69.2% had hearsay information. Local people received only piecemeal and subjective information about their problem.	Da Silva et al. (2002)
Soil-transmitted helminthiasis	Egypt	Mothers	More than adequate knowledge was present on ways to prevent infection. Almost all the respondents considered worms harmful.	Curtale et al. (1998)

Note: KAP, knowledge, attitude, and practice.

TABLE 15.4
Results of a Questionnaire Designed to Investigate the Knowledge of Schistosomiasis Haematobium of Adults in Rural Cross River State, Nigeria. From Table 5 (Useh and Ejezie, 1999). Reproduced from the *Annals of Tropical Medicine and Parasitology* 93, p. 717, with the permission of the Liverpool School of Tropical Medicine.

Knowledge about	Number of Respondents (%)			
	329 Men		251 Women	
Disease (local name)	322	(98)	240	(95)
Causative agent*	50	(15)	20	(8)
Public health status	298	(91)	229	(91)
Vector (snail)*	75	(23)	35	(14)
Source of infection (water)	300	(91)	230	(92)
S. haematobium life history	45	(14)	38	(15)
Blood in urine means disease	295	(90)	224	(89)
Willing to adopt control measures	315	(96)	245	(98)

* Statistically significant difference between the responses.

results of an investigation into knowledge of schistosomiasis hematobium in Nigeria (Table 15.4). Making generalizations, however, may be unwise; there is no universal conclusion to be drawn about health awareness for a particular helminthiasis or how to promote it. In a survey in Ghana, similar to the one carried out in Nigeria (Table 15.4), some residents believed that hematuria, a sign of schistosomiasis hematobium, was caused by the red color of a variety of sugar cane (Aryeetey et al., 1999).

15.3 IMPACT OF HEALTH AWARENESS IN THE CONTROL OF HELMINTHIASIS

Does evidence exist to show that health education campaigns make a significant contribution to the control of helminthiasis? Asaolu and Ofoezie (2003), in line with the policy that health education should be an integral part of control interventions, investigated the value of including health education in seven programs involving several interventions. Of interest was the role of health education alone in the control of *Ascaris lumbricoides* and *Schistosoma mansoni* infections (Table 15.5). Control programs 1 (*Ascaris*) and 7 (*Schistosoma*) depended on health education alone. In program 1, the prevalence and intensity of *Ascaris* infection fell by 26 and 35% respectively while in program 7, the prevalence of *Schistosoma* fell by 18% (Table 15.5). Greater control gains were observed for these and other soil-transmitted helminth infections when health education was included in support of anthelminthic treatment in programs 2, 3, 4, 5, and 6 (Table 15.5).

TABLE 15.5

Changes in the Prevalence and Intensity of Selected Helminth Infections before and after Health Education[1] with or without Chemotherapy[2] Interventions in Different Endemic Regions of the World. From Table 4 (Asaolu and Ofoezie, 2003) with permission from Elsevier.

#	Parasite	Prevalence (%)		Change (%)	Mean Intensity (epg) [c]		Change (%)	Period (years)	Location	Reference
		Pre-control	Post-control		Pre-control	Post-control				
1	Ascaris[a]	49.1	36.1	-26.5	1264	812	35.8	0.5	Indonesia	Hadidjaja et al. (1998)
2	Ascaris[b]	71.4	41.8	-41.5	2435	657	-73.0	0.5	Indonesia	Hadidjaja et al. (1998)
3	Ascaris[b]	17.7	4.4	-75.1	1617	244	-84.9	0.5	Seychelles	Albonico et al. (1996)
4	Trichuris[b]	53.3	27.3	-48.8	782	367	-53.1	0.5	Seychelles	Albonico et al. (1996)
5	Hookworm[b]	6.3	4.2	-33.3	40	27	-32.5	0.5	Seychelles	Albonico et al. (1996)
6	Schistosoma[b]	6.6	0.9	-86.4	NA	NA	NA	4	Mauritius	Dhunputh et al. (1994)
7	Schistosoma[a]	11.0	9.0	-18.2	NA	NA	NA	1	Brazil	Schall (1995)

Note: NA, not available.

Program number.

[1] School or community based awareness program.

[2] Treatment with mebendazole or praziquantel.

[a] Exclusive health education intervention.

[b] In combination with chemotherapy or snail control.

[c] Arithmetic mean.

A major effort to control schistosomiasis japonicum involving health education and health promotion has been made in villages around Lake Poyang in China (Guanghan et al., 2000; Hu et al., 2005). The intervention involved the use of video tapes to train children about the prevention of infection, promotion of understanding about schistosomiasis for women, encouragement in compliance with anthelminthic treatment, and training for men in the prevention of infection. After the program had lasted for 12 years, Hu et al. found that awareness of schistosomiasis had been strengthened and that appropriate knowledge and behavior for avoiding infection had increased. Women had abandoned their practice of washing clothes in water where snails released cercariae and school children showed a reduced frequency of contact with such water. Health education enabled the community to accept the value of anthelminthic treatment. Overall, reinfection rates with *S. japonicum* were found to have fallen, a most encouraging result that had much to do with the community's compliance with anthelminthic treatment (Table 15.6). The program appeared not

TABLE 15.6
The Impact of Health Education on Reinfection Rates with *Schistosoma japonicum* in Villages in the Poyang Lake Area, PR China. The Table Compares Pre- and Post-intervention Results with Those from a Control Village Where Intervention Did Not Take Place. From Table 5 (Hu et al., 2005) with permission from Elsevier.

Village	Study Population	Preintervention		Postintervention		Difference	
		No Examined	No Positive (%)	No Examined	No Positive (%)	χ^2	P-value
B	Schoolchildren (6–15 years)	131	12 (9.2)	127	3 (2.4)	5.44	0.020
	Women (16–60 years)	140	41 (29.3)	141	10 (7.1)	23.29	< 0.001
	Men (16–60 years)	155	29 (18.7)	159	28 (17.6)	0.06	0.800
C	Schoolchildren (6–15 years)	153	22 (14.4)	144	6 (4.2)	9.06	0.003
	Women (16–60 years)	180	27 (15.0)	171	7 (4.1)	11.92	0.001
	Men (16–60 years)	169	26 (15.4)	164	27 (16.5)	0.07	0.788
D	Schoolchildren (6–15 years)	295	36 (12.2)	300	20 (6.7)	5.35	0.016
	Women (16–60 years)	354	34 (9.6)	346	16 (4.6)	6.54	0.012
	Men (16–60 years)	352	38 (10.8)	364	44 (12.1)	0.30	0.124
B′ (control)	Schoolchildren (6–15 years)	151	19 (12.6)	142	18 (12.7)	0.00	0.981
	Women (16–60 years)	171	39 (22.8)	167	39 (23.4)	0.01	0.905
	Men (16–60 years)	183	55 (30.1)	181	46 (25.4)	0.98	0.323

to have been of benefit to the men in the villages (Table 15.6), but their work meant that water contact was inevitable (Hu et al., 2005). No doubt their improved health awareness, however, had helped the women and children in their families to avoid infection and comply with anthelminthic treatment.

Intervention in the form of health education is not always fruitful. Sow et al. (2003) took the opportunity to assess the impact of seven years' of health education about schistosomiasis mansoni in a region of northern Senegal. The intervention began in 1994, a few years after the appearance of the helminthiasis following the construction of the Diama dam in 1986. Would this spell of intensive health education about schistosomiasis mansoni have helped people to protect themselves from infection and disease? The results obtained from answers to a simple questionnaire are shown in Table 15.7. Overall, 20% of children and 37% of adults gave adequate answers to questions about symptoms and transmission. Also 28% of children and 50% of adults gave at least one correct answer to each of both questions (Table 15.7). Sow et al. concluded that this the level of health awareness of the disease in a place where it had become part of daily life and where dedicated intervention had been taking place revealed the limitations of health education in the control of schistosomiasis. Perhaps the expected economic benefits of the dam will gradually increase the level of education in the area so that people will be better placed to be receptive to health education.

TABLE 15.7
Reported Knowledge about Schistosomiasis Mansoni in Ndombo, Northern Senegal (Values Are in Percentages). From Table 1 (Sow et al., 2003) with permission from Blackwell Publishing Ltd.

Reported Knowledge	Children		Adults		All
	Male (151)	Female (89)	Male (138)	Female (188)	(566)
'Ever heard of schistosomiasis'	86.1	68.5	94.2	86.7	85.5
Symptoms					
Adequate*	60.9	53.9	59.4	45.2	54.2
Unclear	15.2	6.7	17.4	16.5	14.8
Incorrect	23.8	39.3	23.2	38.3	30.9
Mode of transmission					
Adequate †	26.5	22.5	63.0	52.7	43.5
Unclear	6.6	1.1	7.2	2.1	4.4
Incorrect	66.9	76.4	29.7	45.3	52.1
Symptoms and mode of transmission combined					
Both adequate	19.9	19.1	40.6	34.0	29.5
Both adequate or unclear	31.8	21.3	59.4	43.6	40.8

* "diarrhea," "abdominal pain," or "bloody stools."
† "From the water" or "from the stream."

15.4 ELEMENTS IN PLANNING FOR HEALTH EDUCATION

Schistosomiasis has featured prominently in operational research intended to improve the implementation of health awareness programs. Decline in efforts to control snail populations and the fact that transmission of the infection depends on water contact has prompted public health workers to focus on health education with the aim of persuading people to voluntarily change behavior patterns (Kloos, 1995). One system for a health education intervention concerning schistosomiasis recognizes ten basic components (WHO, 1990), but these can be adapted for other forms of helminthiasis. The components are:

- title and sponsoring agencies
- description of the target community
- statement of the problem
- list of educational objectives
- description of the means for community involvement
- analysis of the factors that will impede or advance the intervention
- list of appropriate health education strategies
- outline of available resources in relation to those needed for the intervention
- timetable for action
- scheme for monitoring and evaluation

In practice, health education would not be likely to stand alone, but would be a supporting partner in all aspects of the control program.

The plan for health education would be expected to develop each component into a major section. For example, what should be included in the section dealing with health education strategies, methods, and materials? First would be the need to ensure that appropriately trained educators are available to deliver appropriate services. Second would be the need to provide appropriate educational aids such as posters, pamphlets, flipcharts, handbills, audio and video tapes, radio and TV broadcasts, and newspaper articles (Albonico et al., 1996). In the context of health education, appropriate is an important word. Although the eventual purpose will be to stimulate behavioral change, it is unreasonable to promote objectives that are unattainable. Hookworm infections will decline if people wear shoes, but they must have the means of purchase. Water will be safer if boiled, but people must have enough fuel. Food-borne trematodiasis will be reduced if people stop eating raw fish, but food security must not be compromised. And educators should not lose the trust of the community by belittling traditional beliefs or challenging local routines. If a group of women regularly gather at a lake to wash clothes and so expose themselves to infection with schistosomes, educators should be sensitive to the fact that laundry time is part of a social network and friendship.

Another example comes from considering the component covering monitoring and evaluation. In the case of schistosomiasis, arrangements might be made to address the following questions:

1. Is the timetable being adhered to and are planned activities taking place when and as intended?
2. Have people found sustainable ways of reducing their contact with water where transmission occurs?
3. Have latrines been built and are these in use because people understand about the risks of contaminating the water with the eggs of schistosomes?
4. Has participation in screening services increased and is more use being made of health services in the area?
5. Has the morbidity rate declined in response to changed behavior?

Although his review concentrated on schistosomiasis, the literature cited by Kloos (1995) gives an introduction to many aspects of health education as an intervention in the control of helminthiasis.

15.5 HEALTH EDUCATION IN THE SCHOOL SYSTEM

Schools seek to provide an environment that stimulates children to learn and understand. If children have not been exposed to the need for hygienic behavior, their time at school offers an opportunity to make good the omission. If a school has satisfactory water and sanitation facilities the students can learn how to use and maintain them. UNICEF (2005) and its partners are actively promoting hygiene, water, sanitation, and hand-washing in schools. The drive to expand deworming activities in schools (Drake et al., 2002) offers an opportunity to use helminth control as an entry point for health promotion (Taylor et al., 1999). Since children are acknowledged to be more receptive to new information than adults, there is every possibility of their taking health messages back to their families.

Shu et al. (1999) carried out a study with Nigerian children aged from 11 to 17 years to investigate how receptive they were to health education about onchocerciasis. The subjects were tested on their knowledge of onchocerciasis before exposure to educational activities over a period of three months. Nine months later the subjects were tested again and a significantly higher proportion of them ($P < 0.0001$) had acquired knowledge of the disease, its cause, clinical manifestations, diagnosis, treatment, and prevention than before the education. Shu et al. considered that health education for children at school could serve as a useful multiplier resource for the community. A study of similar design in China obtained results to show that primary school children readily learned about urinary schistosomiasis and that a decrease in contact with unsafe water occurred (Yuan et al., 2000). Education in this case was provided through a video and a comic book. Lucien et al. (2003), who studied the impact of health education on urinary schistosomiasis directed at school children in Cameroon, concluded that health education through the framework of school could be adopted as a national policy for urinary schistosomiasis control programs in tropical developing countries. That view could be a recommendation for all governments to consider for all forms of helminthiasis if the policy has not already been adopted.

15.6   COST-EFFECTIVENESS OF HEALTH EDUCATION

How much resource would have to be spent on health education to achieve measurable reductions in morbidity due to helminthiasis? Although few investigations have addressed that question, Mascie-Taylor et al. (2003) have carried out an extensive study of the cost effectiveness of health education directed at soil-transmitted helminth infections in rural Bangladesh. Their project, which ran for 18 months, involved a randomized intervention survey in four communities. All participants were treated with albendazole at the start of the study to ensure that parasitological status was homogeneous. Health education was given to community 1 throughout the study, anthelminthic treatment was given to the index child in each household at 6 and 12 months in community 2, the same treatment regimen was followed plus health education throughout the study in community 3, and no further intervention was offered in community 4. The KAP surveys generated much interesting information about local health awareness that should be of interest elsewhere.

Health education was provided in each of community 1 and community 3 by a team of six Bangladeshi health assistants and a supervisor. Monthly home visits were made, focus groups were arranged, and school sessions were held. The aims were to increase awareness of transmission and morbidity, improve personal hygiene and household hygiene, encourage the wearing of shoes and regular trimming of nails, and stimulate the construction of latrines and drilling of tube wells. Funds were not provided for latrine construction or well drilling.

At the end of the study, Mascie-Taylor et al. found highly significant improvements in knowledge of many aspects of soil-transmitted helminth infections in the communities that had been exposed to health education. Cost-effectiveness was measured by working out how much had been spent on health education to gain an improvement of 1% per household on such subjects as knowledge, water and sanitation, and personal hygiene.

Results are summarized in Table 15.8 in which the relative percentage refers to the difference between the communities with health education and those without.

TABLE 15.8
Soil-Transmitted Helminthiasis: Cost Effectiveness of Health Education in Improving Knowledge, Water and Sanitation Facilities, and Personal Hygiene. From Table 4 (Mascie-Taylor et al., 2003) with permission from Elsevier.

Item	Average Percent Improvement	Relative Average Percent Improvement	Cost of 1% Improvement (USD) Average	Cost of 1% Improvement (USD) Relative
Knowledge	45.0	49.4	0.82	0.75
Water and sanitation	26.6	23.6	1.39	1.57
Personal hygiene	33.6	28.1	1.10	1.32

Note: USD, U.S. Dollars.

Macie-Taylor et al. noted that anthelminthic treatment was the most cost-effective intervention with each USD spent being accompanied by a 3% reduction in prevalence as compared with a 0.3% reduction for the same expenditure on health education. Overall the cost of health education per household was considered to be unlikely to be cost effective at the national level if anything from USD 0.75 to 1.57 would need to be spent per household to gain a 1% improvement in health awareness about soil-transmitted helminth infections. These results from Bangladesh point the way to the development of health education that could cover several topics simultaneously for the same expenditure. For example, what would have been the cost-effective outcome if diarrheal disease, protozoal infections, nutritional needs, and other problems had been included in the program?

REFERENCES

Albonico M, Shamlaye N, Shamlaye C and Savioli L. 1996. Control of intestinal parasitic infections in Seychelles: a comprehensive and sustainable approach. *Bulletin of the World Health Organization* **74**, 577–586.

Aryeetey ME, Aholu C, Wagatsuma Y et al. 1999. Health education and community participation in the control of urinary schistosomiasis in Ghana. *East African medical Journal* **76**, 324–329.

Asaolu SO and Ofoezie IE. 2003. The role of health education and sanitation in the control of helminth infections. *Acta Tropica* **86**, 283–294.

Cairncross S, Shordy K, Zacharia S and Govindan BK. 2005. What causes sustainable changes in hygiene behaviour? A cross-sectional study from Kerala, India. *Social Science and Medicine* **61**, 2212–2220.

Curtale F, Pezzotti P, Sharbini AL et al. 1998. Knowledge, perceptions and behaviours of mothers toward intestinal helminths in Upper Egypt: implications for control. *Health Policy and Planning* **13**, 423–432.

Da Silva RA, de Carvallio ME, Zacharias E et al. 2002. Schistosomiasis mansoni in Bananal (State of Sao Paulo, Brazil): IV. Study on the public awareness of its risks in the Palha District. *Memorias do Instituto Oswaldo Cruz* **97**, 15–18.

Dhunputh J. 1994. Progress in the control of schistosomiasis in Mauritius. *Transactions of The Royal Society of Tropical Medicine and Hygiene* **88**, 507–509.

Drake L, Maier C, Jukes M et al. 2002. School-age children: their nutrition and health. *SCN News* **25**, 4–30.

Eberhard ML, Walker EM, Addis DG and Lammie PJ. 1996. A survey of knowledge, attitudes and perceptions (KAPs) of lymphatic filariasis, elephantiasis and hydrocoele among residents in an endemic area in Haiti. *American Journal of Tropical Medicine and Hygiene* **54**, 299–303.

Guanghan H, Dandan L, Shaoji Z et al. 2000. The role of health education for schistosomiasis control in heavy endemic area of Poyang Lake region, People's Republic of China. *Southeast Asian Journal of Tropical Medicine and Public Health* **31**, 467–472.

Hadidjaja P, Bonwang E, Suyardi MA et al. 1998. The effect of intervention methods on nutritional status and cognitive function of primary school children infected with *Ascaris lumbricoides*. *American Journal of Tropical Medicine and Hygiene* 59, 791–795.

Holland CV, Taren DL, Crompton DWT et al. 1988. Intestinal helminthiasis in relation to the socioeconomic environment of Panamanian children. *Social Science and Medicine* **26**, 209–213.

Hu GH, Jia H, Song KY et al. 2005. The role of health education and health promotion in the control of schistosomiasis: experiences for a 12-year intervention study in the Poyang Lake area. *Acta Tropica* **96**, 232–241.

Kloos H. 1995. Human behavior, health education and schistosomiasis control: a review. *Social Science and Medicine* **40**, 1497–1511.

Lucien KF, Nkwelang G and Ejezie GC. 2003. Health education strategy in the control of urinary schistosomiasis. *Clinical and Laboratory Science* **16**, 137–141.

Mascie-Taylor CGN, Karim R, Karim E et al. 2003. The cost-effectiveness of health education in improving knowledge and awareness about intestinal parasites in rural Bangladesh. *Economics and Human Biology* **1**, 321–330.

Motarjemi Y (ed.). 2000. *Foodborne Disease. A Focus for Health Education.* Geneva: World Health Organization.

Naish S, McCarthy J and Williams GM. 2004. Prevalence and risk factors for soil-transmitted helminth infection in a South Indian fishing village. *Acta Tropica* **91**, 177–187.

Norhayati M, Oothuman P and Fatmah MS. 1998. Some risk factors of *Ascaris* and *Trichuris* infection in Malaysian aborigine (Orang Asli) children. *Medical Journal of Malaysia* **53**, 401–407.

Ogbuokiri JE. 1995. Strategies for improving health of residents in rural Nigeria: cost-effectiveness of a women's health cooperative versus ministry workers in ivermectin (Mectizan) distribution. www.hsph.harvard.edu/takemi/RP104.pdf.

Okyay P, Ertug S, Gultekin B et al. 2004. Intestinal parasites prevalence and related factors in children, a western city sample, Turkey. *BMC Public Health* **4**, 64 (published on line).

Phiri K, Whitty CJ, Graham SM and Ssembatya-Lule G. 2000. Urban/rural differences in prevalence and risk factors for intestinal helminth infection in southern Malawi. *Annals of Tropical Medicine and Parasitology* **94**, 381–387.

Richards F, Klein RE, Gonzales-Peralta C et al. 1991. Knowledge, attitudes and perceptions (KAP) of onchocerciasis: a survey among residents in an endemic area in Guatemala targeted for mass chemotherapy with ivermectin. *Social Science and Medicine* **32**, 1275–1281.

Schall VT. 1995. Health education, public information and communication in schistosomiasis control in Brazil: a brief retrospective and perspectives. *Memorias do Instituto Ozwaldo Cruz* **90**, 229–234.

Shu EN, Okonkwo PO and Onwujekwe EO. 1999. Health education to school children in Okpatu, Nigeria: impact on onchocerciasis-related knowledge. *Public Health* **113**, 215–218.

Sow S, de Vlas SJ, Mbaye A et al. 2003. Low awareness of intestinal schistosomiasis in northern Senegal after 7 years of health education as part of intense control and research activities. *Tropical Medicine and International Health* **8**, 744–749.

Taylor M, Coovadia HM, Kvalsvig JD et al. 1999. Helminth control as an entry point for health-promoting schools in KwaZulu-Natal. *South African Medical Journal* **89**, 273–279.

Traub RJ, Robertson ID, Irwin P et al. 2004. The prevalence, intensities and risk factors associated with geohelminth infection in tea-growing communities in Assam, India. *Tropical Medicine and International Health* **9**, 688–701.

UNICEF. 2003. *The State of the World's Children 2004.* New York: United Nations Children's Fund.

UNICEF. 2004. *The State of the World's Children 2005.* New York: United Nations Children's Fund.

UNICEF. 2005. Water, environment and sanitation. www.unicef.org/wes/index_schools.html.

Useh MF and Ejezie GC. 1999. Modification of behaviour and attitude in the control of schistosomiasis. 1. Observations on water-contact patterns and perception of infection. *Annals of Tropical Medicine and Parasitology* **93**, 711–720.

WHO. 1990. *Health Education in the Control of Schistosomiasis.* Geneva: World Health Organization.

Yuan L, Manderson L, Tempongko MS et al. 2000. The impact of educational videotapes on water contact behaviour of primary school students in the Dongting Lakes region, China. *Tropical Medicine and International Health* **5**, 538–544.

16 Water and Sanitation: Availability, Needs, and Provision

The public health significance of helminthiasis (Table 1.1) will decline and some infections will be eradicated when populations at risk have access to safe water and effective sanitation (Feachem et al., 1983; WHO, 2002). The same generalization applies to the array of water-borne microbial infections (Cotruvo et al., 2004). How easy to state the obvious, how seemingly intractable to achieve it. Surely every government and every agency concerned about health, well being, prosperity, and development should assign the highest priority to providing these fundamental services?

In the context of helminth infections and communicable disease generally, access implies that safe water is available at home or at a reasonable distance from home at all times in the quantity and quality needed for domestic purposes for everybody in the community. Safe means more than having potable water. Water is safe when people are protected from contact with noxious chemicals, carcinogens, pathogenic microbes, and the transmission stages of parasitic infections that may be involved with water. Contact with pathogenic worms may be direct as with schistosome cercariae or indirect as with trematode metacercariae contained in edible fish (Table 16.1). Water is safe when other people and reservoir hosts carrying infectious agents are denied contact with water intended for human use. Safety also addresses protection from the transmission stages of the infectious agents carried by vectors that breed in water. Similar conditions apply to sanitation. Safe sanitation must be accessible, safe at the point of use, comfortable and ventilated, culturally appropriate, amenable to regular cleaning, and connected to a system for the treatment and disposal of excreta. Water and sanitation are always discussed together because they are related determinants of human health. They must never be linked in practice; contamination of water supplies by human and animal waste has to be avoided if helminthiasis is to be controlled.

The risk of acquiring a helminth infection through contact with fresh water is well illustrated by reference to *Fasciola hepatica*. Most courses in parasitology explain that infection with this helminth involves a person swallowing infective metacercariae attached to uncooked, edible water plants. Recently, Mas-Coma (2004) has compiled evidence about the diversity of ways in which metacercariae of *F. hepatica* may be swallowed. Metacercariae may be ingested attached to any one of 15 species of wild freshwater plants that people eat. Metacercariae are regularly ingested with cultivated freshwater plants. Apparently 10% of the green

TABLE 16.1
Unsafe Water: Examples of Human Water Use and Contact and Risks of Exposure to Helminth Infections

Human Water Use	Risk of Helminth Infection	Reference
Domestic and Household Needs		
Drinking water (direct)	*Fasciola hepatica*: swallow free-floating metacercariae	Mas-Coma (2004)
Drinking water (indirect)	*Dracunculus medinensis*: swallow copepods containing L3	Edungbola and Kale (1991)
Water contact (bathing, fishing, laundry, water collection, washing utensils and vegetables, personal hygiene)	*Schistosoma* spp: skin penetration by cercariae	Southgate and Rollinson (1987)
Food Sources and Food Production		
Household fish ponds	*Clonorchis sinensis* and *Opisthorchis viverrini*: swallow metacercariae in raw or undercooked fish	Murrell and Crompton (2006)
Aquatic food plants	*Fasciola hepatica*: swallow metacercariae attached to plants	Mas-Coma (2004)
Rice irrigation schemes	*Wuchereria bancrofti*: increased transmission of L3 as vector population increases	Hunter (1992)

vegetables on sale in a market in Samarkand were found to be carrying metacercariae. Metacercariae have been ingested with five species of terrestrial plants contaminated during irrigation. Metacercariae are not always attached to plants. They have been ingested by drinking contaminated water, by drinking beverages made with contaminated water, by eating vegetables washed in contaminated water, and by using utensils that had been washed in contaminated water. Protecting people from water-borne pathogens is a challenging task.

The gulf between North and South regarding water and sanitation is as wide as the gulf between the strength of their economies. In the North, household water supply and sanitation have become social responsibilities seen as rights and supported by legislation. Enforced regulations govern supply, quality, and purity, while the infrastructure is serviced and replaced when necessary. In the South, about 1.1 billion people do not have access to improved water supplies and 2.4 billion people do not have access to any form of improved sanitation facility (WHO, 2005). For many households in the South, water supply remains an individual responsibility leaving people to walk, collect, and carry their water from streams, ponds, wells, hand pumps, and standpipes. Others may pay for water from privatized companies or buy water of dubious quality from street vendors. In many rural communities, the same water sources serve for personal washing, bathing, animal husbandry, and laundry; analysis and checks for quality and contamination are rarely undertaken except by researchers. Depending on the climate, water supply for rural people may be seasonable, erratic, and unreliable. Communities upstream may deprive their neighbors downstream. Sanitation provision may be nonexistent or deliberately avoided by many households because of its disgusting condition leading to continuous contamination of the environment with pathogens and the transmission stages of helminths.

16.1 NIGHT SOIL

Untreated human excrement (night soil) is a valuable resource that has been used for centuries in Asia to promote vegetable production and enrich the nutrient content of the water in household fish ponds. Night soil is known to have been in use in Japan in the early 8th century. Once urban townships had become established there, night soil was collected from them and transported to the villages for vegetable production (WGGPC, 1998). Unfortunately, this seemingly commendable recycling process also dispersed and recycled pathogens. As recently as the mid-19th century, a similar use for night soil was proposed in England in an attempt to clean urban slums and feed the increasing numbers of workers required by the industrial revolution (Entwisle, 1848). Entwisle advised that water closets and pipes would be needed to support his scheme and so he might have intended to set up some form of sewage treatment before transport to local farms. Attempts to change traditional uses of night soil are unlikely to succeed unless communities understand the need to accept methods for rendering night soil safe from pathogens. If people are denied the use of night soil, environmentally acceptable fertilizers will be required as a replacement. Sustaining food security for poor people has to take higher priority than reducing helminth infections.

16.2 URBANIZATION

In 2000, 47% of the world's population (2.8 billion) was living in urban areas. By 2030, 60% will be living in peri-urban areas with most increase occurring in less-developed countries (Phillips, 1993; PRB, 2005a). Much of this growth is due to the process recognized as urbanization in which poor people seeking opportunities in the informal economy leave rural communities and move into already overcrowded slums that lack the infrastructure to accommodate them (Carty, 1991; Crompton and Savioli, 1993; Stephens, 1996). In developing countries, 60% of urban dwellers do not have access to sanitation, 90% of the sewage that is collected is discharged untreated, and only 30% of the buildings are connected to sewer lines. In Mumbai, the death rate in the slums was found to be double that in the suburbs and in Manila the infant death rate was three times higher than that in the rest of the city (Carty, 1991). Such sanitation as is available is totally unsatisfactory. Grimason et al. (2000) described the state of pit latrines in Blantyre, Malawi. Some had san-plats (concrete platforms) over the squat holes. In others the squat holes were covered with cardboard or iron sheets resulting in the risk of being cut while using the latrine. Most squat holes were fouled with fecal matter. Materials used to surround the latrines and provide some degree of privacy were often stolen. The cloth doors of the latrines were regularly used for wiping hands. There were no facilities for disposal of young children's feces. This is the environment in which soil-transmitted helminth infections (STHs) flourish (Table 16.2; Crompton and Savioli, 1993).

TABLE 16.2
Examples of Socioeconomic Conditions in Urban Slums. From Table 5 (Crompton and Savioli, 1993) with permission from the World Health Organization (www.who.int).

Location	Economic Information	Sanitation	Housing	Water Supply	Reference
Squatter area of Smokey Mountain, Manila, Philippines	Selling recyclable garbage < 3,500 pesos per month for family of 6 (> 99% of families below this poverty line)	2–10% of houses have a latrine	Area 3 m x 3 m. 6 people/house of plywood, plastic, iron sheets, cartons	16 faucets	Auer (1990)
Shanty town of Vila Recreiro, State of São Paulo, Brazil	Average monthly earnings: USD 16 per working person, USD 110 per family	No sewage system	Inadequate	Inadequate	Desai et al. (1980)
Two divisions of shanty town, Coatzacoalcos, Mexico	—	70% defecate in the open air (40% of homes have some facility)	Corrugated iron and concrete blocks, 1–26 people per household (mean 5.5)	Standpipes and drainage canals	Forrester et al. (1988)
Squatter area, Kuala Lumpur, Malaysia	Average household income: < USD 250 per month per family	5–7 families/ latrine: children < 10 years defecate around the home	Houses on stilts over swamps	6 standpipes for 1,200 families	Chai Wee Yan et al. (1978)
Slums in Dhaka, Bangladesh	Work as laborers: c. Rs 94 per family per month	Shallow ditches with temporary fencing	c. 4.5 people per room	Taps and surface water	Khan et al. (1983)

Note: USD, U.S. Dollars; Rs, Rupees.

16.3 WATER: AVAILABILITY AND NEEDS

Only 2.5% of the earth's water is fresh water and most of that is frozen in the icecaps of the Arctic and Antarctic. Less than 1% of the earth's fresh water (about 0.007% of all the earth's water) is directly available for human use, its renewal depending on precipitation (Michigan, 2000). This is surface water found in lakes, rivers, and man-made reservoirs. Water can also be mined from underground aquifers. Water use falls into three main categories: domestic, economic or industrial, and agricultural. Each of these uses is known to be involved in the transmission of helminths to humans (Hunter et al., 1993).

The earth's available fresh water supply is probably similar in amount now as to what it was when hominids appeared a few million years ago. That constancy does not apply to the human population. Two thousand years ago the world's human population was estimated to be 300 million, now it is 6 billion and by 2050 it may have reached 9 billion (PRB, 2005b). The predicted increase of 3 billion more people is expected to occur in Africa, Asia, and Latin America, with Africa at its present growth rate capturing the greatest share (PRB, 2005b). So as the demand for water in the South is increasing, the amount available per capita is decreasing. By the end of the 1980s the annual volume of water usage per capita in Africa was 245 cubic metres (m^3) and 519 m^3 in Asia compared with 1,280 m^3 in Europe, and 1,861 m^3 in North America (Michigan, 2000). Most fresh water is used for agriculture and the pressure on water supply is greater in the South than in the North. Wheat, a staple food in the North, consumes 4,000 m^3 per hectare while rice, a staple food in the South, consumes 7,650 m^3 per hectare (Michigan, 2000). Sixty percent of the world's population depends on rice as its energy source from food with production being dependent on irrigation. As human pressure on water supply increases, the need to make water safe becomes ever greater.

16.4 SANITATION: AVAILABILITY AND NEEDS

Statistics published by UNICEF (2004) show the number of people in the South using improved sources of drinking water and the number using adequate sanitation facilities. Improved and adequate are not defined, but the overall situation for water supply seems to be better than a decade ago. The inadequacy of improved drinking water and sanitation in the world's 50 least developed countries, which include 34 of those that comprise sub-Saharan Africa, is shown in Table 16.3. The provision of adequate sanitation is not keeping pace with that for drinking water, resulting in a daily contamination of human settlements with tens of thousands of tons of fresh feces carrying the transmission stages of pathogenic helminth infections. Measurements of the mean daily fecal discharge by individuals in a range of countries in the south are listed in Table 16.4.

The UNICEF figures indicate that 56% of the population of mainland China (731 million people) do not use adequate sanitation facilities so that 146,200 tons of fresh human feces could be released daily into human communities. This simple calculation is based on a daily fecal production of 200 g per person; that may be an underestimate (Table 16.4). In practice, much of this untreated material in China is

TABLE 16.3

State of the Provision of Drinking Water and Sanitation Facilities in Least Developed Countries (LDCs) and Sub-Saharan Africa (SSA)[1]

	LDCs	SSA
No of countries	50	46
Population (millions)	719	665
Mean U5MR	155	175
No (millions) not using drinking water sources (%)	302 (42)	286 (43)
No (millions) not using adequate sanitation facilities (%)	467 (65)	426 (64)
Daily mass of human stool passed into the environment where sanitation is lacking (tons)[2]	93,400	85,200

Note: U5MR, under five mortality rate.

[1] Based on data published by UNICEF (2004) that recognizes 50 LDCs and 46 countries in SSA. Thirty-four countries in SSA are included in the LDCs.

[2] Based on an assumed average stool production of 200 g per person per day (Crompton and Savioli, 1993; see Table 16.4).

TABLE 16.4

Mean Daily Fecal Production per Person (g) from a Range of Communities. Constructed from Table 1.1 (Feachem et al., 1983).

Community Sampled	No. of Subjects	Mean Wet Fecal Weight (Range)
India, children in New Delhi	36	374 (50–1060)
India, adults in New Delhi	514	311 (19–1505)
Kenya, staff in rural hospital	16	520 (300–> 500)
Malaysia, rural people	28	451 (255–582)
Malaysia, urban people	12	157 (40–300)
South Africa, school children	500	275 (150–350)
Uganda, villagers	15	470 (178–980)

Note: Comparative values from the U.K. and U.S. show that people living there produce less fecal material on average than people in the South. There is evidence that diet and culture affect daily fecal production.

sprayed on to crops to enhance food production and thereby promotes the persistence of ascariasis; a third (> 500 million) of the world's cases of *A. lumbricoides* infection are considered to exist in China (Crompton, 2001).

16.5 PROVISION OF SANITATION

The provision of better sanitation for rural communities should be less problematic than for urban communities because space for construction ought to be available without major disruption to the daily lives of those involved. Kilama (1989) outlined a variety of latrines and conditions for their installation that would seem to protect rural communities from pathogens if used and maintained to acceptable standards. A much more comprehensive document has been published by NSTT (2002). This publication gives technical drawings of different systems, explanations about their operation, costs of installation, and costs of sustained maintenance. Interestingly, unimproved pit toilets, chemical toilets, bucket toilets, and communal block toilets are not recommended and reasons are given. Five factors are of importance in efforts to extend sanitation for communities in need of basic sanitation. First, people must want the provision based on an understanding of the need and the benefits. Second, the sanitation system to be installed must meet local conditions. Third, funds must be obtained to cover the costs of materials, labor, and maintenance. Fourth, the seepage of helminth eggs and other pathogens from excreta to the water supply and elsewhere must be prevented. Fifth, successful sanitation needs enduring organization.

The Ventilated Improved Pit (VIP) latrine or toilet is probably the most effective and practical option for dealing with human excreta, especially in rural communities where the local terrain is amenable for digging. One version of such an on-plot system is illustrated in Figure. 16.1. Waste falls into the pit where the organic material decomposes and liquids percolate into the surrounding soil. In due course, the pit will need to be emptied if a secure lining has been installed or the slab and structure will need to be moved to a new site if not. The design of the VIP latrine causes continuous airflow through the top structure so that gases and smells are carried out and away from the occupant. By maintaining a darkened atmosphere, flies are attracted to the light at the vent pipe where they are trapped by the screen. Experience shows that VIP latrines are better for individual households rather than for groups of households or communities. Cleaning and maintenance is much more likely to occur when ownership and responsibility are clearly defined.

Large-scale sanitation schemes have been attempted in rural settings. Trainer (1989) compared two such schemes, one top-down project in Pakistan that had to be discontinued and another bottom-up project in Nepal that succeeded. A pilot sanitation project covering 188 villages in Punjab Province, Pakistan, was launched with a team of health promoters trained and paid for by UNICEF. The promoters stimulated the people's interest in household latrines, but the program collapsed when UNICEF funding ended and the government at the time could not find funds from its own resources for the project to continue. In 1979 in Nepal, free deworming for ascariasis was introduced as a means of gaining region-wide interest in family planning. On appreciating the health benefits of deworming, however, the villagers formed a management committee and sought advice about how to sustain progress. By the end of 1983, almost every family in a community of 24,000 people was using some form of pit latrine and on average 75% of the costs of latrine construction was provided by the people (Trainer, 1989).

Sanitation on a large scale has also been successfully provided in some poor urban communities. A revolution in sanitation is reported to be in progress in Mumbai to reduce the risks to health of the city's 6.7 million slum dwellers (Chinai, 2002). With a loan to the federal government from the World Bank, toilet blocks for 50 people each have been built in the slum areas of the city (despite the concerns of

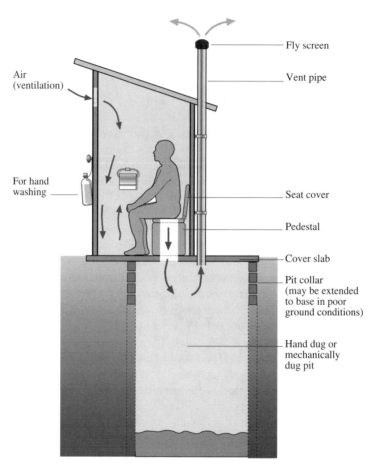

Fly screen

Vent pipe

Air
(ventilation)

For hand
washing

Seat cover

Pedestal

Cover slab

Pit collar
(may be extended
to base in poor
ground conditions)

Hand dug or
mechanically
dug pit

FIGURE 16.1 Ventilated Improved Pit (VIP) latrine or toilet. From NSTT (2002) with permission from the Department of Water Affairs and Forestry, Republic of South Africa.

NSTT, 2002). Each block contains 20 toilets, with separate sections for men and women and with toilets for smaller children. There is a 24-hour supply of water and electricity and the walls are tiled to facilitate easy cleaning. Each block has a paid caretaker. Local people have raised funds for maintenance, another example of bottom-up support being harnessed to sustain health care. The advantages of the scheme include privacy for women, who were restricted to relieving themselves in the open either after nightfall or very early in the morning, safety for children who no longer fear falling into a latrine, a round-the-clock service, cleanliness, and sewage disposal. Ali (1993) documented many of the problems connected with poorly maintained pour-flush pit latrines in another part of India. Problems included disrepair, lack of instruction about maintenance, limited water provision, excreta adhering to squatting slabs, overflow of pit contents in the rainy season, clogged outlets, and so on. The toilet block system in Mumbai seems to have overcome these difficulties. Although universal anthelminthic treatment will rapidly reduce morbidity due to helminthiasis, acceptable sanitation supported by community compliance, is the strategy for a sustained, helminth-free life.

16.6 INTERVENTION OUTCOMES

In addition to the experiences described above, some examples of the impact of improved water supply and sanitation on helminthiasis and infectious disease merit discussion. Esrey et al. (1991) analyzed accounts of 144 studies undertaken to investigate the effects of water supply and sanitation on ascariasis, diarrhea, dracunculiasis, hookworm infection, schistosomiasis, and trachoma. Disease-specific median reduction levels were calculated as a comparative measure of the benefit of the intervention. The median reductions in morbidity attributed to water supply or sanitation were 29% for ascariasis, 26% for diarrhea, 78% for dracunculiasis, 4% for hookworm, 77% for schistosomiasis, and 27% for trachoma. Impressively, overall child mortality was judged to have fallen by 55%.

The Zinder Latrine Promotion Project started in rural Niger in 2002 as an initiative to control trachoma (Carter Center, 2004). By the end of 2003, 2,577 household latrines had been built, thereby catering for about 26,000 people. Although not concerned with helminthiasis, the project is important because it provides insight into how rural people in a low-income country respond to latrines. A year later, the project team assessed latrine use, maintenance, acceptability, and sustainability. Over 90% of adults and 55% of children reported that they always used the latrines. Villagers without latrines had started using those belonging to their neighbors. Separate inspections showed that 86% of the latrines had been used recently. Women were in charge of 75% of the latrines and on average the latrines were cleaned three times weekly. The close proximity of latrines to households was the dominant factor in relation to acceptability. Village leaders said that they would encourage latrine construction; 418 new latrines had been built with local support since the project began.

An extensive study of soil-transmitted helminth infections in children aged 5 to 14 years in relation to community hygiene infrastructure has been carried out by Moraes and Cairncross (2004) and Moraes et al. (2004). The work was done in nine poor urban settlements in Brazil. Three settlements had drainage and sewerage, three had drainage, and three had neither provision. The results of stool examinations showed that the prevalence of infections was least in children living where sewerage was available. The intensity of infection with *Trichuris trichiura* was lowest where sewerage was installed; intensity for *A. lumbricoides* and hookworm was not reduced. In these poor communities, sewerage did not reduce the clustering of infections in households, did not affect individual predisposition to infection, which was more prominent than in the other settlements, and did not weaken the over-dispersed distribution of infections. The main conclusions from the study were (1) sanitation serves to control environmental factors influencing transmission in the public domain and (2) sanitation does not necessarily control household factors influencing transmission. It should be noted that this work was done where poverty and inequality prevail so that many other factors would be expected to affect the lives of separate families.

A smaller-scale study, carried out by Asaolu et al. (2002) in peri-urban communities around Ile-Ife, Nigeria, also compared the effects of hygiene infrastructure on
infection with *A. lumbricoides* in preschool children. Some families in one community had access to flush toilets, pit latrines, and piped water, while families in the
other community did not enjoy these provisions. As expected, the prevalence of
A. lumbricoides infection was much lower in children from the community with
sanitation but, not as expected, the intensity of the infection was much the same in
children from both places. The key point appears to be that if intensity and therefore
morbidity are to be reduced, investment in hygiene infrastructure needs to be sufficient to ensure access for all people in a community and not for a few. The frequency
distribution of numbers of latrines and other hygienic facilities must be across the
board and not be overdispersed. A major effort is being made by UNICEF (2005)
to improve, water supplies and sanitation in schools in places where these services
are inadequate or lacking. School sanitation, supported by hygiene education is
recognized as a potent means of improving health in the community.

16.7 PROSPECTS FOR THE PROVISION OF WATER
SUPPLIES AND SANITATION

One inescapable conclusion emerges from the histories of helminthology, water
provision, and sanitation. Once water provision, sanitation, and sewage treatment
had been installed on an equal footing under law equally to all members of the
community, many helminth infections steadily declined and disappeared. Cairncross
(2003) yet again emphasized the magnitude of the difference in the provision of
sanitation and excreta disposal between the North and the South. He has argued that
a latrine should be seen as a consumer durable to be sold once market research has
been carried out to identify the type of product that poor people might be able to
afford. He readily admitted that some form of subsidy might be needed to make
purchases affordable. Perhaps that approach would stimulate sufficient investment
to generate the funds needed to make appropriate latrines.

Physicians in the North probably never encounter a hookworm or blood fluke
unless a patient happens to have been working abroad or enjoying an adventurous
holiday. Long before safety-tested, efficacious anthelminthic drugs became available,
safe drinking water and effective sanitation cleared helminth and other infections
from the North, a public health triumph on a par with vaccination to eradicate
smallpox. The policy now should be to take every opportunity to improve environmental sanitation. That means intervention to reduce environmental health risks,
including the safe disposal and hygienic management of human and animal excreta,
refuse and waste water; the control of vectors and reservoir hosts; the provision of
safe water; food safety; good housing; domestic hygiene; safe and healthy working
conditions; and basic medical care. Such provision is an intersectoral matter requiring collaboration between numerous government departments in the countries of the
South and partnerships with donors and technical advisors in the north. When the
U.S. government opened the Panama Canal, 5.5% of the allocated budget had been
spent by the Sanitary Department, under the leadership of William Gorgas, on public

health interventions to ensure the availability and productivity of a healthy work force (Watson, 1915). Has there ever been a better investment in economic growth than that expenditure on public health in Panama?

REFERENCES

Ali S. 1993. The sanitation situation: a case study. *Yojana* **37**, 23–25.

Asaolu SO, Ofoezie JE, Odumuyiwa PA et al. 2002. Effect of water supply and sanitation on the prevalence and intensity of *Ascaris lumbricoides* among pre-school age children in Ajebandele and Ifewara, Osun State, Nigeria. *Transactions of the Royal Society of Tropical Medicine and Hygiene* **96**, 600–604.

Auer C. 1990. Health status of children living in a squatter area of Manila, Philippines, with particular emphasis on intestinal parasitoses. *Southeast Asian Journal of Tropical Medicine and Public Health* **21**, 289–300.

Cairncross S. 2003. *International Journal of Environmental Health Research* **13**, 123–131.

Carter Center. 2004. Niger accesses household latrine use, maintenance, and acceptance. *Eye of the Eagle* **5**, pages 1 and 6.

Carty WP. 1991. Towards an urban world. *Earthwatch* **43**, 2–4.

Chai Wee Yan, Fadzrizal bin Ishak, Goh Leng Hee et al. 1978. The problem of soil-transmitted helminths in squatter areas around Kuala Lumpur. *Medical Journal of Malaysia* **33**, 34–43.

Chinai R. 2002. Mumbai slum dwellers' sewage project goes nationwide. *Bulletin of the World Health Organization* **80** [on line], 684–685.

Cotruvo JA, Cotruvo A, Dufour A. et al. (eds.). 2004. *Waterborne Zoonoses. Identification, Causes and Control.* London: IWA Publishing on behalf of the World Health Organization.

Crompton DWT. 2001. *Ascaris* and ascariasis. *Advances in Parasitology* **48**, 285–375.

Crompton DWT and Savioli L. 1993. Intestinal parasitic infections and urbanization. *Bulletin of the World Health Organization* **71**, 1–7.

Desai ID, Garcia Tavares ML, Dutra de Oliveira BS et al. 1980. Food habits and nutritional status of agricultural workers in southern Brazil. *American Journal of Clinical Nutrition* **33**, 702–714.

Edungbola LD and Kale OO. 1991. Guinea-worm disease. *Surgery* **98**, 2351–2354.

Entwisle J. 1848. *A Report of the Sanatory Condition of the Borough of Bolton.* London: Simpkin, Marshall and Co.

Esrey SA, Potash JB, Roberts L and Shiff C. 1991. Effects of improved water supply and sanitation on ascariasis, diarrhoea, dracunculiasis, hookworm infection, schistosomiasis, and trachoma. *Bulletin of the World Health Organization* **69**, 609–621.

Feachem RG, Bradley DJ, Garelick H and Mara D. 1983. Sanitation and Disease: Health aspects of excreta and wastewater management. World Bank Studies in Water Supply and Sanitation No. 3. Chichester, England: John Wiley & Sons.

Forrester JE, Scott ME, Bundy DAP and Golden MHN. 1988. Clustering of *Ascaris lumbricoides* and *Trichuris trichiura* infections in households. *Transactions of the Royal Society of Tropical Medicine and Hygiene* **82**, 282–288.

Grimason AM, Davison K, Tembo KC et al. 2000. Problems associated with the use of pit latrines in Blantyre, Republic of Malawi. *Journal of the Royal Society of Health* **120**, 175–182.

Hunter JM. 1992. Elephantiasis: a disease of development in north east Ghana. *Social Science and Medicine* **35**, 627–645.

Hunter JM, Rey L, Chu KY et al. 1993. *Parasitic Diseases in Water Resources Development.* Geneva: World Health Organization.

Khan MU. 1983. Role of breast-feeding in preventing acquisition of roundworm and hookworm in Dhaka slum children. *Indian Journal of Pediatrics* **50**, 493–495.

Kilama W. 1989. Sanitation in the control of ascariasis. In: *Ascariasis and its Prevention and Control* (eds. DWT Crompton, MC Nesheim and ZS Pawlowski). New York, London and Philadelphia: Taylor & Francis. pp. 289–300.

Mas-Coma S. 2004. Human fascioliasis. In: *Waterborne Zoonoses* (eds. JA Cotruvo, A Cotruvo, A Dufour et al.). London: IWA Publishing on behalf of the World Health Organization. pp. 305–322.

Michigan. 2000. Human appropriation of the world's fresh water supply. University of Michigan. www.globalchange.umich.edu/globalchange2.

Moraes LR and Cairncross S. 2004. Environmental interventions and the pattern of geohelminth infections in Salvador, Brazil. *Parasitology* **129**, 223–232.

Moraes LR, Cancio JA and Cairncross S. 2004. Impact of drainage and sewerage on intestinal nematode infections in poor urban areas in Salvador, Brazil. *Transactions of the Royal Society of Tropical Medicine and Hygiene* **98**, 197–204.

Murrell KD and Crompton DWT. 2006. Foodborne trematode infections and helminths. In *Emerging Foodborne Pathogens* (eds. Y. Motarjemi and M Adams). Cambridge: Woodhead Publishing Ltd.

NSTT. 2002. Sanitation for a Healthy Nation. Sanitation Technology Options. Pretoria, South Africa: Department of Water Affairs and Forestry.

Phillips DR. 1993. Urbanization and human health. *Parasitology* **106**, S93–S107.

PRB. 2005a. Human population fundamentals: fundamentals of growth, population growth and distribution. Largest urban agglomerations. www.prb.org/human_population.

PRB (Population Reference Bureau). 2005b. Human population: fundamentals of growth, population growth and distribution. World population growth. www.prb.org/human_population.

Southgate VR and Rollinson D. 1987. Natural history of transmission and schistosome interactions. In: *The Biology of Schistosomes. From Genes to Latrines* (eds. D Rollinson and AJG Simpson). London and New York: Academic Press. pp. 347–378.

Stephens C. 1996. Urbanization: the implications for health. *African Health* **18**, 14–15.

Trainer ES. 1989. An alternate strategy for promoting ascariasis control. In: *Ascariasis and its Prevention and Control* (eds. DWT Crompton, MC Nesheim and ZS Pawlowski). London, New York and Philadelphia: Taylor & Francis. pp. 321–330.

UNICEF. 2004. *The State of the World's Children 2005.* New York: United Nations Children's Fund.

UNICEF. 2005. Water, sanitation and education. www.unicef.org/wes/index_schools.html.

Watson M. 1915. *Rural Sanitation in the Tropics.* London: John Murray.

WGGPC (Working Group on Global Parasite Control). 1998. *The Global Parasite Control for the 21st Century.* Tokyo: Government of Japan.

WHO. 2002. Prevention and control of schistosomiasis and soil-transmitted helminthiasis. Report of a WHO Expert Committee. *Technical Report Series* 912. Geneva: World Health Organization.

WHO. 2005. Water supply, sanitation and hygiene development. In: *Water Sanitation and Health (WSH).* www.who.int/water_sanitation_health/hygiene/en/index.

17 Helminthiasis and the Millennium Development Goals

Volumes have been written about how to do development and how not to do development. Thousands of people from the North are employed by governments, agencies, and NGOs in the business and industry of development. Ever since the end of World War II, when the U.S. determined to rebuild the world in a direction that understandably suited its national security and economic policies, development has been at the forefront of world affairs. Academics studying development claim that, when stripped down to its bare essentials, development should mean good progress or good change. How can we account for the fact that good progress has not been achieved everywhere despite 50 years of effort? Many authorities in the North are convinced that development is in the doldrums or even stranded in much of sub-Saharan Africa. Somehow much of Africa has been left behind in our runaway world (Giddens, 1999).

17.1 MILLENNIUM DEVELOPMENT GOALS

In September 2000 the member states of the United Nations, in the north and south, unanimously adopted the Millennium Declaration, which committed the international community to work to promote human development. The agenda for this declaration is encapsulated in the Millennium Development Goals (MDGs) which are set out below in a form used by UNICEF (2003). The list does not reflect any order of priority although different governments, UN agencies, and NGOs might be inclined to focus their efforts more toward one or another of the goals rather than all of them:

1. Eradicate extreme poverty and hunger.
2. Achieve universal primary education.
3. Promote gender equality and empower women.
4. Reduce child mortality.
5. Improve maternal health.
6. Combat HIV/AIDS, malaria, and other diseases.
7. Ensure environmental sustainability.
8. Develop a global partnership for development.

Since the MDGs are strongly concerned with development leading to better health, no direct mention of access to safe drinking water or the provision of satisfactory sanitation seems to be an unfortunate omission in this summary of the goals. A more detailed version of goal 7 states that the proportion of people without sustainable access to safe drinking water is to be halved by 2015 (World Bank, 2004). Obviously, permanent reduction in child mortality, improvement in maternal health and successful action against diseases will not be achieved until these fundamental services are generally available.

The goals, which might be better thought of as ideals, challenge the affluent people of the North to accept altruistic action in support of the poor and deprived people of the South. Nowhere is the situation worse or the need greater than in sub-Saharan Africa (UNICEF, 2004). Interestingly, in the UNICEF version of the MDGs, there is no mention of the time by which the goals should be achieved. A wildly optimistic date of 2015 was chosen as the deadline (see World Bank, 2004). With 10 years left to 2015, Gordon Brown, U.K. Chancellor of the Exchequer, expressed his concerns during a speech in Cape Town, South Africa, in January 2005. He said " ... primary education for all will not be delivered as the Millennium Development Goals solemnly promised in 2015 but 2130 — that is 115 years late; the halving of poverty not as the richest countries promised by 2015 but by 2150 — that is 135 years late; and the elimination of avoidable infant deaths not as we the richest nations promised by 2015 but by 2165 — that is 150 years late" (Brown, 2005). So why was 2015 chosen? The date of 2015 serves to concentrate our minds on the urgency of the situation and the responsibility our generation has to strive to make good progress.

17.2 HELMINTHIASIS AND THE G8 AGENDA

At the G8 meeting held in Denver in 1997, Prime Minister Hashimoto of Japan spoke of the importance of parasite control as a public health intervention. In a subsequent report (Hashimoto Initiative, 2000), the case was made for the North (developed countries) to support parasite control efforts being planned or undertaken by the South (developing countries). Suggestions were made about collaboration between North and South over parasite control; understandably the proposals drew heavily on Japan's successful experience in this field of public health during reconstruction and economic growth after the end of World War II. The need to control helminthiasis featured strongly in the report. The need to reinforce anthelminthic treatment with improvements in domestic and community hygiene through the development of a hygiene infrastructure was emphasized just as strongly.

No doubt these views would be endorsed by the presidents and prime ministers who attended the meeting of the G8 in Gleneagles, Scotland, in the summer of 2005. Development, especially with Africa in mind, featured largely in the discussions. The G8 was directed to Africa through the work of the Commission for Africa and powerful advocacy for debt relief and more aid was advanced. Helminthiasis was featured in the report of the Commission presented to those involved with the G8 meeting. On page 191, the report drew attention to the conclusion that helminthiasis causes widespread suffering, reduces economic productivity, and keeps children from school. The report continued that "donors should ensure that there is adequate funding for the prevention and treatment of parasitic diseases and micronutrient deficiency. Governments and global health partnerships should ensure that this is integrated into public health campaigns by 2006" (Commission for Africa, 2005). Delivery and coverage with anthelminthic treatment for those in need worldwide will probably cost about USD 0.2 billion each year for 5 years if morbidity due to helminthiasis is to be reduced significantly. This sum of USD 0.2 billion represents 0.2% of the USD 100 billion needed annually to achieve the MDGs (see Wroe and Doney, 2004).

The evangelistic efforts of Bob Geldof, the Live 8 concerts, and the BBC's concentration on African issues have repeatedly stressed that the North cannot ignore the plight of the poor, powerless, voiceless people of Africa and elsewhere. We cannot say that we did not know. But altruism is a hard approach to sell to wealthy, powerful, and articulate people in a democratic society. If poor nations are to make good progress, political leaders in the North will have to persuade their electorates that there will be a cost. The abolition of agricultural subsidies and the establishment of trading under the same rules and conditions will help the South to make good progress. Genuinely fair competition may be hard to take. The assumption that as the north gets richer some wealth will trickle down to the south seems like wishful thinking; trickle down does not work when the click of a mouse can cause financial upheaval.

When the meeting of the Gleneagles G8 was over, the North had decided to double aid to Africa to reach USD 50 billion by 2010 and to provide debt relief to 18 African countries amounting to USD 40 billion (Webster, 2005). Countries of the North would relieve certain nations from their burden of debt by paying the sums they owed to the World Bank thereby enabling the bank to support new development projects. All caring people in the North would surely welcome these initiatives. At long last Africa's needs have bounded up the development agenda. The G8 did not identify national shares in this exercise of debt relief; details of how much and when were to be left for negotiation. Some of the countries expected to contribute to debt relief are not members of the G8 and must take part in such negotiations (Duncan, 2005). A pledge was made to scrap U.S. and EU agricultural subsidies, but no date was set for this most important measure. Procedural and diplomatic shenanigans and delays must not undermine opportunities to make life better in low-income countries.

17.3 CONTRIBUTION OF DEWORMING TO THE MDGS

In a low-income country where helminthiasis is endemic, deworming is a highly cost-effective public health intervention (World Bank, 1993). Anthelminthic drugs are inexpensive, efficacious, and have an excellent safety record. Deworming needs little or no expenditure on infrastructure, it is readily integrated with other programs, it is popular, and attracts good compliance. The benefits of deworming relate directly to the achieving the MDGs (WHO, 2005). Deworming is accompanied by:

- improvement in birth weights and increased survival during infancy
- improvements in appetite and physical fitness of children
- improvement in the growth and nutritional status of children
- improvement in cognitive development
- increased school attendance
- improvements in maternal health
- increased worker productivity for employers
- increased wage earning for employees
- increased working capacity of unpaid carers for children and households
- increased sense of well-being and self-esteem
- enhancement of existing health care provision

The evidence in support of this list of benefits is to be found in the preceding chapters and in many other sources (WHO, 2005). Deworming brings direct benefits to individuals at all stages in the human life cycle (Figure 17.1) and thereby to entire communities. Deworming is predicted to bring indirect benefits since evidence is emerging to show that worm infections make more subtle contributions to the impact of other diseases including HIV/AIDS, malaria, and TB.

Do you remember Mumbua (Chapter 2) struggling to survive far away from the luxurious surroundings and appointments of Gleneagles? What can be done to offer

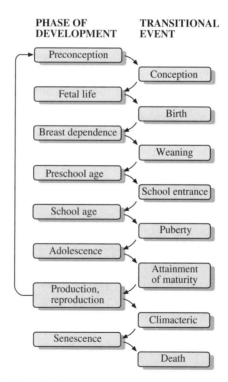

FIGURE 17.1 Events during the human life cycle. From Figure 4.1 (Crompton, 1984) with the author's permission.

Mumbua a better life now and free her and her children from the shackles of poverty and deprivation? Even if worm infections seem to be of low priority compared with other problems that currently beset Africa, deworming would help Mumbua and millions like her, but how will it reach her and who will do it now in a sustainable way?

17.4 DEWORMING — TOP DOWN OR BOTTOM UP?

Although deworming may seem to some to be a minor element in the development agenda, it remains essentially a top-down activity engineered by powerful organizations in the North. Deworming is at risk of being resource driven rather than demand driven. That notion percolates through the pages of this book. Intervention is well intentioned and sincere; we are from the North and we hope to apply our knowledge to help less fortunate people. Furthermore, we have the resources to put our concern into action. Nearly all the basic and operational research results discussed in the book have been carried out by people who were educated, trained, and financed by the North. The public health significance and economic impact of helminthiasis and the technical advice about the design and management of control programs comes from the North. Partnerships for parasite control depend on donations of money, drugs, and expertise from the north. Control is still largely resource driven even though there must be an element of demand. The governments of sovereign states in the South, where helminthiasis is endemic, must first give their agreement before a control scheme begins. Why should an impoverished government commit resources to deworming if the North will do it? What will happen to helminth control if the North has to stop or decides to stop? Can any progress made by control programs be sustained? If sustainability is to depend on economic growth and stability in the South, we had better note the gloomy predictions of the U.K. Chancellor of the Exchequer (Brown, 2005).

If Robert Chambers and like-minded colleagues were to investigate the philosophy and paradigm of contemporary deworming we should expect them to conclude that the uppers own it (see Chambers, 1997). Uppers is a relative term implying the existence of lowers. Uppers are the people who in a given context are dominant or superior to lowers. Lowers are the people who are subordinate and inferior to the uppers. Those of us from the North who work to establish and promote deworming have probably never thought of ourselves as uppers or about this social divide. Our eagerness to do good has overlooked the knowledge, culture and beliefs of those we are so anxious to help. Perhaps we should pay much more attention to the case for facilitating a bottom-up approach to deworming that depends on the empowerment of communities to take charge the management of their own programs. That policy has a chance of relieving Mumbua and the millions like her from the burden of disease due to worm infections.

17.5 DEWORMING THROUGH PARTICIPATORY
RURAL APPRAISAL

Chambers (1997) has undertaken a constructive analysis of approaches to development entitled "*Whose Reality Really Counts? Putting the First Last.*" He is concerned with exploring ways that will enable poor, weak, and vulnerable people to express their realities, to plan, and to act. According to Chambers, that depends on understanding what is meant by reality. Physical reality is the world "out there." Personal reality deals with an individual's perception and interpretation of the physical and social happenings outside herself or himself. Participatory rural appraisal (PRA) is a bottom-up development system that seeks to empower communities in the south to run their own affairs in their way (see Chambers, 1997). PRA depends on the recognition of the importance of personal realities. When we from the north ask can actions match aspirations, we should avoid the imperialism of values in which our actions and aspirations become dominant.

Community deworming can flourish through the PRA system regardless of whether the setting is rural or urban. PRA provides an ideal opportunity for collaboration between communities in need and NGOs or charitable organizations. One example serves to show how the system operates. St Andrew's Clinics for Children (STACC) is a charitable company, registered in Scotland since 1992, with the purpose of developing and supporting basic health care for children in sub-Saharan Africa. STACC raises money to support a variety of health projects in Kenya, Nigeria, Sierra Leone, Uganda, and Tanzania (Zanzibar). All the projects are run by local people, ranging from medically qualified physicians to traditional healers, who serve their communities through fixed and mobile clinics, workshops, hospital care, and deworming programs. STACC sends funds at regular intervals and technical advice when asked. In return STACC receives reports and accounts of expenditure at regular intervals. In Sierra Leone, some health care for children is provided by STACC/Sierra Leone, which is national NGO registered with the government. STACC/SL is running a deworming program that caters for 69,000 primary school children and is expanding. The program, which is taking place up country in places where access is difficult, was established at the request of paramount chiefs and is overseen by District Medical Officers. The influence of personal realities is apparent; if the paramount chief is taking the initiative local people in this particular culture accept and trust deworming with medicine from the north. The unit cost of the program is about USD 0.65 per school child and that includes drugs, transport, gratuities, and every other cost. STACC's contribution to the MDGs is at best a droplet in the ocean, but the North is blessed with thousands of organizations that might be able to support demand-driven development projects perhaps through the PRA approach.

17.6 CONCLUSION

The climate of concern for development generated by the MDGs offers an immediate opportunity to increase efforts to control the morbidity that accompanies helminthiasis. The governments of the UN member states in the South as well as in the North have committed their energies to achieving the MDGs. If the south will promote the demand for helminthiasis control, the North should supply resources for immediate relief and co-operate in the provision of sustained relief. Helminthiasis took a dreadful toll on the health and productivity of poor people in the North for centuries until the industrial revolution produced national wealth. National prosperity ensured that sanitation and clean water supplies could be installed, hygiene regulation was introduced, and permanent relief from helminthiasis followed. All this happened long before satisfactory anthelminthic drugs were available and while health workers were largely ignorant of helminthology. Despite the good intentions expressed in the MDGs, economic growth in the South is bound to be a slow process. Deworming, however, is available now and the cost of its deployment may be as nothing when compared with its economic impact.

REFERENCES

Brown G. 2005. Text of a speech by the Rt Hon Gordon Brown MP, Chancellor of the Exchequer, at the Commission for Africa Meeting, Cape Town, South Africa, 17th January 2005.

Chambers R. 1997. *Whose Reality Counts? Putting the First Last.* London: ITDG Publishing.

Commission for Africa (2005). Our Common Interest. Report of the Commission for Africa. www.commissionforafrica.org.

Crompton DWT. 1984. *Parasites and People.* Basingstoke, Hampshire: Macmillan Publishers Ltd.

Duncan G. 2005. Fears for deal to write off $40bn of Africa's debts. In: *The Times (Business)* 23rd September, 2005.

Giddens A. 1999. *Runaway World. How Globalisation is Reshaping Our Lives.* London: Profile Books Ltd.

Hashimoto R. 2000. Global Parasite Control for the 21st Century. Report of the Hashimoto Initiative Meeting, 27 March 2000.

UNICEF. 2003. *The Millennium Development Goals. They are About Children.* New York: The United Nations Children's Fund (UNICEF).

UNICEF. 2004. *The State of the World's Children 2005. Childhood under Threat.* New York: The United Nations Children's Fund.

Webster P. 2005. Pride, hope and humanity. In: *The Times (World News)*, 9th July 2005. p. 50.

World Bank. 1993 *World Development Report 1993: Investing in Health.* New York: Oxford University Press.

World Bank. 2004. *Miniatlas of Global Development.* Washington DC: The International Bank for Reconstruction and Development.

WHO. 2005. The millennium development goals. Geneva: World Health Organization. WHO/CDS/CPE/PVC/2005.12.

Wroe M and Doney M. 2004. *The Rough Guide to a Better World.* London: Rough Guides Ltd (A project funded by the U.K. Department of International Development).

Glossary of Some Terms and Abbreviations Relating to Human Host-Helminth Interactions and Helminthiasis

AE: Alveolar echinococcosis caused by the development of tumor-like cysts of *Echinococcus multilocularis* in human organs.

AIDS: Acquired immunodeficiency syndrome; disease following HIV infection.

Antibodies: Multifunctional glycoprotein molecules (immunoglobulins, IgG, IgM, IgA, IgD, and IgE) produced by the immune system for the prevention and resolution of infection.

Antigen: Any substance or material that is recognized by the vertebrate immune system thereby stimulating the production of antibody or the activation of immune effector cells.

BBC: British Broadcasting Corporation.

Bilharzia: An obsolete name for schistosomiasis. The disease was named in honor of Thomas Bilharz who described *Schistosoma haematobium* from Egypt in 1850.

Calabar swelling: A transient subcutaneous nodule stimulated by the presence of *Loa loa*.

CCA: Cholangiocarcinoma; malignant cancer of the biliary system.

Cercaria (-ae): A free-living stage in the developmental pattern of a trematode (fluke). Cercariae are released into water from snails and usually develop into metacercariae in or on other aquatic animals and plants. Cercariae of *Schistosoma* spp. penetrate skin during water contact and give rise to schistosomiasis. Schistosomes lack a metacercarial stage.

Community: A group of people living in a particular place or ecological zone.

Culture: A society's shared sets of meanings, beliefs, values, and traditions that help members of the society to make sense of the "world" in which they live.

Cure rate (CR): The number (%) of previously worm-egg-positive subjects found to be egg-negative on examination of stool or urine samples using a standard procedure at a set time after deworming.

CE: Cystic echinococcosis (hydatid disease) caused by the development hydatid cysts of *Echinococcus granulosus* in human organs.

DALY: The "disability-adjusted life year" (DALY) concept is based on the definition, by patients in their particular socioeconomic context, of "disability weights" (ranging from 0 = healthy to 1 = death) for different conditions, and weighting of further years lived with a disability to calculate years lost due to the disability. A major challenge facing public health planners is to place problems in an order of priority for action. The introduction of the DALY is considered to be a useful approach for estimating disease burdens in a quantitative and comparative manner. A WHO Expert Committee, meeting in 2002, drew attention to the fact that some infections, such as schistosome and soil-transmitted helminth infections, frequently occur concurrently in the same person. Moreover, the severity of a helminth infection may be complicated by poor nutrition and other adverse factors. DALYs do not take this into account. Estimates of DALYs should be reviewed regularly and should respond to new information on morbidity and mortality.

DNA: Deoxyribonucleic acid, the primary genetic molecule of life.

Deworming: Use of anthelminthic drugs to reduce and control morbidity. The drugs must have a proven safety record and be of guaranteed quality.

Dioecious: Having separate sexes; male and female worms as separate individuals.

Drug resistance: A genetically transmitted loss of susceptibility to a drug in a worm population that was previously sensitive to the appropriate therapeutic dose.

Dysuria: Erratic, difficult, or painful urination.

Ecological zone in the context of helminthology: A zone reflecting homogeneity in the distribution of a worm species or its intermediate host. This depends on a number of variables such as topography, soil type, altitude, temperature, rainfall, and frost.

Edema: Accumulation of fluid in the body's intercellular spaces causing localized swelling.

Effectiveness: A measure of the effect of an anthelminthic drug under operational conditions.

Efficacy: A measure of the effect of an anthelminthic drug in isolation under ideal conditions.

Egg reduction rate (ERR): The percentage fall in worm-egg counts after deworming based on examination of stool or urine samples using a standard procedure at a set time after deworming.

Elimination: A reduction to zero of the number of new cases of a specific infection in a defined geographical area, as a result of deliberate efforts. Continued intervention or surveillance measures are required.

ELISA: Enzyme-linked immunosorbent assay.

Endemic: When used to describe an infection or disease, endemic means that the infection or disease is habitually present in an area due to the climatic, environmental, social, and biological conditions in the area.

End-organ morbidity: Morbidity caused by schistosomiasis, including hepatic fibrosis, urinary obstruction, and bladder cancer.

Enteral: Pertaining to the gastrointestinal tract.

Environmental sanitation: Interventions to reduce environmental health risks, including the safe disposal and hygienic management of human and animal excreta, refuse, and wastewater; the control of vectors, intermediate hosts, and reservoirs of disease; the provision of safe drinking water; food safety; the provision of housing that is adequate in terms of location, quality of shelter, and indoor living conditions; the provision of facilities for personal and domestic hygiene; and the provision of safe and healthy working conditions.

Eosinophil: Eosinophils are fragile leukocytes (white blood cells) measuring about 10 to 15 μm in diameter and characterized by containing granules (1 μ in diameter) that stain with eosin. In a healthy person, who is free from helminth infection, eosinophils make up a small percentage of circulating leukocytes.

Eosinophilia: Elevated eosinophil count in the circulating blood, often present during a helminth infection.

Epidemiology: Study of the distribution and determinants of infectious agents and diseases.

Eradication: A permanent reduction to zero of the worldwide prevalence of infection caused by a specific agent, as a result of deliberate efforts. Continued intervention or surveillance measures are no longer required.

Etiology: Study of the agents and factors that cause disease.

Food security: A situation that exists when people in a community always have physical, social, and economic access to sufficient, safe, and nutritious food to meets their dietary needs and food preferences for an active, healthy life.

FRF: French franc (currency).

Gross National Income per capita (GNI): The gross national income divided by mid-year population.

Hematuria: Visible blood in urine.

Hemoptysis: The expectoration of blood or blood-stained sputum.

Helminth: A parasitic worm; most species are endoparasitic living inside the host's body.

Helminthiasis: Disease accompanying the course of infection by a species of parasitic worm.

Helminthology: Scientific study of parasitic worms.

HCC: Hepatocarcinoma; malignant cancer of the liver.

Hepatomegaly: Enlargement of the liver.

Hermaphrodite: A worm possessing both male and female gonads.

Hepatosplenomegaly: Combined enlargement of the liver and spleen.

Hookworm anemia: Iron-deficiency anemia related to poor iron status due to gastrointestinal blood loss caused by the feeding activity of the intestinal stages of hookworms.

HIV: Human immunodeficiency virus (see AIDS).

Incidence: The number of new cases of infection in a population in a given period of time.

Intensity (worm burden): The number of worms (measured directly or indirectly) per infected person.

Intervention: Any action with the primary intention of improving health. Intervention incorporates disease-specific actions and integrated care. It covers prevention and health promotion as well as curative care. Intervention also includes intersectoral actions where the purpose is to improve or maintain health.

In vitro: Indicates that a process or reaction takes place out of the body in a test tube or in laboratory equipment.

In vivo: Indicates that a process or reaction takes place inside a living body.

Iron deficiency: Iron deficiency is considered to be insufficient iron supply for the optimal physiological production and function of the body's iron-dependent cellular components. Iron deficiency state is usually recognized by specific biochemical tests to measure variables such as transferrin saturation and serum ferritin.

Iron-deficiency anemia (IDA): Anemia recognized by inadequate iron supply for hemoglobin production.

KAP survey: A series of questions designed to assess people's knowledge, attitudes, and practices concerning an issue such as a helminth infection.

L (larva): In this book L is used as shorthand for a juvenile stage in the developmental pattern of a nematode. Four such stages, L1, L2, L3, and L4, are recognized. During development four molts occur, three between larval stages and one after L4 when adulthood is attained.

Low birth weight (LBW): Defined as a body weight at birth (full term) of less than 2,500 g.

Malaria: A debilitating and often deadly disease caused by infection with *Plasmodium falciparum*, *P. malariae*, *P. ovale*, or *P. vivax* following the blood-feeding activities of female anophiline mosquitoes responsible for transmitting malaria parasites.

MDA: Mass drug administration, a concept of importance in Section 4.8 interventions dealing with lymphatic filariasis (see universal).

Metacercaria (-ae): An encysted stage formed from a cercaria. Metacercariae are involved in the transmission of trematode infections to humans who swallow them in contaminated food.

Metric ton (tonne): One thousand kilograms (kg).

MF (microfilaria, -ae): L1 stage of a filarial nematode. MF may be observed to be sheathed (encased in a thin, extensible egg shell) or unsheathed (free from the egg shell).

Microhematuria: Blood in urine detectable by means of a biochemical test or dipstick.

Morbidity: Clinical or subclinical consequences of infections or diseases that affect an individual's well-being. Changes in morbidity can be measured and used to monitor progress following control interventions.

Morbidity control: Avoidance or reduction of illness caused by worm infections. Morbidity control is achieved by periodically deworming individuals and groups at risk of morbidity.

NCC: Neurocysticercosis caused by the development of cysticerci (single bladder or racemose forms) of *Taenia solium* in the human brain or spinal cord.

NGOs: Nongovernmental organizations that may be national or international in their establishment and sphere of work.

Nutritional status: A person's physiological state depending on the relationship between nutrient requirements and intake and the body's ability to digest, absorb, and use the nutrients.

OCP: Onchocerciasis Control Program in West Africa.

Ovoviviparity: Reproduction in which embryos develop to the L1 stage inside the parent worm's body and are then released.

Parenteral: Pertaining to internal parts of the body other than the alimentary tract.

Pharmacovigilance: The detection, assessment, understanding, and prevention of adverse effects or any other drug-related problem in clinical and public health practice.

PCR (polymerase chain reaction): An *in vitro* method for amplifying specific DNA sequences. PCR generates millions of exact copies, thereby making a genetic analysis of tiny samples of DNA a relatively straightforward process.

Poverty: Socially constructed condition resulting in people being unable to take a full part in human society. Poor people have reduced spending power, lack choice, and are vulnerable to risk from disease, violence, crime, natural disasters, and warfare.

Prevalence: The number (usually expressed as percentage) of individuals in a population estimated to be infected with a particular species of worm at a given time.

Recombinant DNA technology: An umbrella term for a range of techniques for isolating, analyzing, and manipulating DNA *in vitro*.

Risk: Risk is concerned with the assessment of the factors that will increase the chances of an individual acquiring a helminth infection.

Risk group: Those identified as being at risk of morbidity and mortality as a result of a helminth infection.

Sanitation: The provision of access to facilities for the safe collection and disposal of human excreta. Sanitation is usually combined with access to safe drinking water.

School-aged children: Children aged between 6 and 15 years who may or may not be enrolled in school.

Serious adverse event (SAE): Serious adverse event may occur in a person following the administration of a drug for a therapeutic purpose. Any event that is fatal, life-threatening, disabling, or incapacitating or that results in hospitalization, prolongs a hospital stay, or is associated with congenital abnormality, cancer, or overdose (either accidental or intentional). Any experience that the investigator/health worker regards as serious and that may be associated with the drug should be reported as a serious adverse event.

Splenomegaly: Enlargement of the spleen.

Subtle morbidity: Morbidity attributable to either schistosomiasis, soil-transmitted helminthiasis, or other infection that is not normally identified in the clinical case definition for that infection such as anemia, growth impairment, decreased cognitive and work performance, and synergy with other infections.

Sustainable health system: A system having the capacity to function effectively over time with little external output.

TB (tuberculosis): A debilitating and often deadly disease caused by inhalation of *Mycobacterium tuberculosis.*

Under-five morbidity rate (U5MR): Probability of dying between birth and exactly five years of age per 1,000 live births.

UNICEF: United Nations Children's Fund.

URL: Uniform Resource Locator.

USD: U.S. dollar (currency).

Vaccine: A preparation of a pathogenic agent or its antigens that can be administered prophylactically to induce immunity.

Vector: An intermediate host that transports infective agents to definitive hosts.

WHO: World Health Organization, an agency of the United Nations, with regional offices and staff in member states of the UN.

Zoonosis: An infection naturally transmitted between vertebrate animals and humans.

APPENDIX 2

Journals for Helminthology, Helminthiasis, and Control Interventions

We have listed below our selection of biomedical publications that regularly contain research papers and reviews relevant to the themes discussed in this book. We have also included URLs of websites giving access to electronic archives.

American Journal of Tropical Medicine and Hygiene, www.ajtmh.org
Acta Tropica, www.sciencedirect.com
Advances in Parasitology, www.sciencedirect.com
American Journal of Clinical Nutrition, www.ajcn.org
American Journal of Hygiene
Annales de Parasitologie Humaine et Comparée
Annales de la Société Belge de Médecine Tropicale
Annals of Tropical Medicine and Parasitology, www.journalsonline.tandf. co.uk
British Medical Journal, www.bmjjournals.com
Bulletin Société Pathologie Exotique, www.pathexo.fr
Bulletin of the World Health Organization, www.who.int/bulletin
Clinical Microbiology Reviews, www.journals.asm.org
Comparative Parasitology, www.bioone.org
Economics and Human Biology, www.sciencedirect.com
East African Medical Journal, www.ajol.info
Elsevier Parasitology Journals, www.sciencedirect.com
Experimental Parasitology, www.sciencedirect.com
Expert Review of Anti-infective Therapy, www.future-drugs.com/loi/eri
Filaria Journal, www.filariajournal.com
Folia Parasitologia, www.paru.cas.cz
Health and Social Care in the Community, www.blackwell-synergy.com
Helminthologia, www.saske.sk
Helminthological Abstracts, www.cabi-publishing.org
Immunology Today, www.sciencedirect.com
International Journal of Parasitology, www.sciencedirect.com
International Journal of Epidemiology, www.ije.oxfordjournals.org

Japanese Journal of Parasitology
Journal of Helminthology, www.ingentaconnect.com/content/cabi/joh
Journal of Nutrition, www.nutrition.org
Journal of Parasitology, www.bioone.org
Journal of Tropical Medicine and Hygiene
Journal of Infectious Diseases, www.journals.uchicago.edu
Journal of Pediatrics, www.sciencedirect.com
Korean Journal of Parasitology, www.parasitol.or.kr
Lancet, www.sciencedirect.com
Medical Journal of Malaysia, www.mma.org.my
Molecular and Biochemical Parasitology, www.sciencedirect.com
New England Journal of Medicine, www.nejm.org
Nigerian Journal of Parasitology
Parasite, perso.wanadoo.fr/princeps/parasite
Parasitology, www.cambridge.org
Parasitology International, www.sciencedirect.com
Parasitology Today
Parassitologia
Parasite Immunology, www.blackwell-synergy.com
Parasitology Research, www.springerlink.com
Research and Reviews in Parasitology
Reviews of Infectious Diseases
Social Science & Medicine, www.sciencedirect.com
Southeast Asian Journal of Tropical Medicine and Public Health
Transactions of the Royal Society of Tropical Medicine and Hygiene, www.sciencedirect.com
Trends in Immunology, www.sciencedirect.com
Trends in Parasitology, www.sciencedirect.com
Tropenmedizin und Parasitologie
Tropical Diseases Bulletin
Tropical and Geographical Medicine
Tropical Medicine and International Health, www.blackwell-synergy.com
Tropical Medicine and Parasitology
Veterinary Parasitology, www.sciencedirect.com
Zeitschrift fur Parasitenkunde

PubMed is the National Library of Medicine's search service providing millions of citations including public health, helminthology, helminthiasis, and control programs. http://www.ncbi.nlm.nih.gov/entrez/query.fcgi?DB=pubmed

APPENDIX 3

Methods for the Detection of Helminth Eggs in Stool Samples*

1. FECAL CONCENTRATION PROCEDURE — FORMALIN ETHER/ETHYL-ACETATE/GASOLINE

MATERIALS AND REAGENTS

1. Centrifuge, with head and cups to hold 15 ml conical tubes. Sealed buckets must be used.
2. Centrifuge tubes, 15 ml, conical (make a graduation at 10 ml with a grease pencil).
3. Bottles, dispensing or plastic "squeeze," 250 or 500 ml.
4. Wooden applicator sticks, 145 × 2.0 mm.
5. Small beaker, 25, 50, or 100 ml.
6. 400 µm plastic or metal sieve or surgical gauze.
7. Microscope slides (75 × 25 mm).
8. Coverslips.
9. Pipettes, disposable Pasteur, with rubber bulbs.
10. Rubber stoppers for centrifuge tubes.
11. Rack or support for tubes.
12. Formalin (10%).**
13. Ether, ethyl acetate, or, if these solvents are unavailable, gasoline. (**Caution**: Ether is highly volatile and will ignite and explode quickly if there is an open flame or spark nearby. Store open cans or bottles on an open shelf in the coolest part of the laboratory. Do not put opened containers of ether in a refrigerator as fumes escape, build up, and may cause an explosion when the door is opened.) Gasoline can be used as a substitute for ether/ethyl acetate if neither of these two is available.

---

* Reproduced from *Bench Aids for the Diagnosis of Intestinal Helminths* (WHO, 1994) with permission from the World Health Organization (www.who.int).
** For reagent preparations consult the WHO publication, *Basic Laboratory Methods in Medical Parasitology*, 1991, ISBN 92 4 154410 4.

14. Dropping bottles containing:
    saline solution, isotonic (0.85% NaCl).
    Lugol's iodine (1% solution).

PROCEDURE

1. With an applicator stick add 1.0 to 1.5 g
   feces to 10 ml formalin in a centrifuge tube,
   stir, and bring into suspension.
2. Strain suspension through the 400 μm mesh
   sieve or 2 layers of wet surgical gauze directly
   into a different centrifuge tube or into a small
   beaker. Discard the gauze.
3. Add more 10% formalin to the suspension
   in the tube to bring the volume to 10 ml.
4. Add 3.0 ml of ether (or ethyl acetate or gasoline) to the suspension in the tube and mix
   well by putting a rubber stopper in the tube
   and shake vigorously for 10 s.
5. Place the tube with the stopper removed in
   centrifuge; balance the tubes, and centrifuge
   at 400 to 500 × g for 2 to 3 min.
6. Remove the tube from the centrifuge; the
   contents consist of four layers: (a) top layer
   of ether (or ethyl acetate or gasoline), (b) a
   plug of fatty debris that is adherent to the
   wall of the tube, (c) a layer of formalin, and
   (d) sediment (Figure 1).
7. Gently loosen the plug of debris with an
   applicator stick by a spiral movement and
   pour off the top three layers in a single movement, allowing to drain inverted for at least
   5 seconds. When done properly a small
   amount of residual fluid from the walls of
   the tube will flow back onto the sediment
   (Figure 2 and Figure 3).
8. Mix the fluid with the sediment (sometimes
   it is necessary to add a drop of saline to have
   sufficient fluid to suspend the sediment) with
   a disposable glass pipette. Transfer a drop of
   the suspension to a slide for examination
   under a coverslip; an iodine-stained preparation can also be made (Figure 4).
9. Examine the preparations with the 10X
   objective or, if needed for identification,
   higher power objectives of the microscope

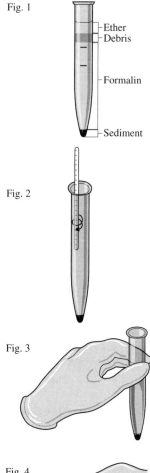

Fig. 1

Ether
Debris

Formalin

Sediment

Fig. 2

Fig. 3

Fig. 4

in a systematic manner so that the entire coverslip area is observed. When organisms or suspicious objects are seen, switch to higher magnification to see more detailed morphology of the material in question.

2   KATO-KATZ TECHNIQUE — CELLOPHANE FECAL THICK SMEAR

MATERIALS AND REAGENTS

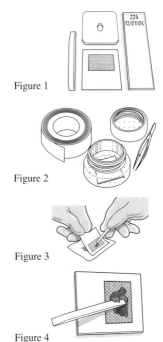

Figure 1

Figure 2

Figure 3

Figure 4

1. Applicator sticks, wooden.
2. Screen, stainless steel, nylon or plastic: 60 to 105 mesh (Figure 1).
3. Template, stainless steel, plastic, or cardboard (Figure 1). Templates of different sizes have been produced in different countries. A 50 mg template will have a hole of 9 mm on a 1 mm thick template; a 41.7 mg a hole of 6 mm on a 1.5 mm thick template; a 20 mg a hole of 6.5 mm on a 0.5 mm thick template. The templates should be standardized in the country and the same size of templates should be used to ensure repeatability and comparability of prevalence and intensity data.
4. Spatula, plastic (Figure 1).
5. Microscope slides (75 × 25 mm).
6. Hydrophillic cellophane, 40 to 50 μm thick, strips 25 × 30 or 25 × 35 mm in size (Figure 2).
7. Flat bottom jar with lid (Figure 2).
8. Forceps.
9. Toilet paper or absorbent tissue.
10. Newspaper.
11. Glycerol-malachite green or glycerol-methylene blue solution. (1 ml of 3% aqueous malachite green or 3% methylene blue is added to 100 ml glycerol and 100 ml distilled water; this solution is mixed well and poured onto the cellophane strips and soaked in this solution in a jar for at least 24 h prior to use.)

PROCEDURE

1. Place a small mound of fecal material on newspaper or scrap paper and press the small screen on top of the fecal material so that some of the feces will be sieved through the screen and accumulate on top of the screen (Figure 3).
2. Scrape the flat-sided spatula across the upper surface of the screen so that the sieved feces accumulate on the spatula (Figure 4).

3. Place template with hole on the center of a microscope slide and add feces from the spatula so that the hole is completely filled (Figure 5). Using the side of the spatula pass over the template to remove excess feces from the edge of the hole (the spatula and screen may be discarded or if carefully washed may be reused again).

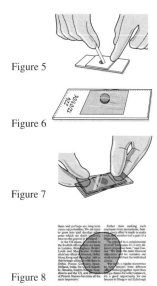

Figure 5

Figure 6

Figure 7

Figure 8

4. Remove the template carefully from the slide so that the cylinder of feces is left completely on the slide.

5. Cover the fecal material with the pre-soaked cellophane strip (Figure 6). The strip must be very wet if the feces are dry and less so with soft feces (if excess glycerol solution is present on upper surface of cellophane wipe the excess with toilet paper). In dry climates excess glycerol will retard but not prevent drying.

6. Invert the microscope slide and firmly press the fecal sample against the hydrophillic cellophane strip on another microscope slide or on a smooth hard surface such as a piece of tile or a flat stone. With this pressure the fecal material will be spread evenly between the microscope slide and the cellophane strip (Figure 7). Newspaper print can be read through the smear after clarification (Figure 8).

7. Carefully remove slide by gently sliding it sideways to avoid separating the cellophane strip or lifting it off. Place the slide on the bench with the cellophane upward. Water evaporates while glycerol clears the feces.

8. For all except hookworm eggs, keep slide for one or more hours at ambient temperature to clear the fecal material prior to examination under the microscope. To significantly expedite clearing and examination, the slide can be placed in a 40°C incubator or kept in direct sunlight for several minutes.

9. *Ascaris* and *Trichuris* eggs will remain visible and recognizable for many months in these preparations. Hookworm eggs clear rapidly and if slides are not examined within 30 to 60 min the eggs will no longer be visible. Schistosome eggs may be recognizable for up to several months but it is preferable in a schistosomiasis endemic area to examine the slide preparations within 24 h.

10. The smear should be examined in a systematic manner and the number of eggs of each species reported. Later multiply by the appropriate number to give the number of eggs per gram of feces (if using a 50 mg template by 20; a 20 mg template by 50; a 41.7 mg template by 24). With high egg counts, to maintain a rigorous approach while reducing reading time, the Stoll quantitative dilution technique with 0.1 N NaOH may be recommended.

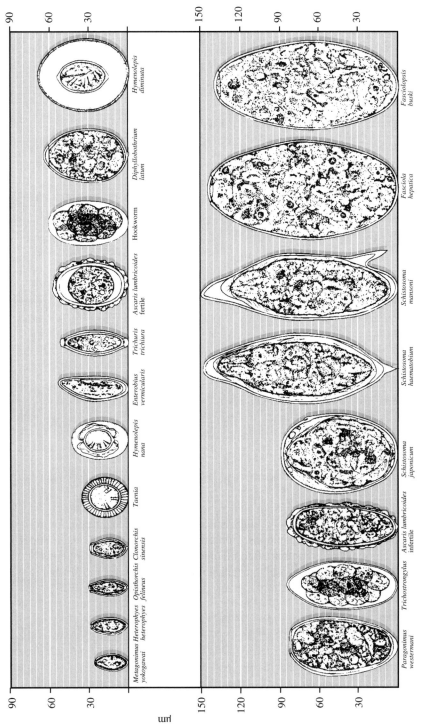

Schistosoma mekongi and Schistosoma intercalatum have been omitted. Eggs of **S.mekongi** measure 51 - 78 µm by 39 - 66 µm; eggs of **S.intercalatum** measure 140 - 200 µm long

APPENDIX 4

A Model Framework for the Control of Morbidity Due to FBT Infections

1. **Priority actions:**
   1.1 Assessment of priorities and selection of cost-effective interventions:
      - Cost of disease burden
      - Cost of treatment and prevention
   1.2 The demographic, epidemiological, and economic information produced should be used for planning and strategy selection.
   1.3 Needs assessment: rapid epidemiological evaluation and choice of interventions:
      - Felt needs of the community and predicted compliance;
      - Types of infection present and their distribution and abundance (check health center records and other sources for data on positive cases)
      - Extent and severity of morbidity
      - Availability and accessibility of primary health care
      - Nature of the environment and ecological features
      - Situation regarding intersectoral collaboration
      - Situation regarding sanitation
      - Levels of managerial and technical support
      - Staff training

2. **Control of FBTs:**
   2.1 Praziquantel is highly efficacious in most cases, but anthelminthic treatment is only one component of disease control.
   2.2 Control of the intermediate host is not a practical option because in many countries freshwater fish, a dietary staple, is imported from many locations. (Snail control may not be practical. It is expensive and there is the risk of disturbing aquatic environments.)
   2.3 Health education, provision of better sanitation, and annual drug treatment have had success in FBT control. It is difficult to change strongly embedded cultural practices that include the eating of uncooked fish.

3. **FBT control: set realistic targets from the following possibilities:**
   3.1 Reduce the prevalence and intensity of FBT infection
   3.2 Elimination of reservoir hosts

3.3 Interrupt the transmission

3.4 Change behavior.

3.5 Disease control

4. **Disease control:**

    4.1 Satisfactory resolution of morbidity is observed after treatment, as shown by significant clinical improvement, normalization of laboratory parameters, and reduction in the frequency of ultrasonographic abnormalities.

    4.2 Monitoring changes in the *Opisthorchis*/cholangiocarcinoma relationship should not be ignored. This point applies to infection with *Clonorchis sinensis*.

5. **Disease control strategies:**

    5.1 Universal drug administration

    5.2 Mass screening and treatment

    5.3 Passive case detection

    5.4 Information, education, communication

    5.5 Reduction of transmission

6. **Repeated universal drug administration: key factors:**

    6.1 Target groups are different in FBT control than in other helminth group

    6.2 How frequent and for how long?

    6.3 Impact on infection control

    6.4 Typically donor dependent

    6.5 Reduction in infection rate

    6.6 Cost of delivery:

    • Training

    • Information campaigns

    • Drugs; availability, cost, and quality

    6.7 Impact on morbidity

7. **Mass screening:**

    7.1 Good impact on people awareness

    7.2 Highly demanding on human resources

8. **Passive case detection (PCD):**

    8.1 PCD is reported as a cost-effective intervention

    8.2 Positive interaction with primary health care

    8.3 "Socialization" of the disease control

    8.4 Combined with health education

9. **Information, education, communication:**

    9.1 What are the objectives?

    • Changing behaviors

    • Increasing awareness

    • Creating needs

        (a) better health services

        (b) sanitation

10. **Anthelminthic drugs:**

    10.1 Price and quality:

- Drug availability may depend largely on the level and duration of donor support
- With the expiry of patents, generic versions of anthelminthic drugs are available at low cost
- Therapeutic value of a drug is affected by variations in content (amount of active ingredient) and purity, disintegration, dissolution, and bio-availability
- Program managers should check drug quality of the drug to avoid fake and counterfeit drugs.

    10.2 Access:

- In its essential medicine strategy, WHO has defined four key elements leading to good access to drugs:

  (a) affordable prices

  (b) reliable supply systems

  (c) sustainable financing

  (d) rational selection and use of drugs

- Despite the current price level of anthelminthic drugs, donor support for helminth control may be needed, particularly in school-based or community-based interventions in the poorest sections of developing countries.
- The "value for money" dimension of helminth control has significantly improved.

11. **Strategies for morbidity reduction:**

    11.1 Attack phase:

- Mass stool examinations and selective treatment

    11.2 Consolidation phase:

- Passive services operated at each level of the health system

    11.3 Cost of service delivery

The high costs of vertical programs for drug administration, in addition to concerns of "sustainability," indicate that integrated approaches capitalizing on existing infrastructures are to be encouraged.

    11.4 The integrated approach has demonstrated that the cost of delivery of treatment is greatly reduced.

    11.5 The delivery cost:

- Cost recovery:

  (a) would people be willing to pay for or contribute to the costs of treatment?

  (b) in many countries the expected answer might be "no," but evidence to substantiate this preconception is limited

  (c) there could be opportunities to propose cost-sharing

12. **Recommended control objectives and strategies:**

    12.1 Ensure access to essential drugs in all health facilities.

    12.2 Morbidity control should be the first objective of community intervention.

    12.3 Where community-wide treatment has achieved low prevalence of infection (with significant decreases in morbidity) a planned, sustainable consolidation phase should be introduced.

12.4 The control strategy must be complemented by improved sanitation, hygiene in aquaculture, and appropriate health education.

13. **Information – education:**
    13.1 Eating raw fish is associated with severe liver disease.
    13.2 Liver disease could lead to liver cancer.
    13.3 Fish should be cooked properly to protect from infection.
    13.4 People should not defecate in the fishponds and waste from domesticated animals should not be added to fishponds.
    13.5 Latrines should be separated from fish ponds.
    13.6 Treatment is safe and effective after diagnosis.
    13.7 Symptoms include: diarrhea, flatulence, fatty-food intolerance, epigastric and right upper quadrant pain, jaundice, fever, hepatomegaly, lassitude, anorexia, emaciation, and edema.
    13.8 If eating raw fish, better to be periodically checked, at least once per year. Go to the health center with a stool sample.

14. **Training:**
    14.1 Health staff in all the primary health care stations in endemic areas should be able to (i) suspect the disease, (ii) recognize trematode eggs with the microscopic examination of a stool sample, and (iii) treat both the infection and the clinical manifestation.

15. **Access to diagnosis and treatment:**
    15.1 Health stations in high endemic areas should be equipped with a microscope, some basic material for stool examination (Kato Katz: Appendix 3), a trained microscopist, and praziquantel.

16. **Measuring effects of control:**
    16.1 Effect on disease and morbidity via clinical surveillance:
        • Ultrasounds indication of disease in liver
        • Organomegaly
        • Other indicators
    16.2 Effect on indices of infection.
    16.3 Effect on human behavior:
        • Contamination measured by infection rate in wild or sentinel snails.

17. **Sustainability:**
    17.1 Integration of control in existing structures and public health interventions, such as school health packages, together with decentralization of decision-making and delivery, are essential in high prevalence areas to ensure commitment and sustainability.
    17.2 Health authorities should recognize FBT control as an integral part of primary health care services.

18. **Conclusions:**

    18.1 The choice of any strategy will depend on the cost-effectiveness ratio (cost per unit of effectiveness achieved) and existing budget constraints.

    18.2 The cost-effectiveness ratio will be affected by a wide variety of economic, epidemiological, demographic, technical, and behavioral factors.

    18.3 A comparative analysis of the cost-effectiveness of treatment should be completed based on parasitological and symptomatic screening.

This framework is based on Annex 4 in *Food-borne Trematode Infections in Asia*, a report of a joint WHO/FAO Workshop held in Viet Nam during November 2002. This version is reproduced with permission from the World Health Organization (www.who.int).

Index

I

M